Applied High-Speed Plate Penetration Dynamics

SOLID MECHANICS AND ITS APPLICATIONS
Volume 132

Series Editor: G.M.L. GLADWELL
Department of Civil Engineering
University of Waterloo
Waterloo, Ontario, Canada N2L 3GI

Aims and Scope of the Series

The fundamental questions arising in mechanics are: *Why?*, *How?*, and *How much?* The aim of this series is to provide lucid accounts written by authoritative researchers giving vision and insight in answering these questions on the subject of mechanics as it relates to solids.

The scope of the series covers the entire spectrum of solid mechanics. Thus it includes the foundation of mechanics; variational formulations; computational mechanics; statics, kinematics and dynamics of rigid and elastic bodies: vibrations of solids and structures; dynamical systems and chaos; the theories of elasticity, plasticity and viscoelasticity; composite materials; rods, beams, shells and membranes; structural control and stability; soils, rocks and geomechanics; fracture; tribology; experimental mechanics; biomechanics and machine design.

The median level of presentation is the first year graduate student. Some texts are monographs defining the current state of the field; others are accessible to final year undergraduates; but essentially the emphasis is on readability and clarity.

For a list of related mechanics titles, see final pages.

Applied High-Speed Plate Penetration Dynamics

by
GABI BEN-DOR

Ben-Gurion University of the Negev, Beer-Sheva, Israel

ANATOLY DUBINSKY

Ben-Gurion University of the Negev, Beer-Sheva, Israel

and

TOV ELPERIN

Ben-Gurion University of the Negev, Beer-Sheva, Israel

Springer

A C.I.P. Catalogue record for this book is available from the Library of Congress.

ISBN-10 1-4020-3452-0 (HB)
ISBN-13 978-1-4020-3452-7 (HB)
ISBN-10 1-4020-4239-6 (e-book)
ISBN-13 978-1-4020-4239-3 (e-book)

Published by Springer,
P.O. Box 17, 3300 AA Dordrecht, The Netherlands.

www.springer.com

Printed on acid-free paper

Printed in the Netherlands.

Contents

Contributing Authors

Gabi Ben-Dor
Anatoly Dubinsky
Tov Elperin

Ben-Gurion University of the Negev, Israel.

Preface

High-speed impact dynamics is of interest in the fundamental sciences, e.g., astrophysics and space sciences, and has a number of important applications in military technologies, homeland security and engineering.

When compared with experiments or numerical simulations, analytical approaches in impact mechanics only seldom yield useful results. However, when successful, analytical approaches allow us to determine general laws that are not only important in themselves but also serve as benchmarks for subsequent numerical simulations and experiments.

The main goal of this monograph is to demonstrate the potential and effectiveness of analytical methods in applied high-speed penetration mechanics for two classes of problem. The first class of problem is shape optimization of impactors penetrating into ductile, concrete and some composite media. The second class of problem comprises investigation of ballistic properties and optimization of multi-layered shields, including spaced and two-component ceramic shields. Despite the massive use of mathematical techniques, the obtained results have a clear engineering meaning and are presented in an easy-to-use form. One of the chapters is devoted solely to some common approximate models, and this is the first time that a comprehensive description of the localized impactor/medium interaction approach is given.

In the monograph the authors present systematically their theoretical results in the field of high-speed impact dynamics obtained during the last decade which only partially appeared in scientific journals and conferences proceedings.

This monograph is written primarily for scientists and engineers working in the field of high-speed impact dynamics and penetration. It can be also used as a textbook in industrial and applied mathematics for demonstrating the applicability of mathematical methods for solving applied problems. Familiarity with basic concept of solid mechanics and mathematics is assumed.

Chapter 1

INTRODUCTION

The main goal of this monograph is the analysis of engineering problems in penetration mechanics and the derivation of solutions that have practical applications. We consider regimes of normal high-speed (but not s hypervelocity) penetration by a rigid impactor that are accompanied by elastic-plastic deformations and brittle failure and when hydrodynamic models are not applicable. The authors aim is to show that despite of the massive use of computer modeling in penetration mechanics – which is facilitated by easy access to powerful computers and multi-purpose codes – the analytical methods that are employed in this book are still useful and retain their importance. The advantage of analytical methods is that they allow us to determine general laws and express them in analytical form. Such results are not only important in themselves, but they can also serve as benchmarks for experiments and numerical simulations. The price of the advantages of the analytical methods is the necessity to use approximate models. These models must not only be sufficiently reliable and adequate, but they must also be relatively simple and convenient. Therefore, in this book the discussion of approximate models receives considerable attention. We do not, however, go into a discussion of the range of applicability and the experimental validation of widely used approximate models, since these discussions can be found in a number of publications, many of which are referred to in our book. Primary emphasis is placed on the association between different approaches for developing approximate models of impactor-shield interaction and on some formal procedures for constructing these models, particularly, the localized interaction approach (LIA). The LIA has received considerable attention (methodological and practical aspects) because of the experience of its application that has accumulated in gasdynamics and because of growing interest in different aspects of LIA in impact dynamics, where many elements of the LIA have been repeatedly rediscovered.

In their exposition of the LIA, the authors use rigorous mathematical proofs and employ numerical simulations only when rigorous mathematical proof is not found. Despite the approximate nature of the basic models, this approach seems reasonable, since it allows us to establish a reliable association between the models and the results obtained on the basis of the models. However, one must keep in mind that results obtained with the approximate models should be used mainly for the qualitative estimation of tendencies and not as quantitative predictions.

This monograph is not intended to serve as a textbook. Similarly, it does not contain a chapter like "Introduction to high-speed impact dynamics", i.e., it is not an initial introduction to this subject. As a basic text, we recommend the monograph (Zukas, 1990), although no special prerequisites are required for understanding the problem under consideration. Our monograph covers the aspects of modeling of high-speed impact that are determined by our research interests and presents the results of our studies. Together with the new material, we include in this monograph some previously published results in revised and expanded form, e.g., we have simplified the proofs, added new examples and calculations, and used a unified style of exposition. The overviews presented in the book are associated directly with the topic of the research. References and information concerning more general problems associated with high-speed plate penetration modeling using approximate models can be found in the studies of Kennedy (1976), Backman and Goldsmith (1978), Jonas and Zukas (1978), Zukas (1982), Brown (1986), Anderson and Bodner (1988), Heuzé (1989), Recht (1990), Zukas and Walters (1990), Abrate (1994), Corbett et al. (1996), Abrate (1998), Teland (1998), Børvik et al. (1998), Kasano (1999), Goldsmith (1999), and Cheeseman and Bogetti (2003).

The monograph is divided into three parts.

Part 1, which comprises *Chapters 2-4*, deals mainly with two universal approximate approaches that are associated with each other and that are widely used in impact dynamics: the LIA and cavity expansion approximations (CEAs).

Chapter 2 is devoted to a discussion of the LIA. In its simplest formulation, it is based on the assumption that the net drag force exerted by the shield on the penetrating impactor can be represented as a sum of independent local drag forces at the contact points. Each local instantaneous drag force is assumed to depend only on the velocity of a penetrator, the angle between the velocity vector and the local normal vector at the impactor's surface, and some global characteristics that remain constant during penetration, e.g., the material properties of the shield. Most of approximate models that are used in practice for describing local interaction between a shield and a high-speed penetrator belong to the class of localized interaction models (LIMs). *Chapter 2* offers, for the first time, a comprehensive exposition of the foundations of localized interaction theory (LIT) applied to modeling of penetration by rigid impactors. Along with the description of the universal model and investigation of general properties of LIMs, this chapter provides a classification of the basic types of LIMs, considers various simplified approaches, and presents formulas for the ballistic limit velocity (BLV) and the depth of penetration (DOP) for impactors of various shapes. LIMs are used in the book as a basis for investigating different problems.

Some features of CEAs are discussed in *Chapter 3*. It is shown that spherical cavity expansion approximations (SCEAs) and the widely used static versions of cylindrical cavity expansion approximations (CCEAs) can be reduced to LIMs. The main emphasis is placed on mathematical formulation of the non-static CCEA model for sharp impactors and impactors with flat bluntness. Notably, in many cases the mathematical derivations are worked down to design formulas.

Whereas *Chapters 2 and 3* deal with the models describing local impactor-shield interaction, *Chapter 4* discusses a power-law relationship between the impact velocity, the residual velocity and the BLV that is widely used for reduction of experimental data. It is demonstrated that some LIMs and their “non-local” generalizations yield such power-law relationships. The accuracy of the power-law approximations is compared with the accuracy of the “basic” power law with an exponent equal to two.

Part 2 of the monograph comprises *Chapters 5-9* dealing with shape optimization of impactors penetrating into different media.

The *Chapter 5* – an introductory chapter – presents an overview of the subject. It also gives a short description of the analytical and numerical methods for solving, in the general case, non-classical variational problems where functionals depend on the integrals of the unknown solution. These methods are not widely known, even through such non-classical problems are often encountered in shape optimization of penetrating impactors (some problems of the latter kind are investigated in *Chapters 6 and 7*).

Optimal impactors having the shape of bodies of revolution with fixed length and radius of the shank and penetrating at the maximum depth into ductile media (metal, soil) and concrete, are considered in *Chapters 6 and 7*, respectively. When the simplest models without friction are employed, the well-known solution of Newton (1687) is obtained once again. More complex models suggested at the Sandia Research Laboratories (Warren and Forrestal, 1998; Forrestal and Tzou, 1997) yield complicated variational problems that are also considered in these chapters. *Chapter 8* contains the solution of a problem of finding the impactor with the minimum BLV that penetrates into fibre-reinforced plastic laminates and has the shape of a body of revolution. In the solution, the model suggested by Wen (2000) and generalized to impactors having an arbitrary shape is used. *Chapter 8* discusses the ballistic properties of three-dimensional penetrating impactors having a star-shaped cross-section. Investigations showed that when solutions in the class of bodies of revolution are considered, optimal penetrating impactors generally have large flat bluntness and a shape close to truncated cones.

In Chapter 9, the well-known in gasdynamics area rules are generalized for the case of penetrating impactors, and a new version of area rules is found. The latter allows us to predict the conditions under which a change of the impactor's shape by a small factor of the order of ε causes changes in the BLV and some other ballistic characteristics by a smaller factor of the order of ε^2.

Part 3 of the monograph, *Chapters 10 and 11,* deals with investigations of the ballistic properties of non-homogeneous (mainly, layered) shields, including spaced shields.

Chapter 10 analyzes the effect on the BLV of the order of the plates and the air gaps in the shield and of separating a monolithic shield into several plates during penetration by a conical impactor. The main emphasis is placed on analytical techniques, although numerical simulations are also used. In *Chapter 10*, it is shown that if any LIM is valid, then the BLV does not change when the initially monolithic shield is replaced by a spaced shield having the same total width of the plates. The latter result does not depend on the widths of the plates or of the air gaps. Analysis of the same problem by means of the CCEA shows that the protective properties of a monolithic shield can be slightly improved by replacing it with a shield with air gaps and that the maximum benefit is achieved when the plates in the spaced shield have the same thickness. The analysis performed yields a criterion that allows us to determine the optimal sequence of the plates in the shield when the shield is composed of plates manufactured from different materials. The magnitude of the gain can, in fact, be quite large.

Two problems of optimization of two-component ceramic armor are solved in *Chapter 11* by applying the model suggested by Florence (1969). The solutions are found in the form of explicit dependencies of the thicknesses of the layers in the optimal armor as functions of the parameters determined by the properties of the shield and the impactor.

In the organization of the monograph, the following set-up is used. Equations are labeled consecutively within each chapter. The number of each equation given as (y.x), where y is a number of the section within the particular chapter (chapters are subdivided into sections) and x is a number of the equation inside the section. As a result formulas with the same equation numbers can be encountered in different chapters. Therefore, formulas in the same chapter are referred to as Eq. (y.x); if the necessity to refer to an equation from another chapter arises, we add the number of the chapter, e.g., Eq. (z-y.x), where z is the number of the chapter. The formulas in *Appendix 2* are denoted as (A2.x), and they are referred to as Eq. (A2.x).

Consecutive labeling is used for figures and tables within each chapter, e.g., *Table z-x* or *Figure z-x*, and they are referred to in the monograph accordingly. The same system is used for numbering and referring to sections, e.g., *Section z-y*, *Section z-y.v.w*, etc.

To unify the system of notations in the monograph, the following methods are used: (1) Summarizing in the tables the numbers of the equations that determine a model or its particular version (this method is often used in *Chapter 2*); (2) Using the same notations to denote parameters that are encountered throughout the monograph, e.g., mass of the impactor, BLV, DOP; (3) Employing notation that characterizes the meaning of the parameter, e.g., μ with a subscript to denote the material properties of a shield; and (4) Presenting a glossary of the notations in *Appendix 1* (in the table). Tensors are not used in the monograph, and a superscript without parentheses always denotes an exponent. To simplify the presentation, we have tried to avoid the use of dimensionless variables.

To master the material presented in the monograph, it is sufficient for the reader to be familiar with basic engineering courses in mathematics and mechanics. However, understanding the proofs and transformations requires a certain mathematical sophistications. The mathematical component of the monograph is intended for researchers in the field of impact dynamics. However, useful information can be obtained simply by skipping the mathematical details and proofs. The authors hope that the theoretical predictions described in the monograph will attract the attention of experimentalists working in the field of penetration mechanics.

PART 1: SOME COMMON MODELS

Chapter 2

LOCALIZED INTERACTION APPROACH

1. BASICS OF THE LOCALIZED INTERACTION APPROACH

Many engineering models for penetration modeling belong to the category of the so-called localized interaction models (LIM) (Bunimovich and Dubinsky, 1995, 1996), in which the integral effect of the interaction between a host medium and a moving projectile is described as a superposition of the independent local interactions of the projectile's surface elements with the medium. Every local interaction is determined by the local geometric and kinematic parameters of the surface element (primarily, by the local velocity of the surface element and the angle between the local surface velocity vector and the local normal vector to the projectile surface) as well as by some global parameters that take into account the integral characteristics of the medium (e.g., hardness, density, etc.). Let us consider different versions of the following unified LIM:

$$d\vec{F} = \begin{cases} [\,\Omega_n(u,v)\vec{n}^0 + \Omega_\tau(u,v)\vec{\tau}^0\,]\,ds \;\; if \;\; 0 < u < 1 \\ \Omega_n(1,v)\vec{n}^0 ds \;\; if \;\; u = 1 \\ 0 \;\; if \;\; u \le 0 \end{cases}, \tag{1.1}$$

$$\vec{\tau}^0 = -(\vec{v}^0 + u\cdot\vec{n}^0)/\sqrt{1-u^2}\,, \tag{1.2}$$

$$u = -\vec{v}^0 \cdot \vec{n}^0 = \cos \hat{\upsilon}\,, \tag{1.3}$$

where (*Figure 2-1*) $d\vec{F}$ is the force acting at the surface element dS of the projectile that is in contact with the host medium, $\vec{n}^0$ and $\vec{\tau}^0$ are the inner normal and tangent vectors at a given location on the projectile's surface, respectively, $\vec{v}^0$ is an unit vector of the surface element velocity of the projectile, $\vec{v}$, and $\hat{\upsilon}$ is the angle between the vector $\vec{n}^0$ and the vector $(-\vec{v}^0)$; the non-negative functions Ω_n and Ω_τ determine the model of the projectile-medium interaction and also depends on the parameters $a_0, a_1, \ldots$ that characterize, mainly, the properties of the host medium. The unit tangent vector $\vec{\tau}^0$ lies in the plane of the vectors $\vec{v}^0$ and $\vec{n}^0$ and is perpendicular to the vector $\vec{n}^0$; its direction is chosen such that $\vec{v}^0 \cdot \vec{\tau}^0 < 0$, i.e., the direction of the friction force is the same as the direction of the vector $\vec{\tau}^0$.

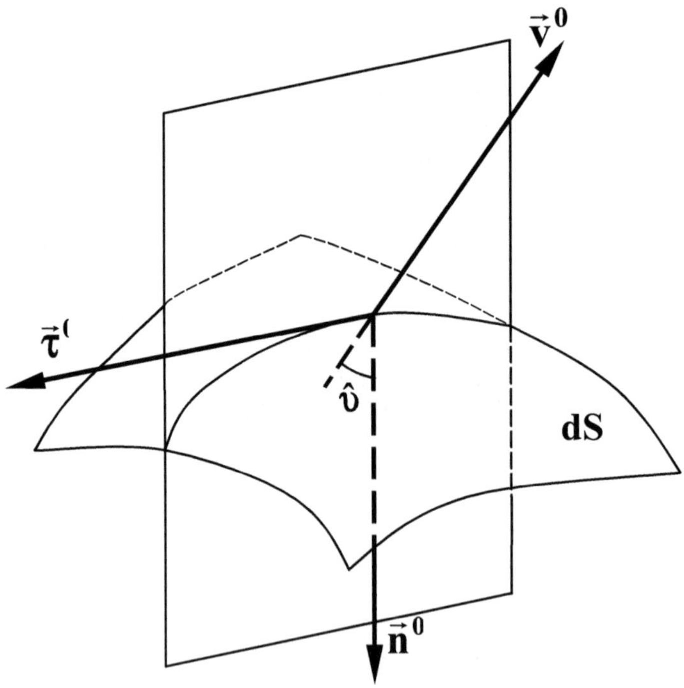

Figure 2-1. Description of the LIM.

The resultant force acting on the body at each instant of time is determined by integrating $d\vec{F}$ over the surface of the impactor-medium contact at the same instant, S. Let us now consider a normal impact (impact velocity is normal to the impacted plate) of a rigid "symmetric body", so that we may assume translational motion of the impactor under the effect of the drag force D, which can be represented in the general case as:

$$D=(-\bar{v}^0)\cdot\iint\limits_{S} d\vec{F}=\iint\limits_{S_{perp}}(-\bar{v}^0)\cdot d\vec{F}+\iint\limits_{S_{lat}}(-\bar{v}^0)\cdot d\vec{F}, \tag{1.4}$$

where S_{perp} is a part of the contact surface S that is normal to the impactor's velocity (as a rule, the flat bluntness) and S_{lat} is the lateral surface of the impactor. Substituting $d\vec{F}$ from Eq. (1.1) into Eq. (1.4), we obtain:

$$D=\Omega_n(1,v)\iint\limits_{S_{perp}} dS+\iint\limits_{S_{lat}}\Omega_0(u,v)dS, \tag{1.5}$$

where

$$\Omega_0(u,v)=u\Omega_n(u,v)+\sqrt{1-u^2}\,\Omega_\tau(u,v). \tag{1.6}$$

2. DESCRIPTION OF NORMAL PENETRATION

From this point onwards, we consider normal penetration and use the following notations (*Figure 2-2a-b*). The coordinate h, the instantaneous depth of the penetration, is defined as the distance between the nose of the impactor and the front surface of the shield, and L is the length of the nose of the impactor. We take into account the possible interaction between the shield and the lateral surface of the impactor only for $0\le h\le L$. The cylindrical coordinates x,ρ,ϑ are associated with the impactor, and the equation $\rho=\Phi(x,\vartheta)$, where Φ is some function, determines the shape of the impactor.

Generally, we consider impactors with flat bluntness and a cylindrical part of length L_0 and assume that this cylindrical part does not interact with the shield. In other words, all the above-mentioned formulas refer to the impactor's nose, which is located between sections $x=0$ and $x=L$. Thus, for example, the notion of a conical penetrator implies a penetrator with a conical nose.

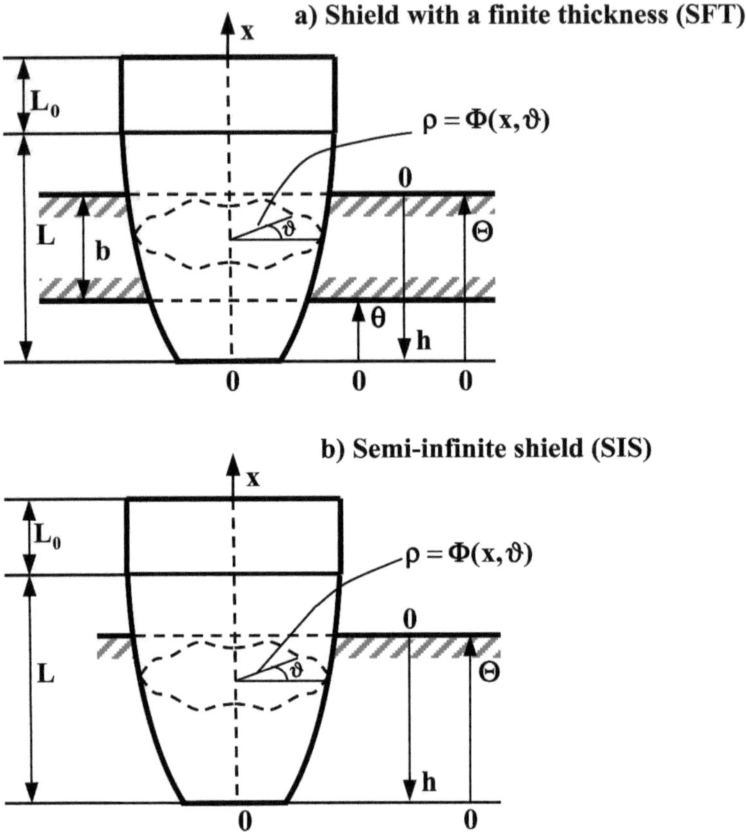

Figure 2-2. The notations.

2.1 Impactor-shield interaction surface

The formalism of the description of the impactor-shield interaction surface in the case of a *semi-infinite shield* (SIS) is illustrated in *Figure 2-3a-b*. Generally, two stages of penetration can be considered. The first stage, entry into the shield, occurs when $0 \le h \le L$ (*Figure 2-3a*). In this case, the flat bluntness of the impactor (if any) and the part of its lateral surface between the sections $x = 0$ and $x = h$ interact with the shield. The second stage (*Figure 2-3b*), i.e., motion inside the shield, is associated with full immersion of the bluntness and of the lateral surface of the impactor into

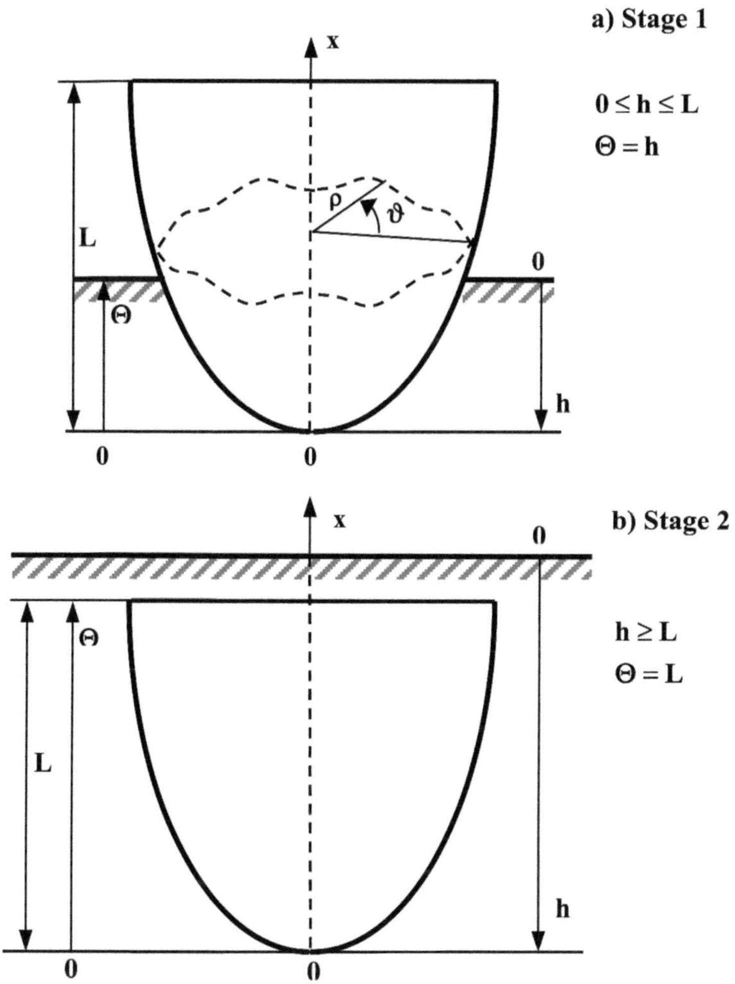

Figure 2-3. Two-stage penetration model for a SIS.

the shield and occurs when $x \geq L$. Therefore, the moving area of the impactor-shield interaction can be described as (*Figure 2-4*):

$$0 \leq x \leq \Theta(h), \qquad (2.1)$$

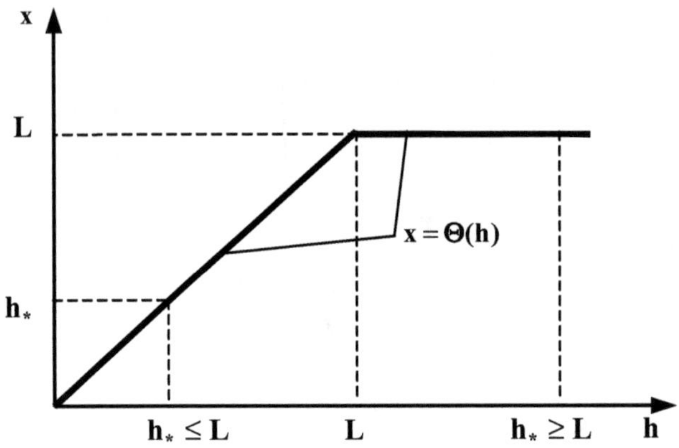

Figure 2-4. Penetration into a SIS. Description of the area of impactor-shield interaction.

where

$$\Theta(h)=\begin{cases} h & if\ 0\le h\le L \\ L & if\ h\ge L \end{cases}. \tag{2.2}$$

In some instances, it is convenient to define function $\Theta(h)$ also for negative values of variable h, assuming that $\Theta(h)=0$ for $h<0$.

In the following exposition, we use a useful elementary identity that is valid for any function $\Xi(x)\ge 0$ and $h_*\ge 0$ (*Figure 2-4*):

$$\int_0^{h_*} dh \int_0^{\Theta(h)} \Xi(x)dx = \int_0^{\Theta(h_*)} (h_*-x)\Xi(x)dx, \tag{2.3}$$

The proof of Eq. (2.3) is based on the change of the order of integration:

$$\int_0^{h_*} dh \int_0^{\Theta(h)} \Xi(x)dx = \begin{cases} \int_0^{h_*} \Xi(x)dx \int_x^{h_*} dh = \int_0^{h_*} (h_*-x)\Xi(x)dx\ if\ h_*\le L \\ \int_0^{L} \Xi(x)dx \int_x^{h_*} dh = \int_0^{L} (h_*-x)\Xi(x)dx\ if\ h_*\ge L \end{cases}. \tag{2.4}$$

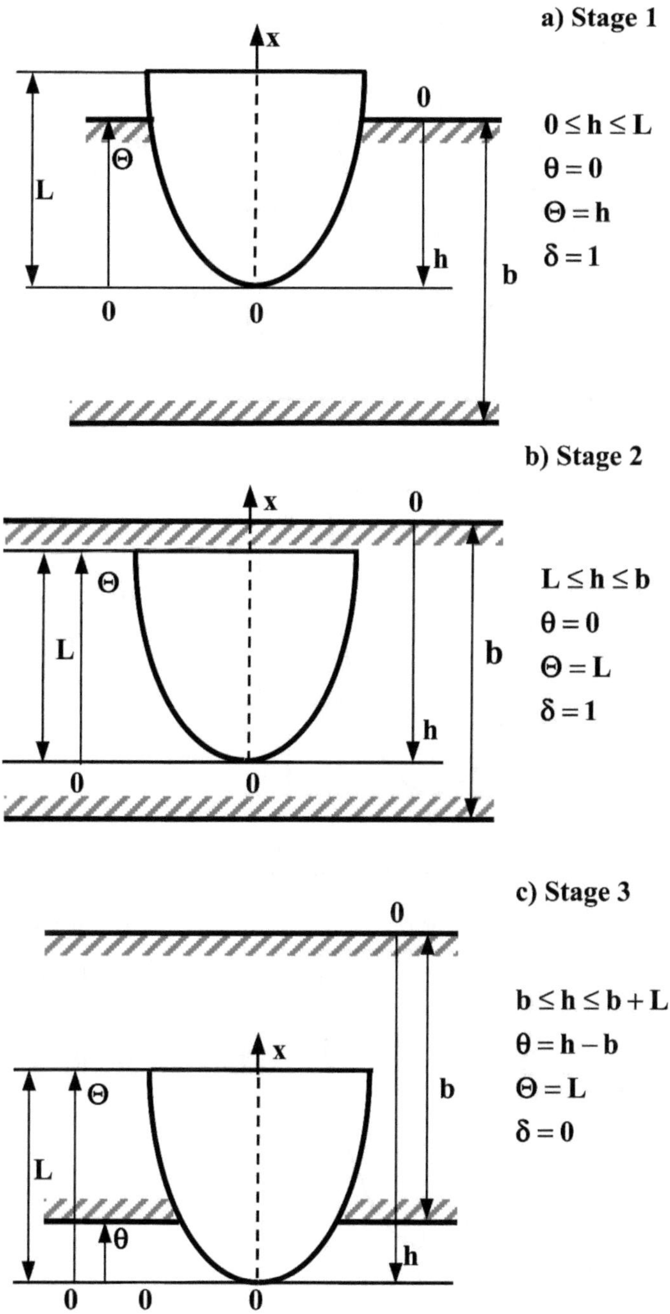

Figure 2-5. Three-stage penetration model for a SFT. The case $L \le b$.

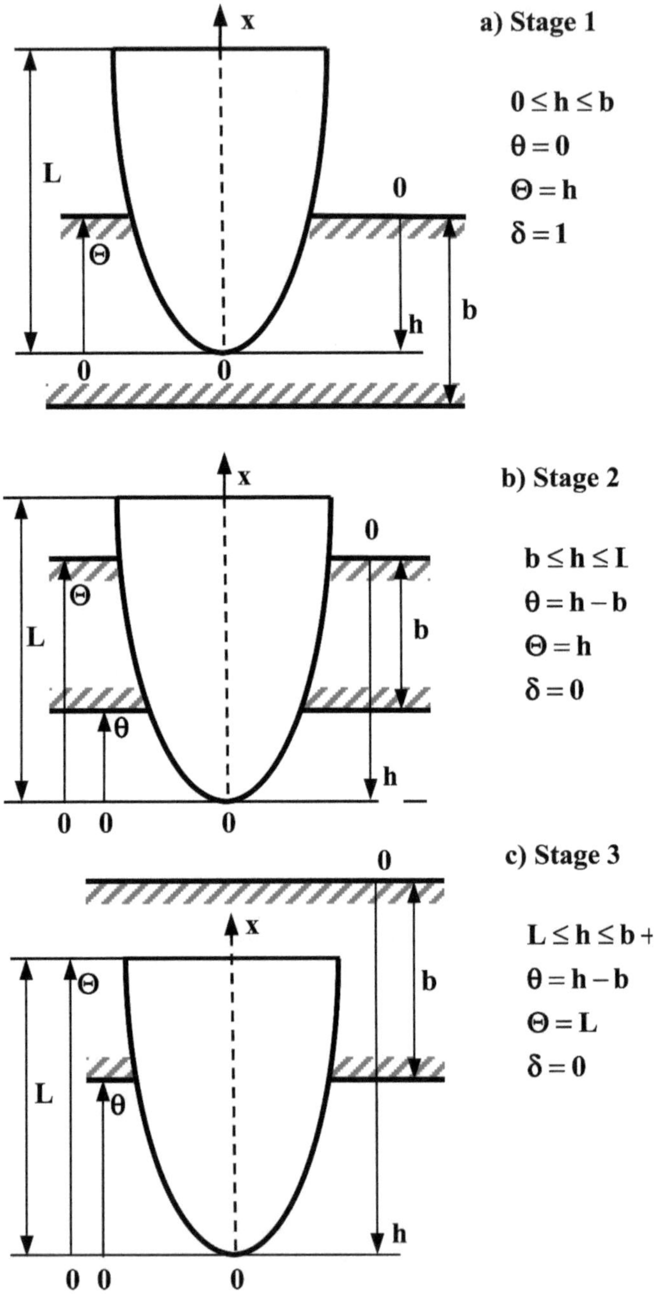

Figure 2-6. Three-stage penetration model for a SFT. The case $L \geq b$.

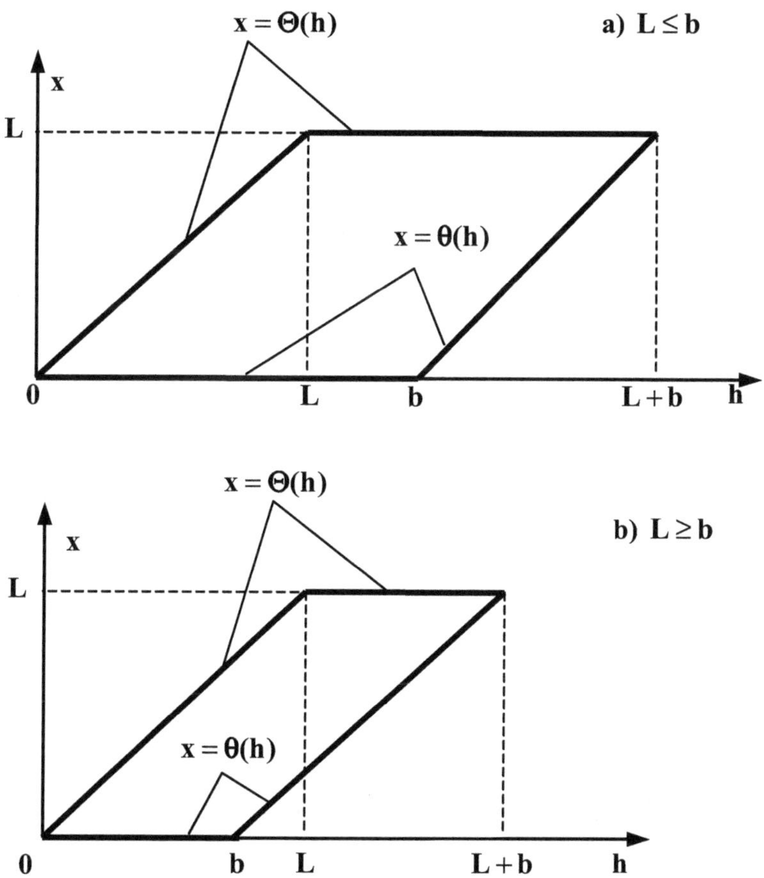

Figure 2-7. Penetration into a SFT. Description of the area of impactor-shield interaction.

Consider a *shield with a finite thickness* (SFT), b. First, let $L \leq b$. We consider a perforation as a three-stage process (*Figure 2-5a-c*). In the first stage (the entry of the impactor into the shield, $0 \leq h \leq L$), the flat bluntness of the impactor (if any) and the part of its lateral surface between the sections $x = 0$ and $x = h$ interact with the shield. In the second stage (the full immersion, $L \leq h \leq b$), the flat bluntness of the impactor (if any) and the entire lateral surface ($0 \leq x \leq L$) interact with the shield. In the third stage (emergence of the impactor from the shield, $b \leq h \leq b + L$), the flat bluntness of the impactor (if any) does not interact with the shield, while the part of the impactor's lateral surface between the sections $x = h - b$ and $x = L$ interactswith the shield. The case $L \geq b$ is illustrated in *Figure 2-6a-c* and can be analyzed in a similar manner to $L \le b$. In both cases, the moving area of the impactor-shield interaction can be described as follows (*Figure 2-7a-b*):

$$\theta(h) \le x \le \Theta(h), \tag{2.5}$$

where function $\Theta(h)$ is defined by Eq. (2.2) and

$$\theta(h) = \begin{cases} 0 & if \quad 0 \le h \le b \\ h - b & if \quad b \le h \le b + L \end{cases} \tag{2.6}$$

In some instances, it is convenient to define functions $\theta(h)$ и $\Theta(h)$ for SFT as follows: $\theta(h) = \Theta(h) = 0$ for $h < 0$ и $\theta(h) = L$ for $h > b + L$. The latter definition implies that $\theta(h) = \Theta(h) = L$ for $h > b + L$.

An identity similar to Eq. (2.3) is valid for some function $\Xi(x) \ge 0$ applied to a SFT (*Figure 2-7a-b*):

$$\int_0^{b+L} dh \int_{\theta(h)}^{\Theta(h)} \Xi(x)dx = \int_0^L \Xi(x)dx \int_{h=x}^{h=x+b} dh = b\int_0^L \Xi(x)dx\,. \tag{2.7}$$

Thus Eq. (2.5) can be used as a unified description of the area of impactor-shield interaction, taking into account that $h \ge 0$ and $\theta = 0$ for a semi-infinite shield and $0 \le h \le b + L$ and $\theta(h)$ is defined by Eq. (2.6) for a SFT.

The model can be simplified if we do not take into account the stage at which penetrator is only partially immersed in the shield. Since such simplification is used frequently, in *Section 2-9* we consider this subclass of the models. However, this simplification may impair the performance of the model. Thus, for example, Li et al. (2004), using experimental data, confirmed the need to take into account the incomplete immersion of the impactor in the shield at the initial stage of penetration, where the length of the impactor and a penetration depth are of the same order.

Sometimes, we shall use other representations of functions $\theta(h)$ and $\Theta(h)$:

$$\theta(h) = x_*(h - b), \quad \Theta(h) = x_*(h) \tag{2.8}$$

and

$$\theta(h)=L\,X_\Diamond\left(\frac{h-b}{L}\right),\quad \Theta(h)=L\,X_\Diamond\left(\frac{h}{L}\right), \tag{2.9}$$

where

$$X_\Diamond(z)=\begin{cases}0 & \text{if} \quad z\le 0\\ z & \text{if} \quad 0\le z\le 1,\\ 1 & \text{if} \quad z\ge 1\end{cases}\quad x_*(z)=\begin{cases}0 & \text{if} \quad z\le 0\\ z & \text{if} \quad 0\le z\le L\,.\\ L & \text{if} \quad z\ge L\end{cases} \tag{2.10}$$

2.2 Drag force. Equation of motion. Depth of penetration and ballistic limit velocity

Now, we can write the expression for the *drag force* using the adopted system of coordinates. Since

$$u(x,\vartheta)=\vec{x}^0\cdot\vec{n}^0=\frac{u_1(x,\vartheta)}{u_0(x,\vartheta)},\quad dS=u_0(x,\vartheta)dxd\vartheta\,, \tag{2.11}$$

$$u_0(x,\vartheta)=\sqrt{\Phi^2(\Phi_x^2+1)+\Phi_\vartheta^2},\quad u_1(x,\vartheta)=\Phi\Phi_x\,, \tag{2.12}$$

Eq. (1.5) can be transformed into the following form:

$$D(h,v)=\Omega_n(1,v)\sigma(0)\delta(h)+\int_{\theta(h)}^{\Theta(h)}dx\int_0^{2\pi}\Omega_0\big(u(x,\vartheta),v\big)u_0(x,\vartheta)d\vartheta, \tag{2.13}$$

where $\sigma(\hat{x})$ is the area of the section of the impactor in the plane $x=\hat{x}$,

$$\sigma(x)=\frac{1}{2}\int_0^{2\pi}\Phi^2(x,\vartheta)d\vartheta\,. \tag{2.14}$$

Function $\delta(h)$ is associated with the description of the resistance of an impactor's nose:

$$\delta(h)=1 \tag{2.15}$$

in the case of penetration into a SIS, and

$$\delta(h)=\begin{cases}0 & if \quad h\leq 0\\ 1 & if \quad 0\leq h\leq b,\\ 0 & if \quad h>b\end{cases} \tag{2.16}$$

when an impactor penetrates into a SFT. Clearly, we do not consider here penetration phenomena that are accompanied by plug formation.

An equation of motion of an impactor with mass m:

$$m\ddot{h}=-D(h,\dot{h}) \tag{2.17}$$

with initial conditions:

$$h(0)=0,\quad \dot{h}(0)=v(0)=v_{imp} \tag{2.18}$$

allows us to determine the depth of penetration h and the velocity of the impactor v as a function of time t and impact velocity v_{imp}.

Since the right-hand side of Eq. (2.17) does not depend on time in the explicit form, the order of this differential equation can be decreased. Considering v as a function h, $v=v(h)$, and taking into account that

$$\ddot{h}=\frac{d^2h}{dt^2}=\frac{dv}{dt}=\frac{dv}{dh}\frac{dh}{dt}=v'(h)v(h) \tag{2.19}$$

we can rewrite Eq. (2.17) as follows:

$$mv\frac{dv}{dh}+D(h,v)=0. \tag{2.20}$$

Let

$$v=V(h;v_{imp}) \tag{2.21}$$

be the solution of Eq. (2.20) with the initial condition:

$$v(0)=v_{imp}. \tag{2.22}$$

Then, the law of motion of the impactor determined as the solution of the differential equation for h with the appropriate initial condition:

$$\frac{dh(t)}{dt} = V(h;v_{imp}), \quad h(0) = 0 \tag{2.23}$$

may be written as:

$$t(h,v_{imp}) = \int_0^h \frac{dz}{V(z;v_{imp})}. \tag{2.24}$$

Equation (2.24) allows us to construct the inverse dependence h vs. t if required. Therefore, Eqs. (2.21) and (2.24) determine function v vs. t in a parametric form.

The *ballistic limit velocity* (BLV), v_{bl}, is usually considered as a characteristic of perforation for a shield with a finite thickness, and it is defined as the initial velocity of the impactor that is required for the impactor to emerge from the shield with zero velocity. Thus, v_{bl} is determined from the equation:

$$V(b+L;v_{bl}) = 0. \tag{2.25}$$

In the case of a semi-infinite shield, the *depth of penetration* (DOP), H, for the known impact velocity, v_{imp}, is determined from the equation:

$$V(H;v_{imp}) = 0. \tag{2.26}$$

Therefore, the general characteristics of the penetration, the BLV and the DOP, can be obtained through solving a first order ordinary differential equation.

3. SUB-CLASS OF MODELS CONSIDERED

From here on, we will consider, as a rule, LIMs of the following type:

$$\Omega_\tau = \mu_{fr}\Omega_n, \tag{3.1}$$

where μ_{fr} is a friction coefficient. If friction between the impactor and the host medium is not taken into account then:

$$\mu_{fr} = 0. \tag{3.2}$$

Taking into account Eqs. (1.6), (3.1) and (2.11)-(2.12), we can rewrite Eq. (2.13) as:

$$D(h,v)=\Omega_n(1,v)\sigma(0)\delta(h)+\int_{\theta(h)}^{\Theta(h)} dx \int_0^{2\pi}\Omega_n(u(x,\vartheta),v)U(x,\vartheta)d\vartheta, \quad (3.3)$$

where

$$U(x,\vartheta)=(u+\mu_{fr}\sqrt{1-u^2}\,)u_0=\Phi\Phi_x+\mu_{fr}\sqrt{\Phi^2+\Phi_\vartheta^2}\,, \quad (3.4)$$

$$u(x,\vartheta)=\frac{\Phi\Phi_x}{\sqrt{\Phi^2(\Phi_x^2+1)+\Phi_\vartheta^2}}\,. \quad (3.5)$$

For purpose of convenience, we have summarized all the required for calculations formulas in *Table 2-1*. All the solutions presented below are derived by applying these general relationships.

Table 2-1. Calculating formulas for the LIM with arbitrary Ω_n and 3-D impactor

Function/variable determined	Equations	Comments
$\sigma(x)$	(2.14)	
$\delta(h)$	$=1$	SIS
	(2.16)	SFT
$\Theta(h)$	(2.2), (2.8), (2.9)	
$\theta(h)$	(2.6), (2.8), (2.9)	SFT
	$=0$	SIS
u,U	(3.4), (3.5)	
$D(h,v)$	(3.3)	
$v=V(h;v_{imp})$	(2.21)	The solution of Eq. (2.20) with initial condition of Eq. (2.22)
v_{bl}	(2.25)	BLV for SFT
H	(2.26)	DOP for SFT

4. CONSIDERED SHAPES OF THE IMPACTOR

Formulas for calculations using the general model determined by Eq. (3.1) and for different shapes of the impactor are summarized in *Table 2-2*.

The volume of the impactor can be represented as:

$$V_{imp} = \int_0^L \sigma(x)dx + \sigma(L)L_0, \tag{4.1}$$

where $\sigma(x)$ is given by Eq. (2.14).

The mass of the impactor m, the volume of the impactor V_{imp} and the density of the material of the impactor γ_{imp} are related by the equation:

$$m = \gamma_{imp} V_{imp}. \tag{4.2}$$

Let us now consider several classes of impactor shapes.

The equation of the lateral surface of a 3-D conical impactor may be written as (*Figure 2-8*):

$$\Phi(x,\vartheta) = (k_0 + kx)\eta(\vartheta), \tag{4.3}$$

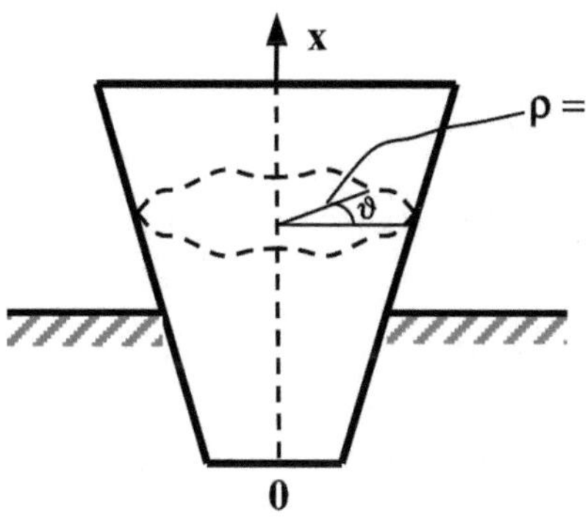

Figure 2-8. 3-D conical impactor.

where function $\eta(\vartheta)$ determines the shape of the impactor's cross section, and k_0 and k are coefficients.

If the impactor has the shape of a body of revolution, then:

$$\Phi = \Phi(x). \tag{4.4}$$

A cone of revolution may be considered as a particular of both case a 3-D conical body and a body of revolution. The equation of its lateral surface is:

$$\Phi(x,\vartheta) = k_0 + kx. \tag{4.5}$$

If an impactor has a flat bluntness then:

$$\sigma(0) = k_0 = r = 0. \tag{4.6}$$

Formulas for $u, U, \sigma(x)$ and $\sigma(0)$ in *Table 2-2* are derived using the corresponding formulas in *Table 2-1* and taking into account the equations describing the surface of the impactor. The following formulas for the drag force D of impactors having different shapes are obtained from Eq. (3.3).

Table 2-2. Description of LIM with arbitrary Ω_n for impactors of different shapes

Function	3-D cone	Body of revolution	Cone of revolution
Φ	$(k_0 + kx)\eta(\vartheta)$	$\Phi(x)$	$k_0 + kx$
u	$k\eta^2 / \sqrt{\eta^2(k^2\eta^2+1)+\eta'^2}$	$\Phi' / \sqrt{\Phi'^2+1}$	$k/\sqrt{k^2+1}$
U	$(k_0 + kx)\tilde{u}(\vartheta)$	$\Phi(\Phi' + \mu_{fr})$	$(k_0 + kx)\tilde{u}$
$\tilde{u}$	$k\eta^2 + \mu_{fr}\sqrt{\eta^2+\eta'^2}$		$k + \mu_{fr}$
$\sigma(x)$	$\frac{1}{2}(k_0 + kx)^2 \int_0^{2\pi} \eta^2(\vartheta)d\vartheta$	$\pi\Phi^2(x)$	$\pi(k_0 + kx)^2$
$\sigma(0)$	$\frac{1}{2}k_0^2 \int_0^{2\pi} \eta^2(\vartheta)d\vartheta$	πr^2	πk_0^2
D	Eqs. (4.10), (4.9), (4.11)	Eq. (4.7)	Eqs. (4.8), (4.9)
v_{bl}, sharp impactor	Eqs. (5.7), (5.5), (4.11)		Eqs. (5.7), (5.5), (4.12)
H, sharp impactor	Eqs. (5.11), (5.5), (4.11)		Eqs. (5.11), (5.5), (4.12)
Comments		$\Phi(0) = r$	$k_0 = r$, $k = R/L = \tan\upsilon$

In the case of *impactor of revolution:*

$$D(h,v)=\pi r^2\Omega_n(1,v)\delta(h)+2\pi\int_{\theta(h)}^{\Theta(h)}\Omega_n\left(\frac{\Phi'}{\sqrt{\Phi'^2+1}},v\right)\Phi(\Phi'+\mu_{fr})dx \tag{4.7}$$

and for a *cone of revolution*:

$$\begin{aligned}D(h,v)&=\pi r^2\Omega_n(1,v)\delta(h)+2\pi\Omega_n\left(\frac{k}{\sqrt{k^2+1}},v\right)(k+\mu_{fr})\hat{D}(h)\\&=\pi r^2\Omega_n(1,v)\delta(h)+2\pi\Omega_n(\sin\upsilon,v)(\tan\upsilon+\mu_{fr})\hat{D}(h),\end{aligned} \tag{4.8}$$

where υ is a half-angle of the apex of the cone and

$$\hat{D}(h)=\int_{\theta(h)}^{\Theta(h)}(k_0+kx)dx=k_0[\Theta(h)-\theta(h)]+\frac{k}{2}[\Theta^2(h)-\theta^2(h)]\,. \tag{4.9}$$

In the case of a *3-D conical impactor*:

$$D(h,v)=\Omega_n(1,v)\delta(h)\sigma(0)+\hat{D}(h)\breve{D}(v)\,, \tag{4.10}$$

where $\hat{D}(h)$ is defined by Eq. (4.9) and

$$\breve{D}(v)=\int_0^{2\pi}\Omega_n(u(\vartheta),v)\tilde{u}(\vartheta)d\vartheta\,. \tag{4.11}$$

For a *cone of revolution,* Eq. (4.11) can be simplified as:

$$\breve{D}(v)=2\pi\Omega_n(\sin\upsilon,v)(\tan\upsilon+\mu_{fr})\,. \tag{4.12}$$

5. SOLUTIONS FOR A THREE-DIMENSIONAL SHARP CONICAL IMPACTOR

Let us now consider a *3-D sharp cone*. Then, Eqs. (4.9) and (4.10) yield:

$$D(h,v)=\hat{D}(h)\breve{D}(v)\,, \tag{5.1}$$

$$\hat{D}(h)=k\int_{\theta(h)}^{\Theta(h)}xdx=\frac{k}{2}[\Theta^2(h)-\theta^2(h)],\qquad(5.2)$$

and the equation of motion of the impactor (Eq. 2.20) may be written as:

$$mv\frac{dv}{dh}=-\hat{D}(h)\breve{D}(v).\qquad(5.3)$$

The solution of this equation with the initial condition $v(0)=v_{imp}$ may be represented in the form:

$$\hat{F}(v_{imp})-\hat{F}(v)=\int_0^h d\tilde{h}\int_{\theta(\tilde{h})}^{\Theta(\tilde{h})}xdx,\qquad(5.4)$$

where

$$\hat{F}(z)=\frac{m}{k}\int_0^z\frac{vdv}{\breve{D}(v)}.\qquad(5.5)$$

Let us now consider shield of finite thickness (SFT) and semi-infinite shield (SIS).

In the case of a SFT, the equation for the BLV can be obtained by substituting $v=0, v_{imp}=v_{bl}, h=b+L$ into Eq. (5.4):

$$\hat{F}(v_{bl})=\int_0^{b+L}dh\int_{\theta(h)}^{\Theta(h)}xdx.\qquad(5.6)$$

Using the identity Eq. (2.7), we can transform the right-hand side of Eq. (5.6) to $bL^2/2$, and this equation becomes $\hat{F}(v_{bl})=bL^2/2$. Since $\hat{F}'(z)>0$ and, consequently, $\hat{F}(z)$ is an increasing function, the latter equation has only one solution:

$$v_{bl}=\hat{F}^{-1}(bL^2/2).\qquad(5.7)$$

In the case of a SIS the equation for the DOP can be obtained by substituting $v=0, h=H, \theta(h)=0$ into Eq. (5.4):

$$\hat{F}(v_{imp}) = \widetilde{F}(H), \tag{5.8}$$

where the right-hand side of this equation can be transformed by using the identity Eq. (2.3):

$$\widetilde{F}(H) = \int_0^H dh \int_0^{\Theta(h)} x dx = \int_0^{\Theta(H)} (H-x) x dx = \frac{1}{6}[3H\Theta^2(H) - 2\Theta^3(H)]$$
$$= \frac{1}{6}\begin{cases} H^3 & \text{if} \quad 0 \le H \le L \\ L^2(3H-2L) & \text{if} \quad H \ge L \end{cases}. \tag{5.9}$$

Considering Eq. (5.8) as a definition of function H vs. v_{imp}, we can differentiate both sides of this equation with respect to v_{imp}:

$$\hat{F}'(v_{imp}) = \widetilde{F}'(H)\frac{dH}{dv_{imp}}, \hat{F}'(v_{imp}) = \frac{d\hat{F}(v_{imp})}{dv_{imp}}, \widetilde{F}'(H) = \frac{d\widetilde{F}(H)}{dH}. \tag{5.10}$$

Since $\hat{F}'(v_{imp}) > 0$ and $\widetilde{F}'(H) > 0$, Eq. (5.10) implies that $dH/dv_{imp} > 0$, i.e., $H(v_{imp})$ is an increasing function, and we can solve Eq. (5.8) with respect to H:

$$H = \begin{cases} \sqrt[3]{6\hat{F}(v_{imp})} & \text{if} \quad 0 \le v_{imp} \le v_{imp0} \\ \frac{2L}{3} + \frac{2\hat{F}(v_{imp})}{L^2} & \text{if} \quad v_{imp} \ge v_{imp0} \end{cases}, \tag{5.11}$$

where v_{imp0} is the impact velocity that provides penetration of the impactor at the depth L:

$$v_{imp0} = \hat{F}^{-1}\left(\widetilde{F}(L)\right). \tag{5.12}$$

For a *sharp cone of revolution,* some equations derived in this section can be simplified using Eq. (4.12).

6. HISTORICAL AND METHODOLOGICAL BACKGROUND OF THE LOCALIZED INTERACTION APPROACH IN IMPACT DYNAMICS

The foundations of the localized interaction approach (LIA) in impact dynamics were developed from three different approaches. In the first - traditional – approach, the study of a corresponding physical phenomenon yielded some types of LIMs. The second approach employs the results obtained in gasdynamics for which the LIA was well developed. The third approach is associated with using the concepts of a localized approach (and not always recognizing them as such) directly in impact dynamics. In this section we consider these approaches in brief.

6.1 Some special localized interaction models

To the best of our knowledge, the first LIM describing bullet-barrier interaction during penetration of a projectile through a plate was suggested by Nishiwaki (1951). This model can be described as a LIM with:

$$\Omega_n = a_0 + a_2 u^2 v^2, \quad \Omega_\tau = \mu_{fr} \Omega_n, \tag{6.1}$$

where a_0 is the "statical contact pressure" and $a_2 = \gamma_{sh}$ is the density of the material of the shield. Using his experimental results for conical bullets and aluminum shields, Nishiwaki (1951) drew the conclusion that a_0 is proportional to the thickness of the perforated plate. He developed a relationship between the impact velocity and the residual velocity of a cone-shaped impactor perforating a SFT, taking into consideration the change in the impactor-plate contact surface during penetration. It is of interest to note that Nishiwaki employed the same model as did Newton (1687). This model is based upon the assumptions that the medium is rarefied and that the interaction of the projectile with the medium occurs through elastic or inelastic (as in this case) collisions of the particles of the medium with a projectile's surface. It later became apparent that this model is also applicable to dense media. Exactly the same situation is true for gasdynamics, for which a similar Newton's model (a_0 – pressure in the undisturbed gas flow) and its modifications are widely used in calculations of the aerodynamic characteristics of supersonic projectiles (Hayes and Probstein, 1959; Chernyi, 1969). Interestingly, Nishiwaki did not refer to Newton's model or to gasdynamics analogy.

Various semi-empirical models were collected and analyzed by Recht (1990). An approach based upon cavity expansion approximations, which is discussed comprehensively in *Chapter 3*, yielded a large number of LIMs. The list of LIMs may be supplemented by adding the models suggested by Landgrov and Sarkisyan (1984) and by Golubev and Medvedkin (2001). The latter model is a modification of the model of Vitman and Stepanov (1959), which takes into account the effect of viscous resistance at the initial stage of penetration. Klepaczko (2001), Jones et al. (2003) and Davis (2003) (see also Klepaczko and Hughes, 2005) proposed the use of concrete shields penetration models with a friction coefficient depending on the local tangential component of the impactor's velocity. The majority of known models employ a second order polynomial of u, v, and the most frequently used models are determined by Eq. (6.1).

A model that is based solely on the dependence between the drag force, D, and the velocity of the impactor, v, and does not consider the influence of the impactor's shape on its resistance or takes it into account through some, usually empirical, coefficients, may be classified as a "degenerate" LIM (DLIM). "Classic" DLIMs postulate a polynomial dependence $D(v)$, and solving the equation of motion in this case reduces to calculating the integral $M(1,1,c_0,c_1,c_2;W)$ in Eq. (A2.01). A brief analysis of these models and references to the early studies can be found in Goldsmith (1960) and Backman and Goldsmith (1978). Heimdahl and Schulz (1986) studied the motion of an impactor for an arbitrary function $D(v)$. A number of investigators have proposed power-law dependences for different media (Mileiko and Sarkisyan, 1981; Mileiko et al., 1994; Forrestal et al., 1984; Forrestal et al., 1986). Various approaches to determine the drag force acting on the body as a function of its velocity and penetration depth were considered by Stone (1994), Zook (1977), Beth (1946), Allen et al. (1957), Bernard (1978), and Dehn (1979, 1986, 1987).

Phenomenological or semi-phenomenological formulas for direct calculations of DOP, BLV or residual velocity seem to have no relation to the problem under consideration. However, they can be useful in developing some particular LIMs (see *Section 2-6.2*). A large number of such models can be found in Heuzé (1989), Backman and Goldsmith (1978), Brown (1986), Corbett et al. (1996), Young (1997), Project THOR (1961), Barr (1990), and Dancygier (1997, 2000). A comparative analysis of a number of known models was performed by Neilson (1985), Børvik et al. (1998), and Teland (1998). The latter survey also briefly summarized the findings of the earlier studies for concrete shields.

6.2 "Tangent cones" approaches

In all the below-described methods, the force at the location of the interaction between the projectile and the host medium is assumed to be equal to the force at the surface of the tangent cone at this location, when the projectile velocity and the host medium are the same in both cases. Different versions of such an approach are known in aerodynamics as "methods of tangent cones" (Chernyi, 1969; Hayes and Probstein, 1959).

It must be noted that such approaches are often accepted intuitively without invoking the theory. As an example, we may refer to the suggested by Bernard and Creighton (1979) "extrapolation to non-normal impact" of the expression for local compressive normal stress on projectile's surface that was employed earlier for normal penetration.

Clearly, the below-described methods alone do not guarantee the validity of the constructed models for non-conical impactors. Development of particular correct procedures for constructing a LIM (including the choice of the type of approximation) is a separate problem that has not yet been addressed in the impact dynamics context.

6.2.1 The version based on the cones resistance measurements

Vitman and Stepanov (1959) investigated experimentally the penetration of conical-nosed impactors with impact velocities up to $1000\, m/s$ into various metal shields. They found that the drag force to shank area ratio exhibits a linear dependence on the impactor's velocity squared provided that the penetration depth exceeds the length of the conical nose. Analysis of these linear correlations for penetrators with different cone angles emptied two interesting observations. The first is that the slope of these lines is approximately the same as the slope of the drag force dependence for a conical projectile moving in a gas flow calculated according to Newton's model. The second is that all the straight lines, obtained in the experiments on penetration into the shields manufactured from the same material intersect at the same point. These two observations imply that the formula for a drag force for a conical impactor may be written as:

$$D_{cone} = \pi R^2 (a_0 + a_2 \sin^2 \upsilon\, v^2), \quad \Omega_\tau = 0, \tag{6.2}$$

where $a_2 = \gamma_{sh}$, a $a_0 = \widetilde{H}_{sh}$ is the "dynamic hardness of the metal for impact velocities of $\widetilde{v} \sim 10\, m/s$". Initially, Vitman and Stepanov (1959) used the following correlation:

$$a_0 = \widetilde{H}_{sh}(v/\widetilde{v})^{\alpha}, \qquad (6.3)$$

but they subsequently decided that the dependence of a_0 on the impactor velocity can be neglected because $0 < \alpha << 1$. Although the development of a LIM describing the interaction between the medium and the impactor's surface was not the direct goal of the study of Vitman and Stepanov (1959), their formula is often referred as a LIM.

Let us now show that under several assumptions Eq. (6.2) implies Eq. (6.1) with $\Omega_\tau = 0$.

Consider the penetration of a sharp straight conical impactor into a SIS without friction for $h \geq L$. Then, under conditions of validity of some LIM and taking into account that $\Theta(h) = L$, $\theta(h) = 0$, $Lk = L\tan\upsilon = R$, Eqs. (4.8) and (4.9) yield the following formula:

$$D = \pi R^2 \Omega_n(\sin\upsilon, v). \qquad (6.4)$$

Comparing Eqs. (6.2) and (6.4) and assuming that a particular LIM without friction is valid, we arrive at the conclusion that this LIM may be expressed as follows:

$$\Omega_n(u,v) = a_0 + a_2 u^2 v^2, \quad \Omega_\tau = 0. \qquad (6.5)$$

Let us demonstrate that such method of construction of LIM can be used in a more general case. During penetration of a sharp straight cone into a SIS (as well as into a SFT), the projection of the lateral surface of the impactor inside the shield upon the plane normal to its direction of motion may be written as:

$$\Delta\sigma(h) = \pi[\Theta(h)\tan\upsilon]^2 - \pi[\theta(h)\tan\upsilon]^2. \qquad (6.6)$$

Then neglecting friction (a method can easily be generalized for a case with friction) and taking into account Eqs. (5.2) and (4.12), we can rewrite Eq. (5.1) as follows:

$$D = \Omega_n(\sin\upsilon, v)\Delta\sigma(h). \qquad (6.7)$$

Let us now assume that penetration experiments performed with different conical impactors and different impact velocities showed that the experimentally measured function $F_D = D_{exp}(\upsilon, h)/\Delta\sigma(h)$, where D_{exp} is the drag force, depends (certainly, in some approximation) on the cone apex half-angle and impactor's velocity at penetration depth h, i.e., $F_D = F_D(\upsilon, v)$. Then, for each penetrator, function F_D will assume the same values for the same penetration velocity, although this velocity in different experiments can be attained at different penetration depths. Then, equating functions $F_D(\upsilon, v)$ and $D/\Delta\sigma(h)$ and using Eq. (6.7), we arrive at the following LIM:

$$\Omega_n(u,v) = F_D(\sin^{-1} u, v), \quad \Omega_\tau = 0. \tag{6.8}$$

Note, that the mere fact that the function F_D depends solely on the penetrator's velocity and does not depend explicitly on the location of the penetrator inside the shield, can be used as a criterion of the "locality" of the model of interaction between the impactor and the shield.

6.2.2 The "local similarity" version

The second method of constructing LIMs is based on the known local model for a sharp straight circular cone. Let us consider the normal penetration of a conical impactor into a shield and assume that functions $\sigma_n(\upsilon, v)$ and $\sigma_\tau(\upsilon, v)$ in the relationship that is similar to Eq. (1.1),

$$d\vec{F} = [\sigma_n(\upsilon, v)\vec{n}^0 + \sigma_\tau(\upsilon, v)\vec{\tau}^0]ds, \quad 0 < \upsilon < \pi/2, \tag{6.9}$$

are known. Then, functions determining the LIM that is suitable for sharp impactors with other shapes are determined as:

$$\Omega_\zeta(u,v) = \sigma_\zeta(\sin^{-1} u, v), \zeta = n, \tau. \tag{6.10}$$

An example of the particular realization of this approach can be found in *Chapter 8* (see also Ben-Dor et al., 2002a,b,c).

6.2.3 The Recht's version

Let us show that the method proposed by Recht (1990) can be considered as a discrete version of the LIT for sharp bodies of revolution. For simplicity, let us neglect friction. Then, Eq. (4.7) for a drag force acting on a sharp impactor of revolution yields:

$$D(h,v) = 2\pi \int_{\theta(h)}^{\Theta(h)} \Omega_n(u,v)\Phi\Phi' dx, \tag{6.11}$$

where $u = u(x) = \Phi' / \sqrt{\Phi'^2 + 1} = \sin\tilde{\upsilon}$, and $\tilde{\upsilon} = \tilde{\upsilon}(x)$ is the angle between the tangent to the generator and the axis of the impactor. The integral in Eq. (6.11) can be approximated by a sum:

$$D(h,v) \approx 2\pi \sum_i \int_{x_i}^{x_{i+1}} \Omega_n(u,v)\Phi\Phi' dx$$
$$\approx \sum_i \sigma_n(\tilde{\upsilon}_i, v) \int_{x_i}^{x_{i+1}} \frac{d(\pi\Phi^2)}{dx} dx \approx \sum_i \sigma_n(\tilde{\upsilon}_i, v) \cdot \delta\sigma(x_i). \tag{6.12}$$

Here, as shown in *Figure 2-9*, the impactor's surface that interacts with the shield is divided into sub-areas using the planes $x = x_i$, where $\theta(h) = x_1 < x_2 < ... < x_i < x_{i+1} < ... < x_N = \Theta(h)$, and $\sigma(x_i)$ is the cross-sectional area of the impactor in the plane $x = x_i, \delta\sigma(x_i) = \sigma(x_{i+1}) - \sigma(x_i), \tilde{\upsilon}_i = \tilde{\upsilon}(x_i)$.

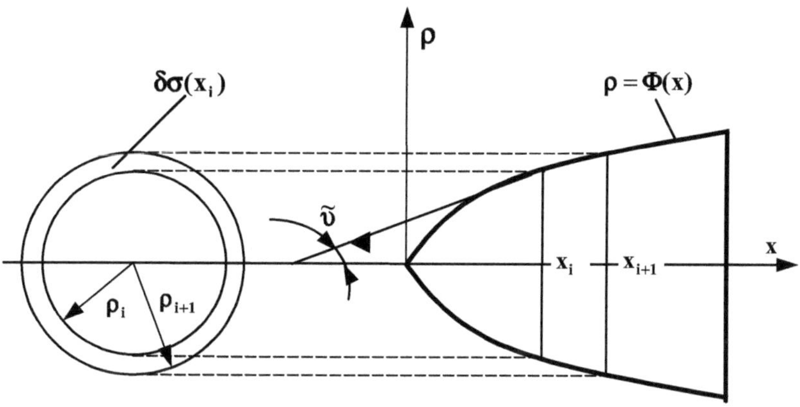

Figure 2-9. Recht's (1990) model.

Thus, if a penetrator is a body of revolution, Eq. (6.12) allows us to calculate the force applied by the shield at any location on the impactor's surface, provided the function $\sigma_n(\upsilon, v)$ is known. This function determines the shield-penetrator interaction force at any location on the surface of a straight circular cone having an apex half-angle υ and velocity v.

6.2.4 The version based on the cones depth of penetration model

A particular LIM can be constructed provided that a function that determines the penetration depth by a circular conical impactor in a SIS as a function of the impact velocity and a cone apex half-angle is known. Let us denote this function by $\hat{H}(\upsilon, v_{imp})$.

Taking into account Eqs. (5.5) and (5.9), and the relationship $k = tan\,\upsilon$ Eq. (5.8) yields:

$$\frac{m}{tan\,\upsilon}\int_0^{v_{imp}}\frac{vdv}{\breve{D}(v)} = \frac{1}{2}\int_0^H \Theta^2 dh, \tag{6.13}$$

Assumption about LIM implies that Eq. (6.13) must be an identity for $H = \hat{H}(\upsilon, v_{imp})$. Differentiating, after this substitution, both sides of this equation with respect to v_{imp}, we obtain:

$$\frac{mv_{imp}}{tan\,\upsilon\,\breve{D}(v_{imp})} = \frac{1}{2}\Theta^2\left(\hat{H}(\upsilon, v_{imp})\right)\frac{\partial\hat{H}}{\partial v_{imp}}. \tag{6.14}$$

Substitution of $\breve{D}$ from Eq. (4.11) yields:

$$\Omega_n\left(sin\,\upsilon, v_{imp}\right) = \frac{mv_{imp}}{\pi\,tan\,\upsilon(tan\,\upsilon + \mu_{fr})}\left[\Theta^2\left(\hat{H}(\upsilon, v_{imp})\right)\frac{\partial\hat{H}}{\partial v_{imp}}\right]^{-1}. \tag{6.15}$$

After the change of variables, $\upsilon = sin^{-1} u,\ v_{imp} = v$, we obtain an expression for function $\Omega_n(u,v)$ that determines the LIM:

$$\Omega_n(u,v) = \frac{mv(1-u^2)}{\pi u(u + \mu_{fr}\sqrt{1-u^2})}\left[\Theta^2\left(\hat{H}(sin^{-1}\upsilon, v)\right)\frac{\partial\hat{H}}{\partial v}\right]^{-1}. \tag{6.16}$$

A similar approach can be used for a SFT, provided that the function $\hat{v}_{bl}(\upsilon, b)$, which determines the dependence of the BLV of a straight circular conical impactor vs. impact velocity, cone apex half-angle and plate thickness, is known.

Equations (5.5) and (5.7) yield an equation similar to Eq. (6.13):

$$\frac{m}{\tan \upsilon} \int_0^{v_{bl}} \frac{v dv}{\breve{D}(v)} = \frac{L^2}{2} \hat{b}(\upsilon, v_{bl}), \tag{6.17}$$

where $\hat{b}(\upsilon, v_{bl})$ is the inverse of $\hat{v}_{bl}(\upsilon, b)$ with respect to the second variable. The latter function does exist, since the BLV is an increasing function of the thickness of the plate, i.e., $\hat{v}_{bl}(\upsilon, \hat{b}(\upsilon, v_{bl})) \equiv v_{bl}$. Then, differentiating Eq. (6.17) with respect to v_{bl}, we obtain:

$$\frac{m v_{bl}}{\tan \upsilon \, \breve{D}(v_{bl})} = \frac{L^2}{2} \frac{\partial \hat{b}(\upsilon, v_{bl})}{\partial v_{bl}}. \tag{6.18}$$

Using the same reasoning as in the case of the SIS, we arrive at the following formula that determines the model:

$$\Omega_n(u, v) = \frac{m v (1 - u^2)}{\pi u (u + \mu_{fr} \sqrt{1 - u^2}) L^2} \left[\frac{\partial \hat{b}(\sin^{-1} \upsilon, v)}{\partial v} \right]^{-1}. \tag{6.19}$$

Note that the idea to restore a model using a known DOP function was used by Eisler et al. (1998) for the simplest degenerate model with $\hat{H}(v_{imp})$, $D = D(v)$, without considering the stage of incomplete immersion of the projectile into the target.

6.3 From aerodynamics to impact dynamics

The conception and development of LIT (primarily for aerodynamic applications) have been discussed in monographs written by Bunimovich and Dubinsky (1995) and Miroshin and Khalidov (2002), as well as in the review of Bunimovich and Dubinsky (1996). Therefore, we will discuss here only briefly the principal points that are important for impact dynamics.

It must be emphasized that the "classic" LIT under discussion here deals mainly with the integral characteristics of a body moving in a fluid. Moreover, in fluid mechanics, the characteristics of the medium and motion of the flying body (including the magnitude and direction of the velocity) are considered to be known. The basic problem here is to determine aerodynamic characteristics of the body for the known values of these parameters.

In 1960s, the problem of rapid calculation of the aerodynamic characteristics of high speed flying bodies over a whole range of flight altitudes became very important. For high altitudes, at which the medium is very rarefied, a simple model could be used to describe a free molecular flow. For low altitudes, at which the atmosphere is relatively dense, various versions of Newton's model were used. The models that worked well for these two extreme situations did indeed have a theoretical foundation and turned out to be LIMs. Theoretical predictions concerning models with the same degree of simplicity for the intermediate range of altitudes were not available, a situation exacerbated by the fact that calculations for these flow conditions were the most problematic.

Barantsev et al. (1969) suggested and validated experimentally an unconventional approach to the problem. They postulated the "locality" of the required model for the intermediate range of flight altitudes and found a class of particular models that satisfied a number of requirements. The most important requirements are the simplicity, a smooth transition to the models for limiting cases of very large and very small degrees of rarefaction, and experimental validation.

Basically, the problem was to develop a unified engineering model applicable to all flight regimes and having parameters that account for the degree of rarefaction. During the first stage of LIT development (as summarized in the monograph by Alekseeva and Barantsev, 1976; see also Barantsev, 1978), these results were received with great interest, not because they employed nontrivial reasoning (from locality to a particular model) but because they constituted a breakthrough in the field of approximate aerodynamic calculations in the intermediate range of altitudes. Interestingly, quite independently, with the same goals and at the same time as Barantsev et al. (1969), local interaction approach suggested Galkin et al., 1977 (these authors claim that their study was performed in 1968). Bunimovich (1973) was the first to note that the significance of the local interaction approach is beyond the scope of rarefied gasdynamics. At present this point of view is generally accepted, and the notions like localized interaction theory, localized interaction models, etc. are widely used (see Bunimovich and Dubinsky, 1995, 1996; Miroshin and Khalidov, 1991, 2002 and references therein). Studies in the framework of the LIT were continued in different directions, such as development of particular LIMs and effective methods for aerodynamic calculations, shape optimization, etc. One of the most important directions of these investigations, which has retained its significance right up to the present, is the development of approaches that employ the locality property and do not dwell on the properties of particular local models (see, e.g., Dubinsky and Elperin, 1997; Yakunina, 2000a,b, 2002). Analysis of the development of LIT allows us to draw the following conclusions:

1) Validity of the "hypothesis of locality" and validity of a particular local model are completely different concepts. On the methodological level, it is often appropriate to consider locality as a fundamental property of an interaction between a projectile and the surrounding medium while a particular local model constitutes only one of a number of possible methods for describing this interaction. The latter approach is especially justified when there are several more or less equivalent models or when there is no any model at all. In the latter case, a hypothesis of locality can be used as one of the instruments for developing a model. Since the hypothesis of locality can be validated experimentally irrespective of the acceptance of a particular local model, a convenient two-stage procedure can be employed in which the hypothesis of locality is validated in the first stage and particular local models are devised in the second stage.

2) LIMs and their subsets have the properties that allow us to develop methods for solving various problems based on locality but not on a particular model.

3) LITs can be used for connecting and mutual cross-fertilization of different fields in mechanics. In the case under consideration, results from aerodynamics may be applied in impact dynamics. It is no accident that many authors using local or similar models in high speed penetration mechanics (HSPM) also have research experience in gasdynamics. A number of monographs (Bunimovich and Dubinsky, 1995; Vedernikov and Shchepanovsly, 1995; Vedernikov et al., 1995; Ostapenko, 1997) have separate sections dealing with application of the approximate models in both fields. However, while in aerodynamics the main field of application of the LIT is calculation and analysis of aerodynamic characteristics of projectiles during their specified motion, in HSPM there are no laws of motion for impactors, and important characteristics, such as BLV and DOP, are determined by solving equations of impactor motion. Seemingly, the direct similarity between HSPM and LIT applies only in the determination of the forces acting upon impactor. However, this connection is deeper than is demonstrated in applications for shape optimization of projectiles (see *Section 5-1*). There are also some tendencies which, at first thought, characterize only HSPM but are known from history of development of LIT. As examples, we can cite to the method suggested by Recht (1990) and the spherical cavity expansion approximation (see *Section 3-1*), which are essentially based on the hypothesis of locality. Using the LIT approach, Ben-Dor et al. (1998a) showed that lack of dependence of BLV in a spaced shield on the thicknesses of the gaps and of the plates (provided the total thickness is kept constant) during penetration by a conical impactor is a consequence of the local character of the penetrator-shield interaction, independently of the choice of a particular local model. Without the use of

the concepts of LIT even such approach to the problem would have been impossible.

Here, we do not intend to reject the application of other methods but only to draw attention to LIT which is a promising method for solving penetration problems already taking root in HSPM. In the exposition that follows, we have adopted LIT viewpoint which is also convenient for better organizing the material. In this chapter, we derived formulas for drag force, DOP and BLV for the basic LIMs used in HSPM. The derived formulas will be used in subsequent chapters.

7. ONE-TERM LOCALIZED INTERACTION MODELS

In this Section, we consider the following model:

$$\Omega_n(u,v) = A_0(u)v^{\alpha}, \quad 0 \le \alpha < 2 \tag{7.1}$$

and particular cases thereof. The most important formulas are summarized in *Table 2-3*. In the following expositions, we substantiate these formulas and some other relationships that follow from the model determined in Eq. (7.1).

Table 2-3. Formulas provided by LIM with $\Omega_n(u,v) = A_0(u)v^{\alpha}$

Function/variable	3-D impactor	3-D conical impactor	Impactor of revolution	Cone of revolution
Φ	$\Phi(x,\vartheta)$	$(k_0 + kx)\eta(\vartheta)$ Sharp cone: $k_0 = 0$	$\Phi(x)$	$k_0 + kx$
δ, θ, Θ	*Table 2-1*	*Table 2-1*	*Table 2-1*	*Table 2-1*
$\sigma, u, U, \tilde{u}$	*Table 2-2*	*Table 2-2*	*Table 2-2*	*Table 2-2*
B_0	Eq. (7.4)	Eq. (7.20)	Eq. (7.32)	
H	Eqs. (7.19), (7.16)	Eqs. (7.25), (7.28) Sharp cone: Eq. (7.30)	Eqs. (7.16), (7.19)	Eqs. (7.25), (7.28) Sharp cone: Eq. (7.30)
$v_{imp\,0}$	Eq. (7.15)	Eq. (7.29) Sharp cone: Eq. (7.31)	Eq. (7.15)	Eq. (7.29) Sharp cone: Eq. (7.31)
F_{init}	Eq. (7.13)	Eq. (7.21)	Eq. (7.33)	Eq. (7.21)
$\tilde{D}_{init}$	Eq. (7.10)	Eq. (7.27)	Eq. (7.34)	Eq. (7.27)
$e_0 \div e_5$		Eqs. (7.24), (7.26)		Eqs. (7.24), (7.26)
C_0		Eq. (7.22)		Eq. (7.35)
v_{bl}	Eq. (7.40)	Eq. (7.40)	Eq. (7.40)	Eq. (7.40)

7.1 General formulas

Equation (3.3) for function $\Omega_n(u,v)$, which is defined by Eq. (7.1), implies that:

$$D(h) = f_0(h)v^{\alpha}, \tag{7.2}$$

where

$$f_0(h) = A_0(1)\sigma(0)\delta(h) + \int_{\theta(h)}^{\Theta(h)} dx \int_0^{2\pi} B_0(x,\vartheta)\,d\vartheta, \tag{7.3}$$

$$B_0(x,\vartheta) = A_0\big(u(x,\vartheta)\big)U(x,\vartheta). \tag{7.4}$$

The equation of motion (Eq. 2.20) can be written as:

$$mv\frac{dv}{dh} + D(h) = 0. \tag{7.5}$$

The solution of this differential equation with the initial conditions given by Eq. (2.22) may be written as:

$$v^{\beta} = v_{imp}^{\beta} - \frac{\beta}{m}F_*(h), \tag{7.6}$$

where

$$F_*(h) = \int_0^h f_0(\hat{h})\,d\hat{h}, \tag{7.7}$$

$$\beta = 2 - \alpha. \tag{7.8}$$

7.2 Semi-infinite shield

7.2.1 3-D impactor

Equations (2.1) and (2.15) are valid in the case of a 3-D impactor penetrating a SIS, and Eq. (7.2) can be written separately for the range $0 \le h \le L$ (initial stage, non-complete immersion of the impactor into the target) and the range $h > L$ (full immersion) in the following form:

$$f_0(h) = \begin{cases} \widetilde{D}_{init}(h) \text{ if } 0 \le h \le L \\ \widetilde{D}_{init}(L) \text{ if } h \ge L \end{cases}, \tag{7.9}$$

where

$$\widetilde{D}_{init}(h) = A_0(1)\sigma(0) + \int_0^h dx \int_0^{2\pi} B_0(x,\vartheta)\, d\vartheta . \tag{7.10}$$

Then, the equation for determination of DOP, H, can be obtained by substituting $h = H$ and $v = 0$ into Eq. (7.6):

$$F(H) = \frac{m}{2} v_{imp}^2 . \tag{7.11}$$

Equations (7.7)-(7.10) allow us to write the following expression for $F(h)$:

$$F(h) = \begin{cases} F_{init}(h) \text{ if } 0 \le h \le L \\ F_{full}(h) \text{ if } h \ge L \end{cases}, \tag{7.12}$$

$$F_{init}(h) = A_0(1)\sigma(0)h + \int_0^h d\hat{h} \int_0^{\hat{h}} dx \int_0^{2\pi} B_0(x,\vartheta) d\vartheta , \tag{7.13}$$

$$F_{full}(h) = \int_0^L \widetilde{D}(\hat{h})d\hat{h} + \int_L^h \widetilde{D}(\hat{h})d\hat{h} = F_{init}(L) + (h-L)\widetilde{D}_{init}(L)$$

$$= A_0(1)\sigma(0)h + \int_0^L d\hat{h} \int_0^{\hat{h}} dx \int_0^{2\pi} B_0(x,\vartheta)d\vartheta \tag{7.14}$$

$$+ (h-L)\int_0^L dx \int_0^{2\pi} B_0(x,\vartheta)d\vartheta .$$

Equations (7.11) and (7.12) show that the impact velocity providing the penetration of the impactor to the depth L, $v_{imp\,0}$, is:

$$v_{imp\,0} = \left[\frac{\beta}{m}F_{init}(L)\right]^{1/\beta} . \tag{7.15}$$

Then, Eq. (7.11) can be written, depending on the impact velocity, in one of two possible forms:

$$F_{init}(H) - \frac{m}{\beta}v_{imp}^{\beta} = 0 \quad \text{if} \quad v_{imp} \le v_{imp\,0} \tag{7.16}$$

or

$$F_{full}(H) - \frac{m}{\beta}v_{imp}^{\beta} = 0 \quad \text{if} \quad v_{imp} \ge v_{imp\,0} \,. \tag{7.17}$$

Since $\Xi(H) = F_{init}(H) - (m/\beta)v_{imp}^{\beta}$ is an increasing function, and $\Xi(0) = -(m/\beta)v_{imp}^{\beta} < 0$ and $\Xi(L) = (m/\beta)(v_{imp\,0}^{\beta} - v_{imp}^{\beta}) \ge 0$, then Eq. (7.16) has only one root, but generally the solution cannot be obtained in a closed form. On the contrary, the exact solution of Eq. (7.17) can be easily written. By substituting $F_{full}(H)$ from Eq. (7.14) and $F_{init}(L)$ from Eq. (7.15), Eq. (7.17) can be written as:

$$\frac{m}{\beta}v_{imp\,0}^{\beta} + (H-L)\widetilde{D}_{init}(L) - \frac{m}{\beta}v_{imp}^{\beta} = 0, \tag{7.18}$$

and the desired solution of Eq. (7.18) is:

$$H = L + \frac{m(v_{imp}^{\beta} - v_{imp\,0}^{\beta})}{\beta \widetilde{D}_{init}(L)}, \quad v_{imp} \geq v_{imp\,0}. \tag{7.20}$$

7.2.2 3-D conical impactor

In the case of a 3-D conical impactor penetrating a SIS, B_0 in Eq. (7.4) may be expressed as:

$$B_0(x,\vartheta) = A_0(u(\vartheta))\widetilde{u}(\vartheta)(k_0 + kx),$$

where $u(\vartheta)$ and $\widetilde{u}(\vartheta)$ are given in *Table 2-2*. Substituting Eq. (7.20) into Eq. (7.13) and calculating the integrals over x and $\hat{h}$, we obtain:

$$F_{init}(h) = A_0(1)\sigma(0)h + C_0 \int_0^h d\hat{h} \int_0^{\hat{h}} (k_0 x + k) dx$$
$$= A_0(1)\sigma(0)h + \frac{C_0 h^2}{6}(kh + 3k_0), \tag{7.21}$$

where $\sigma(0)$ is determined in *Table 2-2* and:

$$C_0 = \int_0^{2\pi} A_0(u(\vartheta))\widetilde{u}(\vartheta) d\vartheta. \tag{7.22}$$

Equation (7.16) thus takes the form:

$$H^3 + e_2 H^2 + e_1 H - e_0 = 0 \tag{7.23}$$

with

$$e_0 = \frac{6 m v_{imp}^{\beta}}{C_0 k \beta}, \; e_1 = \frac{6 A_0(1)\sigma(0)}{C_0 k}, \; e_2 = \frac{3k_0}{k}. \tag{7.24}$$

The solution of this algebraic equation of the third order, i. e., the expression for DOP in the case $v_{imp} \leq v_{imp\,0}$, can be written using Cardano's formulas (Korn and Korn, 1968), and the solution of Eq. (7.16) is:

$$H = \sqrt[3]{\sqrt{e_5} - 0.5 e_4} - \sqrt[3]{\sqrt{e_5} + 0.5 e_4} - e_2/3 \quad if \quad v_{imp} \leq v_{imp\,0}, \tag{7.25}$$

where

$$e_5 = \left(\frac{e_3}{3}\right)^3 + \left(\frac{e_4}{2}\right)^2, e_4 = \frac{2}{27}e_2^3 - \frac{1}{3}e_1 e_2 - e_0, e_3 = e_1 - \frac{1}{3}e_2^2. \quad (7.26)$$

Substituting Eq. (7.20) into Eq. (7.10) and calculating the integrals over $\hat{h}$, we obtain:

$$\widetilde{D}_{init}(h) = A_0(1)\sigma(0) + C_0 h(0.5kh + k_0). \quad (7.27)$$

Using Eqs. (7.15), (7.21) and (7.27), we can rewrite Eq. (7.19) as follows:

$$H = \frac{(2/\beta)mv_{imp}^{\beta} + C_0 L^2 (k_0 + 2/3\,kL)}{2A_0(1)\sigma(0) + C_0 L(2k_0 + kL)}, v_{imp} \geq v_{imp\,0}, \quad (7.28)$$

where

$$v_{imp\,0} = \left\{\frac{L[6A_0(1)\sigma(0) + C_0 L(kL + 3k_0)]\beta}{6m}\right\}^{1/\beta}. \quad (7.29)$$

Equations (7.25) and (7.28) allow us to calculate the DOP of 3-D conical impactor into a SIS. In the case of *a sharp impactor*, $k_0 = 0, \sigma(0) = 0$, and Eqs. (7.25), (7.28) and (7.29) yield:

$$H = \begin{cases} \sqrt[3]{\dfrac{6mv_{imp}^{\beta}}{C_0 k\beta}} & \text{if } v_{imp} \leq v_{imp\,0} \\ \dfrac{2}{3}L + \dfrac{2mv_{imp}^{\beta}}{C_0 kL^2\beta} & \text{if } v_{imp} \geq v_{imp\,0} \end{cases}, \quad (7.30)$$

where

$$v_{imp\,0} = \left(\frac{C_0 kL^3\beta}{6m}\right)^{1/\beta}. \quad (7.31)$$

7.2.3 Impactor of revolution

In the case of an impactor of revolution penetrating a SIS, Eqs. (7.16) and (7.19) for calculating H and Eq. (7.15) for v_{imp0} are valid where:

$$B_0(x) = A_0\left(\frac{\Phi'}{\sqrt{\Phi'^2+1}}\right)\Phi(\Phi'+\mu_{fr}), \tag{7.32}$$

$$F_{init}(h) = \pi[A_0(1)\Phi^2(0)h + 2\int_0^h d\hat{h}\int_0^{\hat{h}} B_0(x)dx], \tag{7.33}$$

$$\widetilde{D}_{init}(h) = \pi[A_0(1)\Phi^2(0) + 2\int_0^h B_0(x)dx]. \tag{7.34}$$

Equations (7.32)-(7.34) are obtained from Eqs. (7.4), (7.10) and (7.13) using *Table 2-2*.

7.2.4 Cone of revolution

In the case of a cone of revolution penetrating a SIS, the formulas for a 3-D conical impactor remain valid, with a simplified Eq. (7.22) for C_0:

$$C_0 = 2\pi A_0\left(\frac{k}{\sqrt{k^2+1}}\right)(k+\mu_{fr}) \tag{7.35}$$

7.3 Shield with a finite thickness

Substituting $v=0, v_{imp}=v_{bl}, h=b+L$ into Eq. (7.6) and taking into account Eq. (7.2), we obtain the equation for determining the BLV for 3-D impactor:

$$\frac{m}{\beta}v_{bl}^{\beta} = A_0(1)\sigma(0)\int_0^{b+L}\delta(h)dh + \int_0^{b+L}dh\int_{\theta(h)}^{\Theta(h)}\widetilde{\psi}(x)dx, \tag{7.36}$$

where

$$\widetilde{\psi}(x) = \int_{0}^{2\pi} B_0(x,\vartheta)d\vartheta, \tag{7.37}$$

and $B_0(x,\vartheta)$ is defined by Eq. (7.4).

Using the definition of $\delta(h)$ given by Eq. (2.16), we calculate the first integral in the left-hand side of Eq. (7.36) as:

$$\int_{0}^{b+L} \delta(h)dh = \int_{0}^{b} \delta(h)dh + \int_{b}^{b+L} \delta(h)dh = b. \tag{7.38}$$

The second integral can be transformed using Eq. (2.7):

$$\int_{0}^{b+L} dh \int_{\theta(h)}^{\Theta(h)} \widetilde{\psi}(x)dx = b\int_{0}^{L} \widetilde{\psi}(x)dx. \tag{7.39}$$

Substituting Eqs. (7.38) and (7.39) into Eq. (7.36) we obtain:

$$v_{bl} = \left[\frac{\beta b}{m} \widetilde{D}_{init}(L) \right]^{1/\beta}, \tag{7.40}$$

where function $\widetilde{D}_{init}(h)$ is defined by Eq. (7.10).

Equation (7.40) also remains valid in the cases of 3-D conical impactors and impactors having the shape of a body of revolution, where function $\widetilde{D}_{init}(h)$ is determined by Eq. (7.27) and Eq. (7.34), respectively.

8. TWO-TERM LOCALIZED INTERACTION MODELS

In this section, we consider the following model:

$$\Omega_n(u,v) = A_2(u)v^2 + A_0(u)v^{\alpha}, \quad 0 \le \alpha < 2. \tag{8.1}$$

The derived final expressions are summarized in *Table 2-4* and *Table 2-5*.

8.1 General solution for a tree-dimensional impactor

Taking into account Eq. (8.1), Eq. (3.3) yields ($i = 0,2$):

$$D(h,v) = f_2(h)v^2 + f_0(h)v^{\alpha}, \tag{8.2}$$

where

$$f_i(h) = A_i(1)\sigma(0)\delta(h) + \int_{\theta(h)}^{\Theta(h)} dx \int_0^{2\pi} A_i(u(x,\vartheta))U(x,\vartheta)d\vartheta. \tag{8.3}$$

The equation of motion (Eq. 2.20) becomes:

$$m\frac{dv}{dh} + f_2(h)v + f_0(h)v^{\alpha-1} = 0. \tag{8.4}$$

This is an ordinary differential equation of the Bernoulli type. Using the change of variable (Kamke, 1959):

$$v = w^{1/\beta}, \tag{8.5}$$

where, as before, $\beta = 2 - \alpha$, Eq. (8.4) can be transformed into a linear ordinary differential equation:

$$\frac{dw}{dh} + \frac{\beta}{m}[f_2(h)w + f_0(h)] = 0. \tag{8.6}$$

The solution of this equation with the initial condition

$$w(0) = v_{imp}^{\beta} \tag{8.9}$$

reads:

$$w(h) = v^{\beta}(h) = \frac{1}{Q(h)}\left[v_{imp}^{\beta} - \frac{\beta}{m}\int_0^h f_0(\tilde{h})Q(\tilde{h})d\tilde{h}\right], \tag{8.8}$$

where

$$Q(h) = exp\left(\frac{\beta}{m}\int_0^h f_2(\zeta)d\zeta\right). \tag{8.9}$$

In the case of a SIS, Eq. (8.8) yields a formula for the DOP, H:

$$\int_0^H f_0(h)Q(h)\,dh = \frac{m}{\beta} v_{imp}^{\beta}, \tag{8.10}$$

where Eq. (2.15) must be taken into account.

8.2 Shield with a finite thickness. General model

8.2.1 3-D impactor

In the case of a SFT, the BLV, v_{bl}, can be calculated from Eq. (8.8) by substituting $h = b + L, v(h) = 0$ and $v_{imp} = v_{bl}$:

$$v_{bl}^{\beta} = \frac{\beta}{m} \int_0^{b+L} f_0(h)Q(h)\,dh. \tag{8.11}$$

This case will be considered below in more detail. Assuming that $v_{imp} \geq v_{bl}$, we can write Eq. (8.8) for the residual velocity, $v_{res} = v(b+L)$:

$$v_{res}^{\beta} = \frac{1}{T}\left[v_{imp}^{\beta} - \frac{\beta}{m} \int_0^{b+L} f_0(h)Q(h)\,dh \right], \tag{8.12}$$

where

$$T = Q(b+L). \tag{8.13}$$

Eqs. (8.11) and (8.12) imply the following simple relationship between the impact velocity, residual velocity and BLV:

$$\left(\frac{v_{imp}}{v_{bl}}\right)^{\beta} - T\left(\frac{v_{res}}{v_{bl}}\right)^{\beta} = 1, \quad v_{imp} \geq v_{bl}. \tag{8.14}$$

8.2.2 3-D conical impactor

Taking into account the formulas from *Table 2-2*, we obtain ($i = 0,2$):

$$f_i(h) = A_i(1)\sigma(0)\delta(h) + C_i \int_{\theta(h)}^{\Theta(h)} (k_0 + kx)dx = \frac{k_0^2}{2} A_i(1)\delta(h) \int_0^{2\pi} \eta^2(\vartheta)d\vartheta$$
$$+ 0.5C_i[\Theta(h) - \theta(h)]\{2k_0 + k[\Theta(h) + \theta(h)]\}, \tag{8.15}$$

where

$$C_i = \int_0^{2\pi} A_i(u(\vartheta))\tilde{u}(\vartheta)d\vartheta, \tag{8.16}$$

and $u(\vartheta)$ and $\tilde{u}(\vartheta)$ are determined by equations in *Table 2-2* for 3-D conical impactor.

For a *cone of revolution*, Eq. (8.16) becomes:

$$C_i = 2\pi A_i\left(\frac{k}{\sqrt{k^2 + 1}}\right)(k + \mu_{fr}). \tag{8.17}$$

In the case of a *sharp 3-D conical impactor*, $k_0 = 0$ and $\sigma(0) = 0$. Then, Eq. (8.15) implies that:

$$f_i(h) = 0.5C_i k[\Theta^2(h) - \theta^2(h)] = C_i k \int_{\theta(h)}^{\Theta(h)} x dx. \tag{8.18}$$

Since

$$\frac{dQ}{dh} = \frac{\beta}{m} f_2(h)Q(h) = \frac{\beta C_2}{mC_0} f_0(h)Q(h), \tag{8.19}$$

Eq. (8.11) can be written as:

$$v_{bl}^{\beta} = \frac{C_0}{C_2} \int_0^{b+L} \frac{dQ}{dh} dh = \frac{C_0}{C_2}[Q(b+L) - Q(0)]. \tag{8.20}$$

Using Eq. (2.7) the expression for $T = Q(b+L)$ can be transformed into the following form:

$$T = exp\left(\frac{\beta C_2 k}{m} \int_0^{b+L} dh \int_{\theta(h)}^{\Theta(h)} x dx \right) = exp\left(\frac{\beta C_2 kbL^2}{2m} \right). \tag{8.21}$$

Then Eq. (8.20) can be written as:

$$v_{bl}^{\beta} = \frac{C_0}{C_2}[T - 1] = \frac{C_0}{C_2}\left[exp\left(\frac{\beta C_2 kbL^2}{2m} \right) - 1 \right]. \tag{8.22}$$

Since Eq. (8.19) implies that

$$\frac{\beta}{m} \int_0^{b+L} f_0(h) Q(h) dh = \frac{C_0}{C_2} \int_0^{b+L} \frac{dQ}{dh} dh = \frac{C_0}{C_2}[T - 1], \tag{8.23}$$

Eq. (8.12) takes the form:

$$v_{res}^{\beta} = \frac{1}{T}\left[v_{imp}^{\beta} - \frac{C_0}{C_2}(T - 1) \right]. \tag{8.24}$$

8.2.3 Impactor of revolution

Taking into account the formulas in *Table 2-2*, Eq. (8.3) yields ($i = 0,2$):

$$\frac{f_i(h)}{\pi} = A_i(1) r^2 \delta(h) + 2 \int_{\theta(h)}^{\Theta(h)} A_i\left(\frac{\Phi'}{\sqrt{\Phi'^2 + 1}} \right) \Phi(\Phi' + \mu_{fr}) dx \tag{8.25}$$

and the other equations for a 3-D impactor remain valid.

Table 2-4. Formulas provided by LIM with $\Omega_n(u,v) = A_2(u)v^2 + A_0(u)v^{\alpha}$

Function/variable	3-D impactor	3-D conical impactor	Impactor of revolution
Φ	$\Phi(x,\vartheta)$	$(k_0 + kx)\eta(\vartheta)$	$\Phi(x)$
	Sharp impactor: $\sigma(0) = 0$	Sharp cone: $k_0 = 0$	Sharp body: $r = 0$
δ, θ, Θ	*Table 2-1*	*Table 2-1*	*Table 2-1*
$\sigma, u, \tilde{u}, U$	*Table 2-2*	*Table 2-2*	*Table 2-2*
H	Eq. (8.10)		
f_0, f_2	Eq. (8.3)	Eq. (8.15)	Eq. (8.25)

Function/variable	3-D impactor	3-D conical impactor	Impactor of revolution
Q	Eq. (8.9)	Eq. (8.9)	Eq. (8.9)
v_{bl}	Eq. (8.11)	Eq. (8.11) Sharp cone: Eq. (8.22)	Eq. (8.11)
v_{res}	Eq. (8.12)	Eq. (8.12) Sharp cone: Eq. (8.24)	Eq. (8.12)
T	Eq. (8.13)	Eq. (8.13) Sharp cone: Eq. (8.21)	Eq. (8.13)
C_0, C_2		Eq. (8.16) Cone of revolution: Eq. (8.17)	

8.3 Generalized "Newton's model"

The most widely used model of the type determined by Eq. (8.1) may be expressed as follows:

$$\Omega_n(u,v) = a_2 u^2 v^2 + a_0 v^\alpha, \tag{8.26}$$

i. e.,

$$A_0(u) = a_0 v^\alpha, \quad A_2(u) = a_2 u^2. \tag{8.27}$$

Although commonly used values of the constants are $\alpha = 0$ and $\beta = 2 - \alpha = 2$, we still consider the case with $0 \le \alpha < 2$, since it does not complicate the obtained formulas or the exposition.

Transition from the model determined by Eq. (8.1) to a particular model, i.e., transition to the model determined by Eq. (8.26), is performed by the choice of f_0 and f_2 in Eq. (8.3) for non-conical strikers or by choosing particular expressions for coefficients C_0 and C_2 in Eq. (8.16) for conical impactors.

In the case of a *3-D impactor*, the following formulas must be used instead of Eq. (8.3):

$$\frac{f_0(h)}{a_0} = \sigma(0)\delta(h) + \int_{\theta(h)}^{\Theta(h)} dx \int_0^{2\pi} (\Phi\Phi_x + \mu_{fr}\sqrt{\Phi^2 + \Phi_\vartheta^2}) d\vartheta, \tag{8.28}$$

$$\frac{f_2(h)}{a_2} = \sigma(0)\delta(h) + \int_{\theta(h)}^{\Theta(h)} dx \int_0^{2\pi} \frac{\Phi^2\Phi_x^2(\Phi\Phi_x + \mu_{fr}\sqrt{\Phi^2 + \Phi_\vartheta^2})}{\Phi^2(\Phi_x^2 + 1) + \Phi_\vartheta^2} d\vartheta. \tag{8.29}$$

For a 3-D conical impactor, expressions for C_i in Eqs. (8.16) assume the following form:

$$C_0 = a_0 \int_0^{2\pi} (k\eta^2 + \mu_{fr}\sqrt{\eta^2 + \eta'^2}\,)d\vartheta, \tag{8.30}$$

$$C_2 = a_2 k^2 \int_0^{2\pi} \frac{\eta^4 (k\eta^2 + \mu_{fr}\sqrt{\eta^2 + \eta'^2}\,)}{\eta^2 (k^2\eta^2 + 1) + \eta'^2} d\vartheta. \tag{8.31}$$

If an impactor's nose has the shape of a *body of revolution*, then instead of Eq. (8.25), the following formulas must be used:

$$\frac{f_0(h)}{\pi a_0} = r^2\,\delta(h) + 2 \int_{\theta(h)}^{\Theta(h)} \Phi(\Phi' + \mu_{fr})dx\,, \tag{8.32}$$

$$\frac{f_2(h)}{\pi a_2} = r^2\,\delta(h) + 2 \int_{\theta(h)}^{\Theta(h)} \frac{\Phi\Phi'^2(\Phi' + \mu_{fr})}{\Phi'^2 + 1} dx\,. \tag{8.33}$$

In the case of a *cone of revolution*, Eqs. (8.30) and (8.31) yield:

$$C_0 = 2\pi a_0 (k + \mu_{fr}), \quad C_2 = \frac{2\pi a_2 k^2 (k + \mu_{fr})}{k^2 + 1}. \tag{8.34}$$

Formulas for considered model are summarized in *Table 2-5*.

Table 2-5. Formulas provided by LIM with $\Omega_n(u,v) = a_2 u^2 v^2 + a_0 v^\alpha$ for SFT

Function/variable	3-D impactor	3-D conical impactor	Impactor of revolution
Φ	$\Phi(x,\vartheta)$ Sharp impactor: $\sigma(0)=0$	$(k_0 + kx)\eta(\vartheta)$ Sharp cone: $k_0 = 0$	$\Phi(x)$ Sharp body: $r = 0$
δ,θ,Θ	*Table 2-1*	*Table 2-1*	*Table 2-1*
$\sigma,u,\tilde{u}$	*Table 2-2*	*Table 2-2*	*Table 2-2*
H	Eq. (8.10)		
f_0, f_2	Eqs. (8.28), (8.29)	Eq. (8.15)	Eqs. (8.32), (8.33)
Q	Eq. (8.9)	Eq. (8.9)	Eq. (8.9)
v_{bl}	Eq. (8.11)	Eq. (8.11) Sharp cone: Eq. (8.22)	Eq. (8.11)

Function/variable	3-D impactor	3-D conical impactor	Impactor of revolution
v_{res}	Eq. (8.12)	Eq. (8.12) Sharp cone: Eq. (8.24)	Eq. (8.12)
T	Eq. (8.13)	Eq. (8.13) Sharp cone: Eq. (8.21)	Eq. (8.13)
C_0, C_2		Eqs. (8.30), (8.31) Cone of revolution: Eq. (8.34)	

9. A CLASS OF AVERAGED LOCALIZED INTERACTION MODELS

The models considered in this section are derived from the initial LIM using averaging over the penetration path, h (but not over the impactor's velocity). Actually, this procedure results in substitution of constant integration limits instead of variable integration limits in the expression for the drag force [$\theta(h)$ is replaced by 0 for SFT and $\Theta(h)$ is replaced by L] and in the appearance of the coefficient in the case of a SFT.

9.1 General localized interaction model for three-dimensional impactor

We start from Eqs. (3.3), (3.4) and (3.5) that determine the drag force of a 3-D impactor D as a function of its instantaneous coordinate h and velocity v.

9.1.1 Shield of finite thickness

Averaging $D(h,v)$ over h $(0 \le h \le b+L)$, we obtain the following expression for the average drag force, D_{av}:

$$D_{av}(v) = \frac{1}{b+L}\int_0^{b+L} D(h,v)dh = \frac{\Omega_n(1,v)\sigma(0)}{b+L}\int_0^{b+L}\delta(h)dh$$
$$+\frac{1}{b+L}\int_0^{b+L}dh\int_{\theta(h)}^{\Theta(h)}dx\left[\int_0^{2\pi}\Omega_n\left(u(x,\vartheta),v\right)U(x,\vartheta)d\vartheta\right], \tag{9.1}$$

where $u(x,\vartheta)$ and $U(x,\vartheta)$ are determined by Eqs. (3.5) and (3.4). Taking into account the definition of the function $\delta(h)$ and using Eq. (2.7), we obtain:

$$D_{av}(v)=\frac{b}{b+L}D_0(v), \tag{9.2}$$

where

$$D_0(v)=\Omega_n(1,v)\sigma(0)+\int_0^L dx\int_0^{2\pi}\Omega_n(u,v)[\Phi\Phi_x+\mu_{fr}\sqrt{\Phi^2+\Phi_\vartheta^2}\,]d\vartheta \tag{9.3}.$$

Since

$$\begin{aligned}\int_0^L dx\int_0^{2\pi}\Phi\Phi_x d\vartheta&=\int_0^L dx\int_0^{2\pi}\frac{1}{2}\cdot\frac{\partial\Phi^2}{\partial x}d\vartheta\\ &=\int_0^L dx\frac{1}{2}\frac{\partial}{\partial x}\int_0^{2\pi}\Phi^2 d\vartheta=\int_0^L\frac{d\sigma(x)}{dx}dx=\sigma(L)-\sigma(0),\end{aligned} \tag{9.4}$$

Eq. (9.3) can be rewritten, after substituting $\sigma(0)$ from Eq. (9.4), in the following form, which is more convenient when optimization problems are investigated:

$$\begin{aligned}&D_0(v)=\Omega_n(1,v)\sigma(L)\\ &+\int_0^L dx\int_0^{2\pi}\{[\Omega_n(u,v)-\Omega_n(1,v)]\Phi\Phi_x+\mu_{fr}\Omega_n(u,v)\sqrt{\Phi^2+\Phi_\vartheta^2}\,\}d\vartheta.\end{aligned} \tag{9.5}$$

The equation of impactor motion (Eq. 2.20) can be simplified:

$$mv\frac{dv}{dh}+\frac{b}{b+L}D_0(v)=0. \tag{9.6}$$

Then

$$m\int_{v_{bl}}^{0}\frac{vdv}{D_0(v)}=-\frac{b}{b+L}\int_0^{b+L}dh, \tag{9.7}$$

and we obtain the equation for determining BLV:

$$\frac{m}{b}\int_0^{v_{bl}}\frac{vdv}{D_0(v)}=1. \tag{9.8}$$

9.1.2 Semi-infinite shield

Averaging $D(h,v)$ over h ($0\leq h\leq H$) where H, as before, is the DOP, we obtain the expression for the average drag force, $\widetilde{D}_{av}$:

$$\widetilde{D}_{av}(v)=\frac{1}{H}\int_0^H D(h,v)dh=\frac{1}{H}\Omega_n(1,v)\sigma(0)\int_0^H dh$$
$$+\frac{1}{H}\int_0^H dh\int_0^{\Theta(h)}dx\left[\int_0^{2\pi}\Omega_n(u,v)Ud\vartheta\right]. \tag{9.9}$$

Using the identity given by Eq. (2.3), we obtain:

$$\widetilde{D}_{av}(v)=\Omega_n(1,v)\sigma(0)+\int_0^{\Theta(H)}\left(1-\frac{x}{H}\right)dx\left[\int_0^{2\pi}\Omega_n(u,v)Ud\vartheta\right]. \tag{9.10}$$

If $H>>L$, then $\Theta(H)=L$, $x<<H$ and the second integral in the right-hand side of Eq. (9.10) can be simplified:

$$\widetilde{D}_{av}(v)=D_0(v). \tag{9.11}$$

The equation of the impactor's motion may be written as:

$$mv\frac{dv}{dh}+D_0(v)=0. \tag{9.12}$$

Then

$$m\int_{v_{imp}}^0\frac{vdv}{D_0(v)}=-\int_0^H dh, \tag{9.13}$$

and we arrive at the following expression for DOP:

$$H = m \int_0^{v_{imp}} \frac{vdv}{D_0(v)} . \tag{9.14}$$

9.2 Three-term localized-interaction models

9.2.1 General three-term model

Here, the term "general three-term model" denotes the following model:

$$\Omega_n(u,v) = A_2(u)v^2 + A_1(u)v + A_0(u) . \tag{9.15}$$

Since we now consider the general *3-D impactor*, $u(x,\vartheta)$ is determined by Eqs. (3.5). Substituting Eq. (9.15) into Eqs. (9.3) and (9.5), we obtain:

$$D_0(v) = B_2 v^2 + B_1 v + B_0 \tag{9.16}$$

where $B_i \ (i = 0,1,2)$ are determined by one of two possible formulas, depending on which equation, Eq. (9.3) or Eq. (9.5), is used:

$$B_i = A_i(1)\sigma(0) + \int_0^L dx \int_0^{2\pi} A_i(u)[\Phi\Phi_x + \mu_{fr} A_i(u)\sqrt{\Phi^2 + \Phi_\vartheta^2}\,]d\vartheta$$

$$= A_i(1)\sigma(L) + \int_0^L dx \int_0^{2\pi} \{[A_i(u) - A_i(1)]\Phi\Phi_x + \mu_{fr} A_i(u)\sqrt{\Phi^2 + \Phi_\vartheta^2}\,\}d\vartheta . \tag{9.17}$$

Then the BLV and the DOP can be found from Eq. (9.8) and Eq. (9.14), respectively:

$$\frac{m}{b} M(1,1,B_0,B_1,B_2;v_{bl}) = 1 , \tag{9.18}$$

$$H = mM(1,1,B_0,B_1,B_2;v_{imp}) , \tag{9.19}$$

where

$$M(1,1,B_0,B_1,B_2;W) = \int_0^W \frac{\zeta d\zeta}{B_2\zeta^2 + B_1\zeta + B_0} . \tag{9.20}$$

The latter integral can be expressed through elementary functions, and the corresponding formulas are presented in *Appendix 2*.

The choice of the type of striker affects only the formulas for B_i, and these formulas can be simply obtained using *Table 2-2*. In particular, in the case of an *impactor of revolution*:

$$\frac{B_i}{\pi} = A_i(1)r^2 + 2\int_0^L A_i(u)\Phi(\Phi' + \mu_{fr})dx$$
$$= A_i(1)R^2 + 2\int_0^L \Phi\{[A_i(u) - A_i(1)]\Phi' + \mu_{fr}A_i(u)\}dx, \tag{9.21}$$

where $u = \Phi' / \sqrt{\Phi'^2 + 1}$.

9.2.2 Three-term sub-model

For this model:

$$\Omega_n(u,v) = a_2u^2v^2 + a_1uv + a_0, \tag{9.22}$$

i. e., when $A_i(u) = a_iu^i, i = 0,1,2$, Eq. (9.17) for a *3-D impactor* implies:

$$\frac{B_i}{a_i} = \sigma(0) + \int_0^L dx \int_0^{2\pi} \frac{\Phi^i\Phi_x^i(\Phi\Phi_x + \mu_{fr}\sqrt{\Phi^2 + \Phi_\vartheta^2})}{[\Phi^2(\Phi_x^2 + 1) + \Phi_\vartheta^2]^{i/2}} d\vartheta$$
$$= \sigma(L) + \int_0^L dx \int_0^{2\pi} \Phi\Phi_x \left\{ \frac{\Phi^{i-1}\Phi_x^{i-1}(\Phi\Phi_x + \mu_{fr}\sqrt{\Phi^2 + \Phi_\vartheta^2})}{[\Phi^2(\Phi_x^2 + 1) + \Phi_\vartheta^2]^{i/2}} - 1 \right\} d\vartheta. \tag{9.23}$$

In particular, the formula for B_0 is quite simple:

$$\frac{B_0}{a_0} = \sigma(L) + \mu_{fr} \int_0^L dx \int_0^{2\pi} \sqrt{\Phi^2 + \Phi_\vartheta^2}\, d\vartheta. \tag{9.24}$$

In the case of an *impactor of revolution*:

$$\frac{B_i}{\pi a_i} = r^2 + 2\int_0^L \frac{\Phi\Phi'^i(\Phi' + \mu_{fr})}{(\Phi'^2 + 1)^{i/2}} dx$$
$$= R^2 + 2\int_0^L \Phi\Phi'\left[\frac{\Phi'^{i-1}(\Phi' + \mu_{fr})}{(\Phi'^2 + 1)^{i/2}} - 1\right]dx \tag{9.25}$$

with

$$\frac{B_0}{\pi a_0} = R^2 + 2\mu_{fr}\int_0^L \Phi dx. \tag{9.26}$$

The formulas associated with the three-term model are summarized in *Table 2-6*.

Table 2-6. Formulas provided by three-term averaged LIMs

Function / Variable	$\Omega_n(u,v) = A_2(u)v^2 + A_1(u)v + A_0(u)$		$\Omega_n(u,v) = a_2u^2v^2 + a_1uv + a_0$	
	3-D impactor	Impactor of revolution	3-D impactor	Impactor of revolution
Φ	$\Phi(x,\vartheta)$ Sharp impactor: $\sigma(0)=0$	$\Phi(x)$ Sharp body: $r=0$	$\Phi(x,\vartheta)$ Sharp impactor: $\sigma(0)=0$	$\Phi(x)$ Sharp body: $r=0$
u	Eq. (3.5)	*Table 2-2*	Eq. (3.5)	*Table 2-2*
$\sigma(x)$	*Table 2-1*	*Table 2-2*	*Table 2-1*	*Table 2-2*
B_i, $i=0,1,2$	Eq. (9.17)	Eq. (9.21)	Eqs. (9.23), (9.24)	Eqs. (9.25), (9.26)
v_{bl}	Eq. (9.18)	Eq. (9.18)	Eq. (9.18)	Eq. (9.18)
H	Eq. (9.19)	Eq. (9.19)	Eq. (9.19)	Eq. (9.19)
$M(...)$	Eq. (9.20)	Eq. (9.20)	Eq. (9.20)	Eq. (9.20)

9.3 Two-term localized-interaction models

9.3.1 General two-terms model

Let us consider the model:

$$\Omega_n(u,v) = A_2(u)v^2 + A_0(u) \tag{9.27}$$

and *3-D impactor*. In this case, Eq. (9.17) for B_0 and B_2 and Eqs. (9.16) and (9.18)-(9.19) with $B_1 = 0$, are valid. The integral in Eq. (9.20) can be simplified:

$$M(1,1,B_0,0,B_2;W) = \int_0^W \frac{\zeta d\zeta}{B_2\zeta^2 + B_0}, \tag{9.28}$$

and the result is given by Eq. (A2.14). Then, Eqs. (9.18) and (9.19) yield:

$$v_{bl} = \sqrt{\frac{B_0}{B_2}\left[exp\left(\frac{2bB_2}{m}\right) - 1\right]}, \tag{9.29}$$

$$H = \frac{m}{2B_2} ln\left(1 + \frac{B_2}{B_0} v_{imp}^2\right). \tag{9.30}$$

In the case of an *impactor of revolution*, B_0 and B_2 are determined by Eq. (9.21).

9.3.2 Two-term sub-model

In this case, the model is determined by the following formula:

$$\Omega_n(u,v) = a_2 u^2 v^2 + a_0, \tag{9.31}$$

i. e., $A_i(u) = a_i u^i, i = 0,2$ in Eq. (9.27).

For a *3-D impactor*, Eqs. (9.29) and (9.30) remain valid, with B_0 and B_2 determined by Eqs. (9.23)-(9.24).

In the case of an *impactor of revolution,* B_0 and B_2 are determined by Eqs. (9.25)-(9.26).

Formulas associated with the two-terms model are summarized in *Table 2-7.*

Table 2-7. Formulas provided by two-term averaged LIMs

Function/ variable	$\Omega_n(u,v) = A_2(u)v^2 + A_0(u)$		$\Omega_n(u,v) = a_2 u^2 v^2 + a_0$	
	3-D impactor	Impactor of revolution	3-D impactor	Impactor of revolution
Φ	$\Phi(x,\vartheta)$	$\Phi(x)$	$\Phi(x,\vartheta)$	$\Phi(x)$
	Sharp impactor: $\sigma(0)=0$	Sharp body: $r=0$	Sharp impactor: $\sigma(0)=0$	Sharp body: $r=0$
u	Eq. (3.5)	*Table 2-2*	Eq. (3.5)	*Table 2-2*

Function/ variable	$\Omega_n(u,v) = A_2(u)v^2 + A_0(u)$		$\Omega_n(u,v) = a_2 u^2 v^2 + a_0$	
	3-D impactor	Impactor of revolution	3-D impactor	Impactor of revolution
$\sigma(x)$	*Table 2-1*	*Table 2-2*	*Table 2-1*	*Table 2-2*
B_0, B_2	Eq. (9.17)	Eq. (9.21)	Eqs. (9.23), (9.24)	Eqs. (9.25), (9.26)
v_{bl}	Eq. (9.29)	Eq. (9.29)	Eq. (9.29)	Eq. (9.29)
H	Eq. (9.30)	Eq. (9.30)	Eq. (9.30)	Eq. (9.30)

Chapter 3

CAVITY EXPANSION APPROXIMATIONS

1. INTRODUCTION TO THE CAVITY EXPANSION APPROXIMATIONS

1.1 Spherical cavity expansion approximation

The spherical cavity expansion approximation (SCEA) in a quasi-static version is widely used for constructing impactor-shield interaction models for SISs. In these models, expansion of a spherically symmetrical cavity from a zero initial radius at a constant velocity is considered by means of some continuum mechanics model of the material. Let the solution of this problem be represented in the form:

$$p = \widehat{\omega}(\vec{a}; \dot{y}), \tag{1.1}$$

where p is the stress at the boundary of the cavity, y is the radius of the hole, and $\widehat{\omega}$ is some function. Then, it is assumed that the normal stress at the surface of the impactor moving in the same medium is given by the following formula:

$$\sigma_n = \widehat{\omega}(\vec{a}; uv). \tag{1.2}$$

The latter assumption implies that the normal stress caused by host medium-penetrator interaction at the impactor's surface at some location moving with the instantaneous normal velocity $v_n = v \cos \hat{\nu} = uv$ is equal to the stress at the boundary of the cavity that expands with constant velocity v_n. Clearly, a quasi-static SCEA implies a LIM with $\Omega_n = \widehat{\omega}(\vec{a}; uv)$ in Eq. (2-1.1). Friction between impactor and shield can be taken into account by using Eq. (2-3.1).

1.2 Cylindrical cavity expansion approximation

Another widely used approach is known as the cylindrical cavity expansion approximation (CCEA) (model, method, etc.). Sometimes other names are used, e.g., the method of plane sections (Sagomonyan, 1960; Rakhmatulin et al., 1964) and the disks model (Yankelevsky and Adin, 1980). The CCEA has been applied to modeling penetration into SISs and perforation of SFTs. In this approach, normal penetration of a slender body of revolution is usually considered, and it is assumed that particles of the material of the shield move in a radial direction during penetration by the impactor. The shield is divided into infinitely thin layers, and in each layer a cavity expansion caused by the moving impactor is modeled. This approach facilitates calculating the stress at the boundary of the hole in each layer and, consequently, the force acting at the penetrator at each location on the impactor's lateral surface. The common technique for applying the CCEA to penetration mechanics may be described as follows. The solution of a "dynamical problem" of hole expansion with time is usually represented for each layer as:

$$p = \breve{\omega}(\vec{a}; y, \dot{y}, \ddot{y}) . \tag{1.3}$$

Let $\rho = \Phi(x)$ be the equation of the surface of the impactor (a body of revolution). Then, for the layer with the coordinate ξ (see *Figure 3-1*), the condition that the boundary of the hole coincides with the boundary of the impactor (Rakhmatulin et al., 1964) reads:

$$y = \Phi(h - \xi) . \tag{1.4}$$

Differentiation of Eq. (1.4) yields:

$$\dot{y} = \Phi'(x)\dot{h}, \ddot{y} = \Phi''(x)\dot{h}^2 + \Phi'(x)\ddot{h}, \quad x = h - \xi . \tag{1.5}$$

The expression for the normal stress at the surface of the impactor, $\sigma_n = p$, is obtained by substituting $y, \dot{y}$ and $\ddot{y}$ from Eqs. (1.4) and (1.5) into Eq. (1.3). Since σ_n is a function of $x, \dot{h} = v$, and $\ddot{h} = v\,dv/dh$, the equation

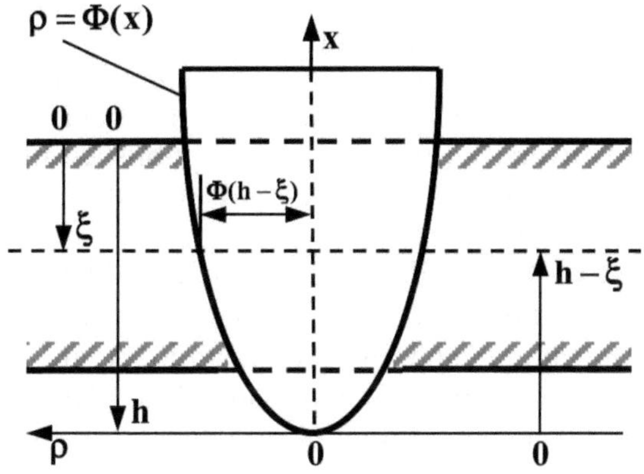

Figure 3-1. The cylindrical cavity expansion approximation.

of motion of an impactor in the case of CCEA is similar in structure to Eq. (2-2.20).

In the case of a "quasi-static" model—when Eq. (1.1) instead of Eq. (1.3) is assumed to be valid—the CCEA yields:

$$\sigma_n = \hat{\omega}(\vec{a};\Phi' v) = \hat{\omega}(\vec{a};\frac{uv}{\sqrt{1-u^2}}) \equiv \tilde{\omega}(\vec{a};u,v), \quad (1.6)$$

i.e., the CCEA is reduced to a LIM with $\Omega_n = \tilde{\omega}$.

In the following exposition, we comprehensively discuss applications of the CCEA to particular models.

1.3 Cavity expansion approximation models in penetration mechanics

Bishop et al. (1945) pioneered the application of cavity expansion models in penetration mechanics. They obtained solutions describing the quasi-static expansion of cylindrical and spherical cavities in an infinite medium from zero initial radius and used these solutions to determine the forces acting at a conical impactor. A survey of the state-of-the-art up to the late 1950s concerning the dynamic expansion of cavities in solids was written by Hopkins (1960). Useful information on this topic is summarized in the monograph of Yu (2000), which consists of two parts, namely, "Fundamental Solutions" and "Geotechnical Applications". Cavity expansion models applied to penetration mechanics have been described and analyzed by Teland (1999) and Satapathy (1997). Recent studies directly associated with the application of cavity expansion methods in modeling ballistic impact are surveyed below. The most intensive research in this field has been conducted at Moscow State University (the results were published mainly in Russian) and Sandia Research Laboratories, and, as was noted by Isbell et al. (1992), some of the results obtained are similar.

Investigations that were performed at Moscow State University in the 1950s and at the beginning of 1960s in the field of soil dynamics were summarized in the monograph by Rakhmatulin et al. (1964), in which penetration modeling occupies an important place. The authors described dynamic solutions for the expansion of cylindrical and spherical cavities in soil. The problem of impactor penetration into a SIS (soil) was solved by applying the developed cylindrical cavity expansion models. Solutions were found for the DOP, which took into account incomplete immersion of the impactor in the shield at the initial stage of penetration.

The results of subsequent investigations, mainly associated with the development and application of the cavity expansion approach (CEA) were summarized in the monographs of Sagomonyan (1974, 1988). A dynamic CCEA was used to model perforation of a metal plate and a brittle plate by a sharp impactor; a technique for obtaining an analytical solution for the BLV was outlined, and some calculations for cone-nosed impactors were performed. In an investigation of the penetration of an impactor with plane bluntness into an elastic-plastic shield with plug formation, the penetration phenomenon was considered as two simultaneous processes, namely, expansion of the cavity in the shield and motion of the plug. A model was developed to describe the expansion of a cylindrical cavity inside an elastic-plastic medium, starting from non-zero radius. Sagomonyan (1975) noted that the CCEA may be generalized to layered shields.

Extensive studies of CEAs in penetration dynamics were performed at the Sandia Research Laboratories by Forrestal and his colleagues. They proposed a large variety of spherical and cylindrical cavity expansion models for materials with different mechanical properties. The distinct features of their approach are described below. Simple "quasi-static" two-term or three-term models for engineering applications were developed by using theoretical studies of cavity expansion problems, numerical simulations, and experimental investigations. The friction coefficient was often introduced into a model to account for the tangent component of the impactor-shield interaction force. A limited number of impactor nose shapes (cone, sphere and ogive) were considered comprehensively, although the approach is applicable to a wide class of bodies of revolution. Forrestal and coauthors did not consider separately the stage of penetration with incomplete immersion of the impactor when normal penetration was modeled.

Two-term SCEAs,

$$\hat{\omega}(\vec{a};\dot{y}) = a_0 + a_2\dot{y}^2, \tag{1.7}$$

have been developed for concrete shields (Luk and Forrestal, 1987, 1989), shields manufactured from elastic-plastic materials (Forrestal and Luk, 1988; Forrestal et al., 1995), strain-hardening materials (Luk et al., 1991; Forrestal et al., 1981b, 1991; Forrestal and Luk, 1992a; Piekutowski et al., 1996) and soils (Forrestal and Luk, 1992b). Three-term SCEAs,

$$\hat{\omega}(\vec{a};\dot{y}) = a_0 + a_1\dot{y} + a_2\dot{y}^2, \tag{1.8}$$

have been proposed for concrete shields (Forrestal and Tzou, 1997) and metal shields taking into account strain hardening, compressibility, and strain-rate effects (Warren and Forrestal, 1998). The coefficients a_i in Eqs. (1.7) and (1.8) depend, generally, on the mechanical properties of the material of the shield and are determined experimentally or obtained from numerical simulations.

Development of the approaches based on CEA accounts for some additional features of penetration. In the studies of Littlefield et al. (1997), Partom (1996), and Teland and Sjøl (2000), cavity expansion models were modified to account for the finite size of the shield in the direction normal to the direction of penetration. Warren and Poormon (2001), Warren et al. (2001) and Longcope et al. (1999) took into account the influence of the free surface. For this purpose, in the developed spherical cavity expansion models it was assumed that radial stress vanishes at some sphere that represents the "free surface". A spherical cavity expansion technique was also developed by Macek and Duffey (2000) to take into account near-surface effects and layering.

Cylindrical cavity expansion models have been developed for geological shields (Forrestal et al., 1981a, 1987; Forrestal and Longcope, 1982; Forrestal, 1983, 1986; Longcope and Forrestal, 1981, 1983; Norwood and Sears, 1982), metal shields (Forrestal et al., 1981b, 1990; Rosenberg and Forrestal, 1988; Luk and Amos, 1991; Warren, 1999) and concrete shields (Forrestal et al., 1988a).

Some two-term CCEAs and SCEAs were summarized and compared by Forrestal et al. (1988b)—using experimental data—for conical-, spherical and ogive-nosed projectiles. Brown et al. (2003) and Teland (2002) described computer tools which are based on some cavity expansion models. Analysis of various CEAs using experimental data and benchmark calculations were undertaken by a number of researchers (Sjøl and Teland, 2000; Sjøl et al., 2002; Teland and Moxnes, 2003; Børvik et al., 2004). Forrestal and Longcope (1990), Satapathy and Bless (1996, 2000), Kartuzov et al.(1999, 2002), Satapathy (2001) and Mastilovic and Krajcinovic (1999a,b) also applied cavity expansion analysis to brittle materials. Aptukov (1991a,b) and Aptukov et al. (1992) took into account the influence of the free surface, "spherical layering" of the medium, and temperature effects, and obtained the solution for the one-dimensional problem of the expansion of a spherical cavity into a compressible elastic-plastic medium. Some engineering approximations of dynamic spherical cavity expansion solutions in an elastic-plastic medium applied to penetration problems have been obtained. Kravchenko et al. (1994) used a combined approach, including the "method of plane sections" and cylindrical (spherical) cavity expansion models, in modeling penetration by a solid body of revolution into soils and rocks. Bashurov et al. (1994) used a three-term SCEA, given by Eq. (1.8), in modeling penetration into concrete, metal, ice and geological media.

Luk and Forrestal (1987, 1989) developed a model to estimate the DOPs and the forces acting at the surface of ogive- and spherical-nose projectiles penetrating into a concrete SIS. The model was based on a quasi-static SCEA. Forrestal et al. (1988a) proposed a cylindrical cavity expansion model and an iterative procedure for determining its parameters. Forrestal et al. (1994) suggested the following two-stage penetration model. In the first stage of the penetration, the resistance force was given by:

$$D=\tilde{c}\,h, \quad 0\leq h\leq \kappa R, \tag{1.9}$$

where $\tilde{c}$ is a constant and $\kappa=4$. In the second stage ($h\geq \kappa R$), it was assumed that the spherical cavity expansion model (Forrestal and Luk, 1992b) described by Eq. (1.7) is valid, with $a_0=s\mu_8$, $a_2=\gamma_{sh}$, where μ_8 was considered as a mechanical characteristic of concrete similar to the unconfined compressive strength (for details see Forrestal et al., 2003), and dimensionless empirical constant s could be calculated from experimental data. The constant $\tilde{c}$ was determined by using the condition of continuity of the resistance force at $h=\kappa R$. This model yielded an explicit solution for the DOP. More recently, Forrestal et al. (1996) and Frew et al. (1998) showed that s can be considered as a function of μ_8, and they plotted the corresponding curve. Frew et al. (1998) suggested the approximation $s=82.6\,\mu_8^{-0.544}$ (μ_8 in MPa), but Forrestal et al. (2003) preferred to interpret

a_0 as "the measure of the shield resistance". Forrestal and Tzou (1997) used approximation of the type described by Eq. (1.8) for the second step of the penetration.

Although Forrestal and Luk (1992b) developed their model for ogive-shaped impactors, the model may easily be generalized to arbitrary bodies of revolution and projectiles with plane bluntness (see *Chapter 2*). The corresponding formulas for cone, truncated-ogive and segmental-spherical noses were derived by Chen and Li (2002). Qian et al. (2000) introduced an empirical constant to take into account the truncation effect of the ogive-nose projectile. In some studies (Qian et al., 2000; Gomez and Shukla, 2001), it was proposed that κ in Eq. (1.9) should be considered as an empirical constant. Teland and Sjøl (2004) suggested their version of the model with $\kappa = L/R$ and conducted a number of experiments that validated application of two-step models to describe penetration of flat-nosed projectiles into a concrete shield. Li and Tong (2003), Chen et al. (2004), Chen and Li (2003) generalized the model, taking plug formation into account.

An elastic-cracked model based on the SCEA was developed by Xu et al. (1997). Yankelevsky (1997) suggested a two-stage model for concrete slab penetration/perforation in which the disks model (Yankelevsky and Adin, 1980) was used. Gomez and Shukla (2001) extended the model of Forrestal and Luk (1992b) to multiple impacts. To this end, an empirical coefficient that is a function of the number of impacts was introduced into the formula derived by Forrestal and Luk (1992b). On the basis of the same model, Choudhury et al. (2002) and Siddiqui et al. (2002) derived expressions for the DOP in a buried shield and applied sensitivity analysis to study the influence of various random parameters on projectile reliability and shield safety. Chen and Li (2002) and Li and Chen (2002, 2003) performed a dimensional analysis of analytical perforation models for concrete, metal and soil and concluded that two dimensionless parameters would suffice to describe the DOP with reasonable accuracy.

Frew et al. (2000) and Forrestal and Hanchak (2002) proposed that the model of Forrestal and Luk (1992b) could be applied to a limestone shield.

Modeling of penetration into a shield with a pre-drilled cavity has also attracted the attention of researchers in the field. Yankelevsky (1983a) developed a model based upon his own version of CCEA. To determine the DOP, Murphy (1984), Folsom (1987) and Mostert (1999) modified models developed for homogeneous shields. A better substantiated approach takes into account the influence of the pre-drilled cavity on the contact surface between an impactor and a shield during penetration. The area and the shape of this contact surface affect the drag force. Teland (2001a,b) suggested an SCEA-based model and noted that a similar approach had been developed independently by Szendrei (2000). Although all these studies used some particular models describing impactor-barrier interaction, the LIT can be generalized to describe this situation.

Clearly, the majority of CEA models used are LIMs, and methods used for LIMs can be directly applied in these cases. In this chapter, we also discuss CCEA models of the type determined by Eq. (1.3) that are not reducible to LIMs and the applications of these models.

2. CYLINDRICAL CAVITY EXPANSION APPROXIMATION MODEL FOR A BLUNT IMPACTOR AGAINST AN ELASTIC-PLASTIC SHIELD

2.1 Introduction

The procedure of constructing the penetration model that employs the CCE concept involves several steps, namely, modeling of the cavity expansion (step 1), formulation of the equation of motion of the impactor (step 2), and determining the desired characteristics (the BLV, the maximum DOP) (step 3). Clearly, the models developed in step 1 with the purpose of obtaining analytical formulas for the stress at the wall of the cavity are approximate. Further simplifications are often introduced in step 2 to obtain a linear differential equation with constant coefficients for the squared relative velocity of the impactor. Such equation is obtained by assuming that the impactor is completely immersed into the host medium. However, analytical solutions of the equation in step 3 can often be determined in the integral form for CCE models without using the simplifications described above and/or some additional approximations. Moreover, for conical sharp impactors, these solutions can be expressed through elementary functions, which is very convenient for the analysis and applications.

To illustrate the procedure of constructing a penetration model on the basis of CCEA, we have chosen the case in which: *(i)* CCEA is not simplified, so as to represent it as a LIM, *(ii)* non-complete immersion into the host medium at corresponding stages is taken into account, *(iii)* impactors (bodies of revolution) with an arbitrary generator are considered, *(iv)* impactors with a flat bluntness are allowed. The latter condition implies a problem that is more involved than that for sharp impactors, because more complicated solutions must be used in step 1 and because the resistance of the flat part of the impactor's nose and plugging (for a shield with a finite thickness) must be taken into account. This problem was formulated and applied to shields with a finite thickness by Sagomonyan (1988), who considered hole expansion from the initial non-zero radius (step 1) and outlined the procedure for modeling the motion of the impactor. Here, we have extended this approach, and have determined the closed-form solutions for the BLV by using the hole expansion model without any additional simplifications. On the basis of these solutions, we will later investigate some problems associated with spaced shields.

2.2 Cavity expansion equation

Consider the high-speed normal penetration of a rigid body of revolution into a shield with thickness b. The notations are shown in *Figure 3-2a-b*.

As usual, the coordinate h, the depth of the penetration, is defined as the distance between the nose of the impactor and the front surface of the target. The coordinates x, ρ are associated with the impactor, function $\rho = \Phi(x)$ determines generator of the nose of the impactor having bluntness with the radius r. The impactor may also have a cylindrical part that does not make a contribution to its resistance. The part of the lateral surface of the nose of the impactor that interacts with the target can be described by Eq. (2-2.5).

Following Sagomonyan (1988), we consider an elastic-plastic shield and use a CCEA that implies the solution for the hole expansion from time $t = 0$:

$$p = \begin{cases} p_{elast} & if \quad r \le y \le (\mu_1 + 1)r \\ p_{plast} & if \quad y \ge (\mu_1 + 1)r \end{cases}, \tag{2.1}$$

where

$$p_{elast} = \frac{\mu_9}{(1+\mu_{10})r}(y - r), \tag{2.2}$$

$$p_{plast} = 0.5\gamma_{sh}\varphi_2(y)\dot{y}^2 + 0.5\gamma_{sh}\varphi_1(y)y\ddot{y} + \mu_{11}\varphi_0(y), \tag{2.3}$$

$$\varphi_0(y) = \varphi_1(y) + 1, \tag{2.4}$$

$$\varphi_1(y) = ln\left[\frac{(\mu_2 + 1)(y^2 - r^2)}{y^2}\right], \tag{2.5}$$

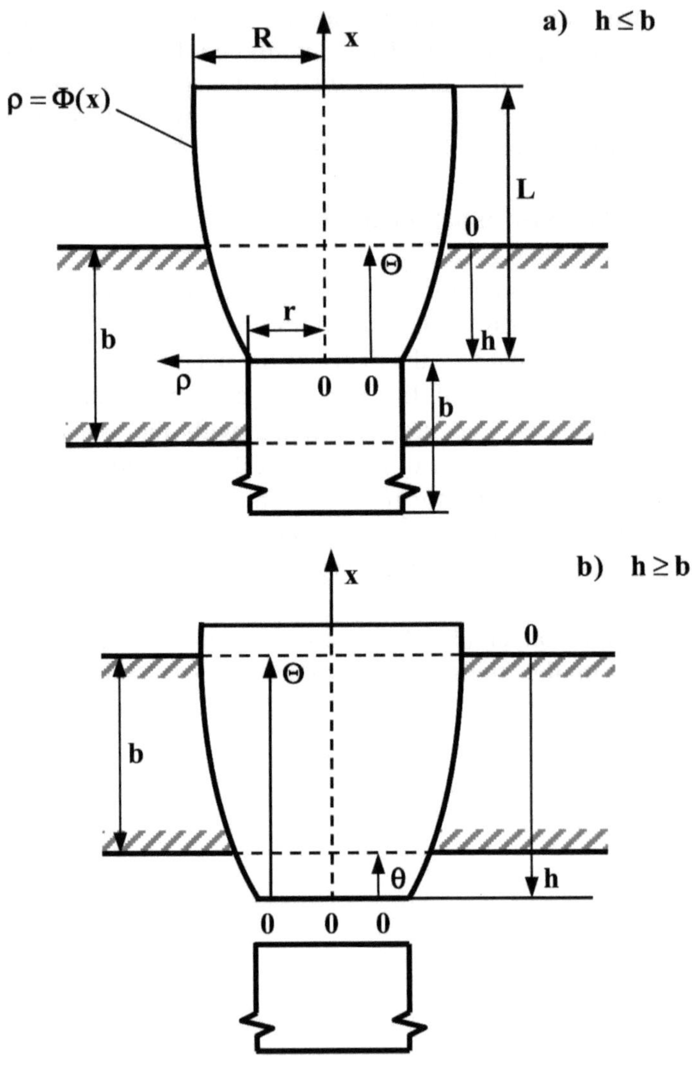

Figure 3-2. Coordinates and notations. The first stage (a) and the second stage (b) of penetration.

$$\varphi_2(y)=\varphi_1(y)-\frac{\mu_2}{\mu_2+1}+\frac{r^2}{(\mu_2+1)(y^2-r^2)}, \tag{2.6}$$

$$\mu_1=\frac{(1+\mu_{10})\mu_{11}}{\mu_9}, \quad \mu_2=\frac{1}{\mu_1(\mu_1+2)}, \tag{2.7}$$

y is the radius of the hole that varies from the initial radius r, p is, as before, a stress at the boundary of the cavity, p_{elast} and p_{plast} are associated with the elastic and plastic regimes at the boundary of the hole, respectively; $\mu_9, \mu_{10}, \mu_{11}$ and γ_{sh} are Young's modulus, Poisson's ratio, shear strength and mass density of the material of the shield, respectively. In order to simplify the algebra, friction is neglected.

2.3 Penetration model

Equations (2.1)-(2.3) can be rewritten using Eqs. (1.4) and (1.5) for $\sigma_n = p$ as follows:

$$\sigma_n=\begin{cases}\sigma_{n\,elast} & if \quad r\le\Phi(x)\le(\mu_1+1)r\\ \sigma_{n\,plast} & if \quad \Phi(x)\ge(\mu_1+1)r\end{cases}, \tag{2.8}$$

$$\sigma_{n\,elast}=\frac{\mu_9}{(1+\mu_{10})r}[\Phi(x)-r], \tag{2.9}$$

$$\begin{aligned}\sigma_{n\,plast}&=0.5\gamma_{sh}[\varphi_2(\Phi(x))\Phi'^2+\varphi_1(\Phi(x))\Phi\Phi'']\dot{h}^2\\&+0.5\gamma_{sh}\varphi_1(\Phi(x))\Phi\Phi'\ddot{h}+\mu_{11}\varphi_0(\Phi(x)).\end{aligned} \tag{2.10}$$

Then, the expression for the drag force acting at the impactor's lateral surface, D_{imp}, reads:

$$D_{imp}(h,\dot{h},\ddot{h})=2\pi\int_{\theta(h)}^{\Theta(h)}\sigma_n\Phi\Phi' dx. \tag{2.11}$$

For a blunt impactor, we use a simplified model of plugging (see, e.g., Sagomonyan, 1988; Holt et al., 1993, Chen and Davies, 1997; Goldsmith and Finnegan, 1971) assuming that the cylindrical plug of the radius r, height b and mass

$$m_{plug}=\pi\gamma_{sh}r^2 b \tag{2.12}$$

is formed at the beginning of penetration, and that the plug moves together with the impactor, while $h \le b$ (the first stage, see *Figure 3-2a*). The formula for the resistance force associated with the plug during this stage of the penetration reads:

$$D_{plug} = 2\pi\mu_{11} r(b-h). \tag{2.13}$$

When $h = b$, the impactor and the plug break apart, and the impactor with mass m moves under the action of force D_{imp} (the second stage, see *Figure 3-2b*). Therefore, equations describing the motion of the impactor read:

$$(m + m_{plug})\ddot{h} + D_{imp}(h,\dot{h},\ddot{h}) + D_{plug}(h) = 0, \quad 0 \le h < b, \tag{2.14}$$

$$m\ddot{h} + D_{imp}(h,\dot{h},\ddot{h}) = 0, \quad b \le h \le b + L. \tag{2.15}$$

Considering h as an independent variable

$$w = v^2 = \dot{h}^2 \tag{2.16}$$

and taking into account Eqs. (2.12) and (2.13) and identity:

$$\ddot{h} = \frac{d}{dt}\sqrt{w} = \frac{d}{dh}\sqrt{w}\frac{dh}{dt} = \frac{1}{2\sqrt{w}}\frac{dw}{dh}\sqrt{w} = \frac{1}{2}\frac{dw}{dh}, \tag{2.17}$$

we can rewrite Eqs. (2.14) and (2.15) as follows:

$$\frac{1}{2}[m + \pi\gamma_{sh} r^2 b]\frac{dw}{dh} + D_{imp}\left(h,\sqrt{w},\frac{1}{2}\frac{dw}{dh}\right) + 2\pi\mu_{11} r(b-h) = 0, 0 \le h < b, \tag{2.18}$$

$$\frac{1}{2}m\frac{dw}{dh} + D_{imp}\left(h,\sqrt{w},\frac{1}{2}\frac{dw}{dh}\right) = 0, \; b \le h < b + L. \tag{2.19}$$

Let $w = w_2(h)$ be the solution of Eq. (2.19) with the boundary condition $w(b+L) = 0$ and $w = w_1(h)$ be the solution of Eq. (2.18) with the boundary condition $w_1(b) = w_2(b)$. Then the BLV is determined as $v_{bl}^2 = w_{bl} = w_1(0)$.

Equations. (2.18) and (2.19) can be replaced by one equation:

$$\frac{1}{2}[m+\pi\gamma_{sh}r^2 b\delta(h)]\frac{dw}{dh}+D_{imp}\left(h,\sqrt{w},\frac{1}{2}\frac{dw}{dh}\right)+2\pi\mu_{11}r(b-h)\delta(h)=0\,, \tag{2.20}$$

where function $\delta(h)$ is determined by Eq. (2-2.16). The BLV is determined as $v_{bl}^2 = w_{bl} = w(0)$, where $w = w(h)$ is the solution of Eq. (2.20) with the boundary condition $w(b+L)=0$. Strictly speaking, the correctness of this change of variables calls for proof, because $\delta(h)$ is a non-smooth function. In the cases considered below, such proof can be obtained directly by comparing the solutions.

3. SOLUTION FOR AN ARBITRARY BODY OF REVOLUTION

3.1 Elastic response

Equation (2.8) shows that the elastic mode of deformation of the shield at the boundary of the hole occurs only if the following condition is satisfied:

$$(\mu_1+1)r \geq R \equiv \Phi(L)\ . \tag{3.1}$$

Since $\sigma_n = \sigma_{n\,elast}$, Eqs. (2.9) and (2.11) imply that:

$$D_{imp} = \frac{2\pi\mu_9}{(1+\mu_{10})r}\int_{\theta(h)}^{\Theta(h)}\varphi(x)dx, \quad \varphi(x)=[\Phi(x)-r]\Phi\Phi'. \tag{3.2}$$

Then Eq. (2.20) can be represented in the following form:

$$dw = -\widetilde{\Psi}(h)dh\,, \tag{3.3}$$

where

$$\widetilde{\Psi}(h)=4\pi\frac{\dfrac{\mu_9}{(1+\mu_{10})r}\displaystyle\int_{\theta(h)}^{\Theta(h)}\varphi(x)dx+\mu_{11}r(b-h)\delta(h)}{m+\pi\gamma_{sh}r^2 b\delta(h)}\,. \tag{3.4}$$

Integrating Eq. (3.3) over w from $w=0$ to $w=w_{bl}=w(0)$ and over h from $h=b+L$ to $h=0$, we obtain:

$$w_{bl}=e_6\int_0^b dh\int_{\theta(h)}^{\Theta(h)} dx\varphi(x)+\frac{2}{b^2}e_8\int_0^b (b-h)dh+e_7\int_b^{b+L} dh\int_{\theta(h)}^{\Theta(h)} dx\varphi(x), \quad (3.5)$$

where

$$e_6=\frac{4\pi\mu_9}{(1+\mu_{10})(m+\pi\gamma_{sh}r^2b)r}, e_7=\frac{4\pi\mu_9}{(1+\mu_{10})mr}, e_8=\frac{2\pi b^2\mu_{11}r}{m+\pi\gamma_{sh}r^2b}. \quad (3.6)$$

Introducing the function

$$\psi(z)=\int_0^z dh\int_{\theta(h)}^{\Theta(h)} dx\varphi(x) \quad (3.7)$$

and calculating the second integral in the right-hand side of Eq. (3.5), we may rewrite the latter equation as:

$$\begin{aligned} w_{bl}&=e_6\psi(b)+e_8+e_7[\psi(b+L)-\psi(b)] \\ &=e_8+e_7\psi(b+L)-(e_7-e_6)\psi(b). \end{aligned} \quad (3.8)$$

The expression for the function $\psi(z)$ can be simplified by changing the order of integration in the integrals. If $b\le L$, (see *Figure 3-3a*) then:

$$\psi(b)=\int_0^b dx\varphi(x)\int_x^b dh=\int_0^b (b-x)\varphi(x)dx . \quad (3.9)$$

If $b\ge L$ (see *Figure 3-3b*), then:

$$\psi(b)=\int_0^L dx\varphi(x)\int_x^b dh=\int_0^L (b-x)\varphi(x)dx . \quad (3.10)$$

Finally, we obtain (see the identity given by Eq. 2-2.7):

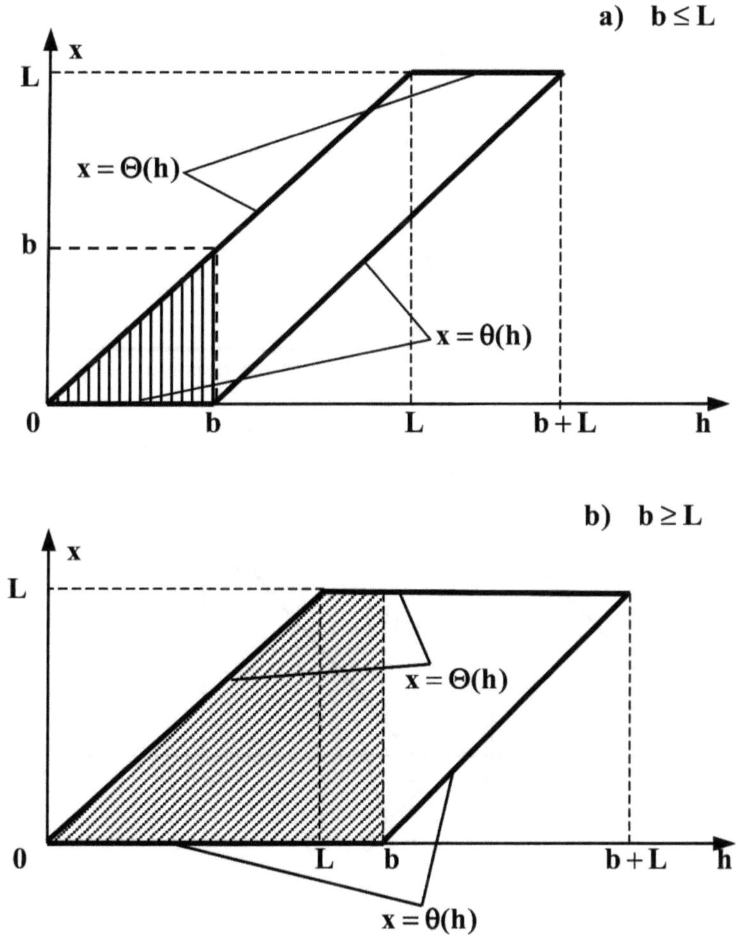

Figure 3-3. Possible structures of the domain on h,x plane in description of impactor-shield interaction.

$$\psi(b+L) = \int_0^L dx\varphi(x) \int_x^{b+x} dh = b\int_0^L \varphi(x)dx \ . \tag{3.11}$$

Therefore,

$$w_{bl} = e_8 + e_7 b\int_0^L \varphi(x)dx - (e_7 - e_6)\int_0^{min(b,L)} (b-x)\varphi(x)dx \, . \tag{3.12}$$

3.2 Elastic-plastic response

When $(\mu_1 + 1)r < R$, Eq. (2.8) implies that both elastic and plastic regimes occur during penetration. The corresponding areas on the plane h,x are shown in *Figure 3-4*.

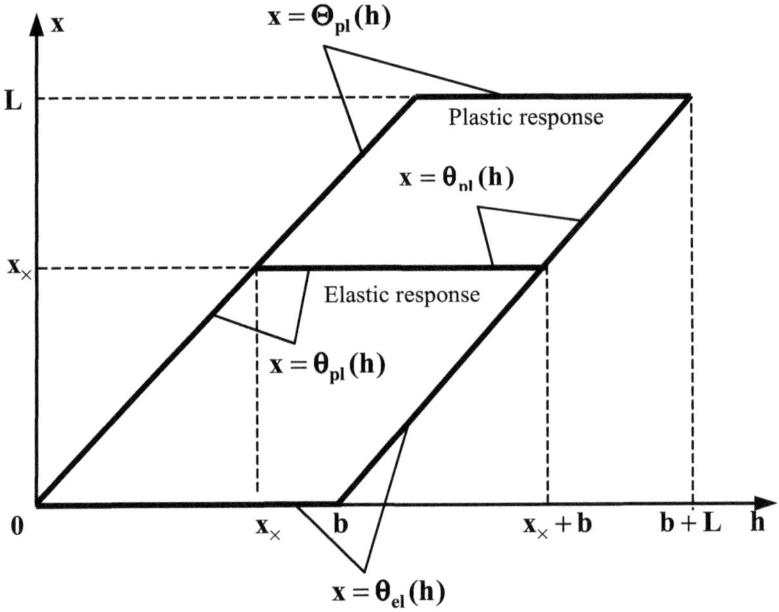

Figure 3-4. Subdomains of impactor-shield interaction on h,x plane in the case of elastic-plastic regime.

The domain of the elastic region is bounded by the lines $x = 0$, $x = x_\times$, $x = h$ and $x = h - b$, where

$$x_\times = \Phi^{-1}((\mu_1 + 1)r). \tag{3.13}$$

This domain can be described as follows:

$$\theta_{el}(h) \le x \le \Theta_{el}(h), \quad 0 \le h \le b + x_\times, \tag{3.14}$$

where

$$\theta_{el}(h)=x_{\times}X_{\Diamond}\left(\frac{h-b}{x_{\times}}\right),\quad \Theta_{el}(h)=x_{\times}X_{\Diamond}\left(\frac{h}{x_{\times}}\right) \tag{3.15}$$

and function $X_{\Diamond}(z)$ is defined by Eq. (2-2.10).

The domain of the plastic region is bounded by the lines $x=x_{\times}$, $x=L$, $x=h$ and $x=h-b$, and can be described as follows:

$$\theta_{pl}(h)\le x\le \Theta_{pl}(h),\quad x_{\times}\le h\le b+L, \tag{3.16}$$

where

$$\begin{aligned}\theta_{pl}(h)&=x_{\times}+(L-x_{\times})X_{\Diamond}\left(\frac{h-x_{\times}-b}{L-x_{\times}}\right),\\ \Theta_{pl}(h)&=(L-x_{\times})X_{\Diamond}\left(\frac{h-x_{\times}}{L-x_{\times}}\right).\end{aligned} \tag{3.17}$$

Functions $\theta_{pl}(h)$, $\Theta_{pl}(h)$, $\theta_{el}(h)$ and $\Theta_{el}(h)$ are formally defined for $-\infty<h<\infty$, and $\theta_{el}(h)=\Theta_{el}(h)$ outside the interval $(0,b+x_{\times})$ and $\theta_{pl}(h)=\Theta_{pl}(h)$ outside the interval $(x_{\times},b+L)$. Therefore, the following expression for the drag force acting at the impactor's lateral surface D_{imp} is valid for $0\le h\le b+L$:

$$D_{imp}=2\pi\left[\int_{\theta_{el}(h)}^{\Theta_{el}(h)}\sigma_{n\,elast}\Phi\Phi' dx+\int_{\theta_{pl}(h)}^{\Theta_{pl}(h)}\sigma_{n\,plast}\Phi\Phi' dx\right]. \tag{3.18}$$

Substituting $\sigma_{n\,elast}$ and $\sigma_{n\,plast}$ from Eqs. (2.9) and (2.10) into Eq. (3.18) and taking into account Eqs. (2.16) and (2.17), we obtain:

$$D_{imp}=g_0(h)+g_2(h)w+g_1(h)\frac{dw}{dh}, \tag{3.19}$$

where

$$g_0(h)=2\pi\left[\frac{\mu_9}{(1+\mu_{10})r}\int\limits_{\theta_{el}(h)}^{\Theta_{el}(h)}(\Phi-r)\Phi\Phi' dx+\mu_{11}\int\limits_{\theta_{pl}(h)}^{\Theta_{pl}(h)}\varphi_0(\Phi)\Phi\Phi' dx\right], \quad (3.20)$$

$$g_1(h)=\frac{\pi\gamma_{sh}}{2}\int\limits_{\theta_{pl}(h)}^{\Theta_{pl}(h)}\varphi_1(\Phi)\Phi^2\Phi'^2 dx, \quad (3.21)$$

$$g_2(h)=\pi\gamma_{sh}\int\limits_{\theta_{pl}(h)}^{\Theta_{pl}(h)}[\varphi_2(\Phi)\Phi'^2+\varphi_1(\Phi)\Phi\Phi'']\Phi\Phi' dx. \quad (3.22)$$

Then equation of motion of the impactor, Eq. (2.20), can be written in the following form:

$$\frac{dw}{dh}+\alpha(h)w+\beta(h)=0, \quad (3.23)$$

where

$$\alpha(h)=\frac{2g_2(h)}{\zeta(h)},\ \beta(h)=\frac{2[g_0(h)+2\pi\mu_{11}r(b-h)\delta(h)]}{\zeta(h)},$$
$$\zeta(h)=m+\pi\gamma_{sh}r^2 b\delta(h)+2g_1(h). \quad (3.24)$$

Equation (3.23) implies the following expression for the BLV:

$$w_{bl}=\int\limits_0^{b+L}dh\beta(h)exp\left(\int\limits_0^h\alpha(\tilde{h})d\tilde{h}\right). \quad (3.25)$$

This equation can be transformed by taking into account the special features of the stages of penetration. To this end, let us rewrite the integral in Eq. (3.25) as:

$$
\begin{aligned}
w_{bl} &= \int_0^b dh\beta(h)\exp\left(\int_0^h \alpha(\tilde{h})d\tilde{h}\right) \\
&+ \int_b^{b+L} dh\beta(h)\exp\left(\int_0^b \alpha(\tilde{h})d\tilde{h} + \int_b^h \alpha(\tilde{h})d\tilde{h}\right) \\
&= \int_0^b dh\beta_1(h)\exp\left(\int_0^h \alpha_1(\tilde{h})d\tilde{h}\right) \\
&+ \exp\left(\int_0^b \alpha_1(\tilde{h})d\tilde{h}\right) \times \int_b^{b+L} dh\beta_2(h)\exp\left(\int_b^h \alpha_2(\tilde{h})d\tilde{h}\right),
\end{aligned}
\tag{3.26}
$$

where $\alpha_i(h) = \alpha(h)$ and $\beta_i(h) = \beta(h)$ for the stage with the number $i\,(i = 1,2)$, namely:

$$
\begin{aligned}
&\alpha_1(h) = \frac{2g_2(h)}{\zeta_1(h)}, \beta_1(h) = \frac{2[g_0(h) + 2\pi\mu_{11}r(b-h)]}{\zeta_1(h)}, \\
&\zeta_1(h) = m + \pi\gamma_{sh}r^2 b + 2g_1(h), \\
&\alpha_2(h) = \frac{2g_2(h)}{\zeta_2(h)}, \beta_2(h) = \frac{2g_0(h)}{\zeta_1(h)}, \zeta_2(h) = m + 2g_1(h).
\end{aligned}
\tag{3.27}
$$

3.3 Plastic response (sharp impactor)

When $r = 0$, i.e, a sharp impactor is considered, only the plastic regime occurs during penetration and a plug is not formed. Then

$$
D_{imp} = 2\pi \int_{\theta(h)}^{\Theta(h)} \sigma_{n\,plast}\Phi\Phi' dx, \tag{3.28}
$$

where a plastic response is described by Eq. (2.10) with constant φ_0, φ_1 and φ_2 in Eqs. (2.4)-(2.5):

$$
\varphi_0 = \varphi_1 + 1, \varphi_1 = \ln(\mu_2 + 1), \; \varphi_2 = \varphi_1 - \frac{\mu_2}{\mu_2 + 1}. \tag{3.29}
$$

For purposes of convenience, we consider the solution of the CCEA problem written in a more general form:

$$p_{plast} = a_2\dot{y}^2 + a_1 y\ddot{y} + a_0 . \tag{3.30}$$

Then Eq. (2.10) can be rewritten as:

$$\sigma_{n\ plast} = (a_2\Phi'^2 + a_1\Phi\Phi'')\dot{h}^2 + a_1\Phi\Phi'\ddot{h} + a_0 . \tag{3.31}$$

Clearly, Eqs. (3.30) and (3.31) coincide with Eqs. (2.3) and (2.10), respectively, when

$$a_2 = 0.5\gamma_{sh}\varphi_2, \quad a_1 = 0.5\gamma_{sh}\varphi_1, \quad a_0 = \mu_{11}\varphi_0 . \tag{3.32}$$

Substituting $\sigma_{n\ plast}$ from Eq. (3.31) into Eq. (3.28), we arrive at Eq. (3.19) with

$$g_0(h) = 2\pi a_0 \int_{\theta(h)}^{\Theta(h)} \Phi\Phi' dx = \pi a_0 \left[\Phi^2(\Theta(h)) - \Phi^2(\theta(h))\right], \tag{3.33}$$

$$g_1(h) = \pi a_1 \int_{\theta(h)}^{\Theta(h)} \Phi^2\Phi'^2 dx , \tag{3.34}$$

$$g_2(h) = 2\pi \left[a_2 \int_{\theta(h)}^{\Theta(h)} \Phi\Phi'^3 dx + a_1 \int_{\theta(h)}^{\Theta(h)} \Phi\Phi'\Phi'' dx \right]. \tag{3.35}$$

The BLV is determined by using Eq. (3.25) with

$$\alpha(h) = \frac{2g_2(h)}{\zeta(h)}, \beta(h) = \frac{2g_0(h)}{\zeta(h)}, \zeta(h) = m + 2g_1(h) . \tag{3.36}$$

4. SOLUTIONS FOR SOME PARTICULAR IMPACTOR'S SHAPES

4.1 Truncated cone. General formulas

The equation of the generator for the lateral surface of the cone with flat bluntness reads:

$$\Phi(x) = r + kx. \tag{4.1}$$

Consequently, $\varphi(x) = k^2 x(r + kx)$, where function $\varphi(x)$ is defined by Eq. (3.2). Calculating the integrals in Eqs. (3.12) we can determine the expression for w_{bl} in the case of *elastic response* in the following form:

$$w_{bl} = \begin{cases} \dfrac{k^2 b}{12}[b^2(2r + kb)e_6 + (6rL^2 + 4kL^3 - 2rb^2 - kb^3)e_7] \\ + e_8 \quad if \quad b \le L \\ \\ \dfrac{k^2 L^2}{12}[(6br + 4kbL - 4rL - 3kL^2)e_6 + L(4r + 3kL)e_7] \\ + e_8 \quad if \quad b \ge L \end{cases} . \tag{4.2}$$

In the case of an *elastic-plastic regime,* integrals in Eqs. (3.20)-(3.22) can be calculated and the expressions for functions $g_i\,(i = 0,1,2)$ can be simplified:

$$\frac{1}{2\pi} g_0(h) = \frac{\mu_9 k^2}{(1+\mu_{10})r} F_0(h) + \mu_{11}[ln(\mu_2 + 1) + 1]F_1(h) + \mu_{11}F_2, \tag{4.3}$$

$$\frac{2}{\pi \gamma_{sh} k} g_1(h) = ln(\mu_2 + 1)F_5(h) + F_4(h), \tag{4.4}$$

$$\frac{1}{\pi \gamma_{sh} k^2} g_2(h) = [ln(\mu_2 + 1) - \frac{\mu_2}{\mu_2 + 1}]F_1(h) + F_2 + \frac{1}{\mu_2 + 1} F_3, \tag{4.5}$$

where

$$\begin{aligned} &F_0(h) = f_0(\Theta_{el}(h)) - f_0(\theta_{el}(h)), \\ &f_0(x) = \int x(r + kx)dx = \frac{1}{6} x^2 (3r + 2kx), \end{aligned} \tag{4.6}$$

$$F_i(h) = f_i(\Theta_{pl}(h)) - f_i(\theta_{pl}(h)), \quad i = 1,2,3,4,5 \tag{4.7}$$

$$f_1(x) = \int \Phi d\Phi = \frac{1}{2}\Phi^2, \quad f_5(x) = \int \Phi^2 d\Phi = \frac{1}{3}\Phi^3, \tag{4.8}$$

$$f_2(x)=\int \Phi\, ln\left(\frac{\Phi^2-r^2}{\Phi^2}\right)d\Phi=\frac{1}{2}\begin{bmatrix}(\Phi^2-r^2)ln(\Phi^2-r^2)\\-2\Phi^2\,ln\,\Phi+r^2\end{bmatrix}, \tag{4.9}$$

$$f_3(x)=r^2\int\frac{\Phi}{\Phi^2-r^2}d\Phi=\frac{r^2}{2}ln(\Phi^2-r^2), \tag{4.10}$$

$$f_4(x)=\int \Phi^2\, ln\left(\frac{\Phi^2-r^2}{\Phi^2}\right)d\Phi=\frac{1}{3}\begin{bmatrix}\Phi^3\,ln(\Phi^2-r^2)-2\Phi^3\,ln\,\Phi\\+r^3\,ln\left(\frac{\Phi+r}{\Phi-r}\right)-2r^2\Phi\end{bmatrix}. \tag{4.11}$$

4.2 Sharp cone. Shield with a finite thickness

In the case of a sharp cone (a plastic regime), $\Phi(x)=kx$ and formulas from *Section 3-4.1* are simplified, namely:

$$g_0(h)=\pi a_0 k^2[\Theta^2(h)-\theta^2(h)], \tag{4.12}$$

$$g_1(h)=\frac{\pi a_1 k^4}{3}[\Theta^3(h)-\theta^3(h)], \tag{4.13}$$

$$g_2(h)=\pi a_2 k^4[\Theta^2(h)-\theta^2(h)], \tag{4.14}$$

where $\varphi_0, \varphi_1, \varphi_2$ are defined by Eq. (3.29), while Eqs. (3.25) and (3.36) remain valid. Taking into account Eqs. (4.12) and (4.14), Eq. (3.36) yields:

$$\beta(h)=\frac{g_0(h)}{g_2(h)}\alpha(h)=\lambda_0\alpha(h), \tag{4.15}$$

where

$$\lambda_0=\frac{\chi}{k^2},\quad \chi=\frac{a_0}{a_2}. \tag{4.16}$$

The equation of motion of the impactor (Eq. 3.23) can be written as:

$$\frac{dw}{dh}+(w+\lambda_0)\alpha(h)=0 \tag{4.17}$$

or

$$\frac{dw}{w+\lambda_0} = -\alpha(h)dh. \tag{4.18}$$

Integrating Eq. (4.18) over w from $w = w_{res} = w(b+L)$ to $w = w_{imp} = w(0)$ and over h from $h = b+L$ to $h = 0$, we obtain the relationship between the impact velocity and the residual velocity of the impactor:

$$w_{imp} - \hat{T}w_{res} = \lambda_0(\hat{T}-1), \quad w_{imp} \geq \lambda_0(\hat{T}-1), \tag{4.19}$$

where

$$\hat{T} = exp\left(\int_0^{b+L} \alpha(h)dh\right). \tag{4.20}$$

The expression for the BLV can be obtained from Eq. (4.19) at $w_{res} = 0$:

$$w_{bl} = v_{bl}^2 = \lambda_0(\hat{T}-1). \tag{4.21}$$

In the case considered (a sharp cone), the integral in Eq. (4.20) can be in terms of elementary functions. For this purpose, it is convenient to use dimensionless variables and to represent the integral in the form:

$$\int_0^{b+L} \alpha(h)dh = \frac{3a_2}{a_1}\bar{I}, \quad \bar{I} = \int_0^{\bar{b}+1} \frac{\bar{\Theta}^2(\bar{h}) - \bar{\theta}^2(\bar{h})}{\bar{m}_{imp} + \bar{\Theta}^3(\bar{h}) - \bar{\theta}^3(\bar{h})} d\bar{h}, \tag{4.22}$$

where

$$\bar{m}_{imp} = \frac{3m}{2\pi a_1 k^4 L^3}, \quad \bar{h} = \frac{h}{L}, \quad \bar{b} = \frac{b}{L} \tag{4.23}$$

$$\bar{\Theta}(\bar{h}) = \begin{cases} 0 & if \quad \bar{h} \leq 0 \\ \bar{h} & if \quad 0 \leq \bar{h} \leq 1 \\ 1 & if \quad \bar{h} \geq 1 \end{cases}, \quad \bar{\theta}(\bar{h}) = \begin{cases} 0 & if \quad \bar{h} \leq \bar{b} \\ \bar{h} - \bar{b} & if \quad \bar{b} \leq \bar{h} \leq \bar{b}+1 \\ 1 & if \quad \bar{h} \geq \bar{b}+1 \end{cases}. \tag{4.24}$$

Calculating the integral $\bar{I}$ for $\bar{b} \leq 1$ and for $\bar{b} \geq 1$ (see *Figure 3-5a-b*),

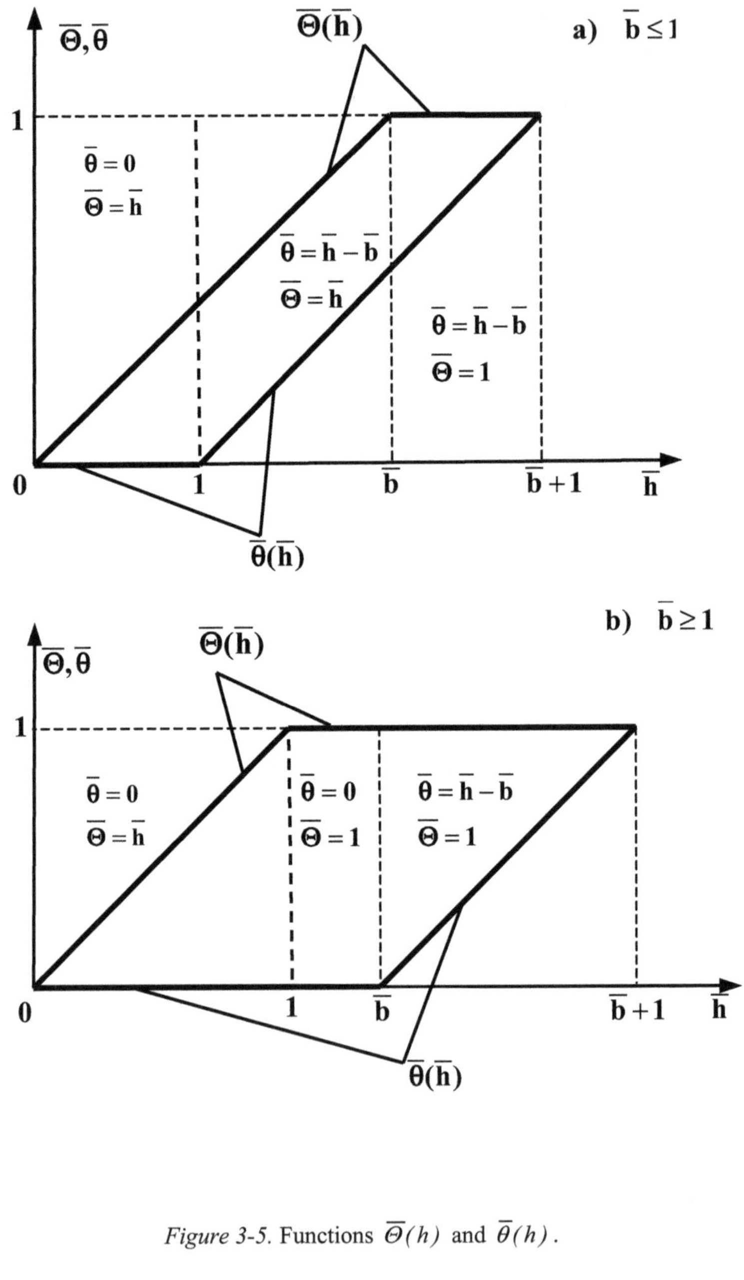

Figure 3-5. Functions $\overline{\Theta}(h)$ and $\overline{\theta}(h)$.

we obtain:

$$\bar{I}=\begin{cases}\int_0^{\bar{b}}\frac{\bar{h}^2 d\bar{h}}{\overline{m}_{imp}+\bar{h}^3}+\int_{\bar{b}}^{1}\frac{\bar{h}^2-(\bar{h}-\bar{b})^2}{\overline{m}_{imp}+\bar{h}^3-(\bar{h}-\bar{b})^3}d\bar{h} \\ +\int_1^{\bar{b}+1}\frac{1-(\bar{h}-\bar{b})^2}{\overline{m}_{imp}+1-(\bar{h}-\bar{b})^3}d\bar{h}=\int_1^{\bar{b}+1}\frac{d\bar{h}}{\overline{m}_{imp}+1-(\bar{h}-\bar{b})^3} & if\ \bar{b}\le 1 \\ \int_0^{1}\frac{\bar{h}^2 d\bar{h}}{\overline{m}_{imp}+\bar{h}^3}+\int_1^{\bar{b}}\frac{d\bar{h}}{\overline{m}_{imp}+1}+\int_{\bar{b}}^{\bar{b}+1}\frac{1-(\bar{h}-\bar{b})^2}{\overline{m}_{imp}+1-(\bar{h}-\bar{b})^3}d\bar{h} \\ =\frac{\bar{b}-1}{\overline{m}_{imp}+1}+\int_0^{1}\frac{d\bar{h}}{\overline{m}_{imp}-\bar{h}^3} & if\ \bar{b}\ge 1\end{cases} \tag{4.25}$$

Introducing the function:

$$\Psi(z)=\int_0^{z}\frac{dt}{1-t^3}=\frac{1}{\sqrt{3}}tan^{-1}\left(\frac{2z+1}{\sqrt{3}}\right)-\frac{1}{6}ln\left[\frac{(z-1)^2}{z^2+z+1}\right]-\frac{\pi}{6\sqrt{3}} \tag{4.26}$$

we can represent the expression for $\bar{I}$ as:

$$\bar{I}=\lambda_1^2 G(\lambda_1,\bar{b}), \tag{4.27}$$

where

$$G(\lambda_1,z)=\begin{cases}\Psi(\lambda_1)-\Psi(\lambda_1(1-z)) & if \quad z\le 1 \\ \Psi(\lambda_1)+\lambda_1(z-1) & if \quad z\ge 1\end{cases}, \tag{4.28}$$

$$\lambda_1=\frac{1}{\sqrt[3]{\overline{m}_{imp}+1}}<1. \tag{4.29}$$

Thus, the relationship between the impact velocity and the residual velocity of the impactor and the formula for the BLV are determined by Eqs. (4.19) and (4.21), respectively, with

$$\hat{T}=exp\left(\frac{3a_2}{a_1}\bar{I}\right), \tag{4.30}$$

where $\bar{I}$ can be calculated using the explicit formulas given by Eqs. (4.27)-(4.29).

4.3 Sharp cone. Semi-infinite shield

In the case of a SIS, Eqs. (4.12)-(4.18) remain valid with $\Theta(h)$ defined by Eq. (2-2.2) and $\theta(h)=0$. To obtain a relationship for determining the DOP, H, we integrate Eq. (4.18) over w from $w=0$ to $w=w_{imp}=w(0)$ and over h from $h=H$ to $h=0$. After some algebra we obtain:

$$ln\left(1+\frac{w_{imp}}{\lambda_0}\right)=\bar{J}(\bar{H}), \tag{4.31}$$

where

$$\bar{J}(\bar{H})=\int_0^{\bar{H}}\frac{\bar{\Theta}^2(\bar{h})}{\bar{m}_{imp}+\bar{\Theta}^3(\bar{h})}d\bar{h}, \quad \bar{H}=\frac{H}{L}, \tag{4.32}$$

and $\bar{\Theta}(\bar{h})$ is determined by Eq. (4.24).

Let us calculate the integral $\bar{J}(\bar{H})$ (function $\bar{\Theta}(\bar{h})$ is shown in *Figure 3-5a-b*):

$$\bar{J}(\bar{H})=\begin{cases}\displaystyle\int_0^{\bar{H}}\frac{\bar{h}^2 d\bar{h}}{\bar{m}_{imp}+\bar{h}^3}=\frac{1}{3}ln\left(1+\frac{\bar{H}^3}{\bar{m}_{imp}}\right) & if \quad \bar{H}\le 1\\ \displaystyle\int_0^{1}\frac{\bar{h}^2 d\bar{h}}{\bar{m}_{imp}+\bar{h}^3}+\int_1^{\bar{H}}\frac{d\bar{h}}{\bar{m}_{imp}+1} & \\ \displaystyle=\frac{1}{3}ln\left(1+\frac{1}{\bar{m}_{imp}}\right)+\frac{\bar{H}-1}{\bar{m}_{imp}+1} & if \quad \bar{H}\ge 1\end{cases}. \tag{4.33}$$

Then Eq. (4.31) allows us to determine $\bar{H}$ (Ben-Dor et al., 2000d):

$$\overline{H}=\begin{cases}\sqrt[3]{\overline{m}_{imp}\left[\left(1+\dfrac{w_{imp}}{\lambda_0}\right)^{\frac{a_1}{a_2}}-1\right]} & if \quad w_{imp}\le w_{imp\,0}\\ 1+(\overline{m}_{imp}+1)\ln\left[\left(1+\dfrac{w_{imp}}{\lambda_0}\right)^{\frac{a_1}{3a_2}}\sqrt[3]{\dfrac{\overline{m}_{imp}}{\overline{m}_{imp}+1}}\right] & \\ & if \quad w_{imp}\ge w_{imp\,0}\end{cases}, \tag{4.34}$$

where

$$w_{imp\,0}=\lambda_0\left[\left(1+\frac{1}{\overline{m}_{imp}}\right)^{\frac{a_2}{a_1}}-1\right]. \tag{4.35}$$

4.4 Cylinder

In this case, the initial energy of the impactor is expended solely for plugging. Then, $D_{imp}=0$ in Eq. (2.20), and the equation of motion of the impactor for $0\le h\le b$ reads:

$$\frac{dw}{dh}+\lambda_{19}(b-h)=0,\quad \lambda_{19}=\frac{4\pi\mu_{11}r}{m+\pi\gamma_{sh}r^2b}\,. \tag{4.36}$$

Integrating the equation $dw=-\lambda_{19}(b-h)dh$ over w from $w=0$ to $w=w_{bl}$ and over h from $h=b$ to $h=0$, we obtain the formula for the BLV:

$$w_{bl}=v_{bl}^2=\lambda_{19}b^2/2\,. \tag{4.37}$$

Chapter 4

POWER-LAW RELATIONSHIPS BETWEEN IMPACT, RESIDUAL AND BALLISTIC LIMIT VELOCITY

1. INTRODUCTION

The following formula:

$$v_{imp}^{\beta} - a_{\Diamond} v_{res}^{\beta} = v_{bl}^{\beta}, \quad v_{imp} \geq v_{bl} \tag{1.1}$$

where v_{imp}, v_{res} and v_{bl} are the impact velocity, the residual velocity and the BLV, respectively, and $v_{res} = 0$ for $v_{imp} < v_{bl}$, is used widely in impact dynamics. Lambert and Jonas (1976) and Lambert (1978) proposed this formula as a unified relation for processing ballistic impact data, where $a_{\Diamond}$ and β (and, actually, v_{bl}) are the coefficients of approximation. Many empirical and semi-empirical models can be represented in the form of Eq. (1.1), particularly the models based on the conservation of energy and/or momentum (Zukas, 1982; Recht and Ipson, 1963; Ipson and Recht, 1975; Hetherington and Rajagopalan, 1991; Hetherington, 1992b; Hetherington, 1996; Giere, 1964). Among the most known models of this type is the model suggested by Recht and Ipson (1963), which can be justified by assuming constant energy absorbed by a shield. This model may be represented in the form given by Eq. (1.1) as:

$$\beta = 2, \quad a_{\Diamond} = m_{res}/m_{imp}, \tag{1.2}$$

where m_{imp} is the impact (initial) mass of the impactor, m_{res} is the residual mass of the impactor, including the mass of material expelled from the target by a blunt impactor. Mileiko and Sarkisyan (1981) and Mileiko et al. (1994) (see also Mileiko, 1997) demonstrated that a solution of the equation of motion of the impactor yields Eq. (1.1), with $a_\Diamond = 1$, when a power-law dependence between the impactor's drag force and its velocity is valid. Nixdorff (1983; 1984a,b; 1987a) showed that under certain assumptions the theory of Awerbuch and Bodner (Awerbuch, 1970; Awerbuch and Bodner, 1974) implies Eq. (1.1). The model determined by Eq. (2-8.1) also has the same property (see Eq. 2-8.14). Although other relationships between v_{res}, v_{imp} and v_{bl} can be used (see, e.g., Anderson et al., 1999), at present Eq. (1.1) is considered as the method of choice for the reduction of experimental data. Therefore, it would useful to understand the cause of the efficiency of correlations of this type. One way to explain the efficiency of these correlations is to construct physically realistic models of penetration that imply Eq. (1.1). This approach is described below (see Ben-Dor et al., 1998b; 2001b; 2002d).

2. A CLASS OF GENERALIZED LOCALIZED INTERACTION MODELS

Let us consider the penetration model that is based on the following assumptions (the notations are shown in *Figure 4-1*):

1) During penetration, the impactor can change its shape and accumulate and/or lose mass (even if the impactor's nose can be deformed, we measure the coordinate h from the front side of the shield to the point of the impactor having the coordinate $h = 0$ at the beginning of its motion). The rate of change of mass is a function of the DOP, and it does not depend upon the impactor's velocity, i.e.,

$$m = m(h),\ m^- = m^-(h),\ m^+ = m^+(h)\,, \tag{2.1}$$

where m is the mass of the impactor, and m^+ and m^- are mass accumulation and mass loss from the beginning of motion, respectively. It is assumed that

$$v^+ = 0, \quad v^- = v\,, \tag{2.2}$$

where v^+ is the velocity of the accumulated particles and v^- is the velocity of the lost particles.

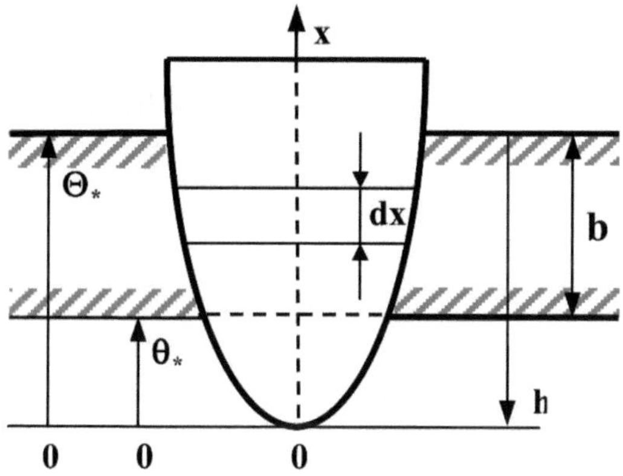

Figure 4-1. Notations for the generalized model.

2) The expression for the differential drag force dD acting on the impactor's surface element between the sections x and $x+dx$ (see *Figure 4-1*) reads:

$$dD = [\,P_2(x,h)v^2 + P_0(x,h)v^{\alpha}\,]dx, \tag{2.3}$$

where P_0 and P_2 are non-negative functions and parameter $0 \le \alpha < 2$. Clearly, this is more general model than the model considered in *Section 2-8*.

3) The part of the impactor interacting with the shield depends only on the DOP, i.e.,

$$x = \theta_*(h), \quad x = \Theta_*(h). \tag{2.4}$$

Impactor-shield interaction begins when $h=0$ and is completed when $h=b+L$. The functions $\theta_*(h)$ and $\Theta_*(h)$ have the meaning of $\theta(h)$ and $\Theta(h)$, correspondingly, but we do not specify their particular forms.

4) The impactor can change its shape during penetration. Generally, the functions $m, m^-, m^+, G_0, G_2, \theta_*, \Theta_*$ depend on the instantaneous shape of the impactor, i.e., they account for deformation, accumulation and loss of mass of the impactor.

The equation of motion of a body with a variable mass can be written as (see, e.g., Avallone and Baumeister, 1996; Corben and Stehle, 1994):

$$m\frac{dv}{dt}+(v-v^{+})\frac{dm^{+}}{dt}+(v-v^{-})\frac{dm^{-}}{dt}=-D(h,v)\,. \tag{2.5}$$

The expression for the drag force can be obtained from Eqs. (2.3) and (2.4):

$$D(h,v)=\hat{f}_2(h)v^2+\hat{f}_0(h)v^{\alpha}\,, \tag{2.6}$$

where

$$\hat{f}_i(h)=\int_{\theta_*(h)}^{\Theta_*(h)}P_i(x,h)dx,\quad i=0,2, \tag{2.7}$$

The model also includes the equation of mass balance:

$$m(h)=m_{imp}+m^{-}(h)+m^{+}(h)\,, \tag{2.8}$$

where $m_{imp}=m(0)$ and $m_{res}=m(b+L)$.

After substituting D given by Eq. (2.6), v^{+} and v^{-} given by Eq. (2.2) and using change of variables $d/dt=v\,d/dh$, we may write Eq. (2.5) as:

$$m\frac{dv}{dh}+f_2(h)v+f_0(h)v^{\alpha-1}=0\,, \tag{2.9}$$

where

$$f_2(h)=\frac{1}{m(h)}\left[\frac{dm^{+}}{dh}+\hat{f}_2(h)\right],\quad f_0(h)=\frac{\hat{f}_0(h)}{m(h)}\,. \tag{2.10}$$

Equation (2.9) coincides with Eq. (2-8.4), and we can immediately write relationship between impact velocity, residual velocity and BLV that is similar to Eq. (2-8.14) in the form of Eq. (1.1) with:

$$a_{\Diamond}=exp\left(\beta\int_0^{b+L}f_2(h)\,dh\right)=exp\left(\beta\int_0^{b+L}\left[\frac{dm^{+}}{dh}+\hat{f}_2(h)\right]\frac{dh}{m(h)}\right), \tag{2.11}$$

$$\beta=2-\alpha\,. \tag{2.12}$$

Differentiation of Eq. (2.8) yields:

$$\frac{1}{m(h)}\frac{dm^+}{dh}=\frac{1}{m(h)}\left[\frac{dm}{dh}-\frac{dm^-}{dh}\right]=\frac{d\,ln(m(h))}{dh}-\frac{1}{m(h)}\frac{dm^-}{dh}. \qquad (2.13)$$

Then Eq. (2.11) can be written in the following form:

$$a_\Diamond=\left(\frac{m_{res}}{m_{imp}}\right)^\beta exp\left(\beta\int_0^{b+L}\left[\hat{f}_2(h)-\frac{dm^-}{dh}\right]\frac{dh}{m(h)}\right). \qquad (2.14)$$

Therefore, the above-suggested model implies a power-law dependence between the impact, the residual and the ballistic limit velocities of the impactor given by Eq. (1.1). Note that the parameter $a_\Diamond$ does not depend upon the function $\hat{f}_0(h)$.

Let us consider the case when $P_2(x,h)=0$ in Eq. (2.3) and, hence, $\hat{f}_2(h)=0$. If the impactor only accumulates mass ($m^-=0$), then Eq. (2.14) implies that:

$$a_\Diamond=\left(\frac{m_{res}}{m_{imp}}\right)^\beta. \qquad (2.15)$$

If the impactor only loses mass ($m^+=0$), then Eq. (2.11) implies that:

$$a_\Diamond=1. \qquad (2.16)$$

In both the cases, the parameter $a_\Diamond$ is independent of the history of mass change that is determined by the functions m^+ and m^-.

3. A BENCHMARK ANALYSIS OF POWER-LAW APPROXIMATIONS

It is of interest to compare the performance of the most justified version of the model with $\beta=2$ and versions of the model with $\beta\neq 2$.

Let us rewrite Eq. (1.1) as:

$$v_{res}=c_\Diamond(v_{imp}^\beta-v_{bl}^\beta)^{1/\beta},\quad v_{imp}\geq v_{bl},\quad c_\Diamond=(1/a_\Diamond)^{1/\beta}, \qquad (3.1)$$

and assume that this is an "exact" model. Our immediate task is to determine the best approximation of the "exact" model in the form:

$$\hat{v}_{res} = \hat{c}(v_{imp}^2 - \hat{v}_{bl}^2)^{1/2}, \quad v_{imp} \geq \hat{v}_{bl} \tag{3.2}$$

for $\hat{v}_{bl} \leq v_{imp} \leq 2\hat{v}_{bl}$ where $\hat{c}$ and $\hat{v}_{bl}$ are the unknown coefficients of the model with $\beta = 2$ ($\hat{v}_{bl}$ is interpreted as the BLV), and $\hat{v}_{res}$ is the residual velocity calculated using the model with $\beta = 2$.

The difference between these models can be estimated by means of the following norm:

$$\varepsilon = \frac{1}{c_{\Diamond} v_{bl}^2} \int_{v_{bl}}^{2v_{bl}} | v_{res} - \hat{v}_{res} | \, dv_{imp} . \tag{3.3}$$

Substituting v_{res} and $\hat{v}_{res}$ from Eqs. (3.1) and (3.2) and changing variable $v_{imp} = z\, v_{bl}$, we may rewrite Eq. (3.3) as:

$$\varepsilon = \int_{1}^{2} |(z^{\beta} - 1)^{1/\beta} - \lambda_c (z^2 - \lambda_v^2)^{1/2} | \, dz , \tag{3.4}$$

where

$$\lambda_c = \hat{c}/c_{\Diamond}, \quad \lambda_v = \hat{v}_{bl}/v_{bl} . \tag{3.5}$$

The values of the parameters λ_c and λ_v can be found by minimizing ε, taking into account the conditions $\lambda_c > 0$ and $0 < \lambda_v \leq 1$. The results of the calculations are shown in *Table 4-1*, where the difference in the results obtained using these models is estimated by the following parameter:

$$\tilde{\varepsilon} = \frac{1}{c_{\Diamond} v_{bl}} (v_{res} - \hat{v}_{res}) = (z^{\beta} - 1)^{1/\beta} - \lambda_c (z^2 - \lambda_v^2)^{1/2} . \tag{3.6}$$

Table 4-1 shows that both Eqs. (3.1) and (3.2) provide estimates of the residual velocity with almost the same accuracy. In the neighborhood of the BLV where the residual velocity changes rapidly, neither of these approximations is reliable.

Table 4-1. Estimation of the difference in the models, $\tilde{\varepsilon}$

Z	β					
	2.1	2.2	2.3	2.4	2.5	2.6
1.0	-0.144	-0.250	-0.291	-0.328	-0.362	-0.444
1.1	-0.001	-0.018	-0.016	-0.015	-0.014	-0.026
1.2	0.004	-0.002	0.004	0.009	0.012	0.009
1.3	0.004	0.002	0.010	0.015	0.018	0.018
1.4	0.003	0.003	0.010	0.015	0.017	0.018
1.5	0.001	0.002	0.008	0.012	0.013	0.014
1.6	-0.001	0.0	0.005	0.008	0.008	0.008
1.7	-0.003	-0.003	0.001	0.002	0.001	0.0
1.8	-0.005	-0.006	-0.003	-0.003	-0.006	-0.008
1.9	-0.008	-0.009	-0.008	-0.009	-0.013	-0.017
2.0	-0.010	-0.013	-0.013	-0.016	-0.021	-0.026
λ_c	1.020	1.030	1.040	1.050	1.060	1.070
λ_v	0.990	0.970	0.960	0.950	0.093	0.910

The results of calculations using the experimental data of Børvik et al. (2002) are presented in *Table 4-2* (blunt cylinder) and in *Table 4-3* (cylinder with hemispherical nose). In these tables the magnitude of the error $\varepsilon_i^{[2]}$ is determined as:

$$\varepsilon_i^{[2]} = \hat{c}_\Diamond (v_{imp,i}^2 - \hat{v}_{bl}^2)^{1/2} - v_{res,i}, \tag{3.7}$$

Table 4-2. Errors of approximation of the experimental data of Børvik et al. (2002) for a blunt-nosed cylinder

	Impact velocity (m/s)					
	303.5	285.4	244.2	224.7	200.4	189.6
$\varepsilon^{[\beta]}$	1.1	1.5	4.6	-1.6	0.2	0.4
$\varepsilon^{[2]}$	0.2	-0.0	2.0	-4.3	-0.5	4.8

Table 4-3. Errors of approximation of the experimental data of Børvik et al. (2002) for a hemispherical-nosed cylinder

	Impact velocity (m/s)				
	452.0	420.6	362.9	326.7	300.0
$\varepsilon^{[\beta]}$	-5.2	2.6	-2.2	6.6	-6.3
$\varepsilon^{[2]}$	1.0	4.1	-8.3	-0.6	0.9

where $v_{imp,i}$ and $v_{res,i}$ are the impact velocity and the residual velocities in the i-th experiment, respectively; and parameters $\hat{c}$ and $\hat{v}_{bl}$ minimize the sum $\sum | \varepsilon_i^{[2]} |$. The values of the errors $\varepsilon_i^{[\beta]}$ are calculated as:

$$\varepsilon_i^{[\beta]} = c_\lozenge (v_{imp,i}^{\beta} - v_{bl}^{\beta})^{1/\beta} - v_{res,i} , \quad (3.8)$$

where the parameters $c_\lozenge, \beta, v_{bl}$ in these models were obtained by Børvik et al. (2002) using the method of least squares. The above-described results clearly demonstrate that of the considered models have approximately the same accuracy.

PART 2: SHAPE OPTIMIZATION OF IMPACTORS

Chapter 5

TOWARDS SHAPE OPTIMIZATION OF IMPACTORS

1. INTRODUCTION

In this chapter, the notion of shape optimization of impactors implies the choice of the best impactor from an infinite number of possible solutions. Emphasis is placed on variational problems that permit analytical solution. Certainly, the latter requirement does not prevent us from using numerical calculations for determining a particular solution. To study such problems, it is feasible to employ only those models that allow us to continuously vary the impactor's shape, i.e., the models describing impactor-shield interaction for a wide class of impactor shapes (e.g., convex bodies of revolution). The models used must also be sufficiently simple to allow determination of the target functional in a relatively compact form. Sometimes, instead of a direct criterion, e.g., the BLV or DOP, an indirect optimization criterion was used with the goal of circumventing the latter requirement. Such approach was used in the early studies of impactor shape optimization.

Kucher (1967) optimized the penetrator's shape using as the criterion the "dynamic work" from Thomson's theory (Thomson, 1955) for thin plates. Nixdorff (1987b) compared the efficiency of conical, different power-law and ogival heads and found that there exist impactors that are superior to "Kucher's optimum head", which was determined by solving the corresponding variational problem. This paradox was explained (Ben-Dor et al., 2001a) with reference to the correct solution (Miele, 1962, 1965) of the mathematically similar variational problem in hypersonic aerodynamics, namely, determining a thin head with minimum drag by means of the Newton-Busemann model for a projectile-medium interaction. In solving this problem in gasdynamics, investigators also encountered serious difficulties, which were reflected in corresponding sections in the monographs of Hayes and Probstein (1959) and Miele (1965). Note that this situation is not the only example of a problem of shape optimization that turns out to be more complicated than it seems at first sight.

The first studies of shape optimization of non-thin impactors also used indirect optimization criteria. Gendugov et al. (1984), Bunimovich and Yakunina (1987a,b; 1989), Ostapenko and Yakunina (1992), and Ostapenko (1997) determined the shapes of 3-D bodies with a minimum "shape factor" that is equivalent to the minimum resistance during the motion of an impactor inside a dense medium with constant velocity. We note (Ben-Dor et al., 1997c, 2001a) that using the two-term LIM yields minimum BLV and/or maximum DOP solutions that coincide with the solution obtained by applying the above-mentioned indirect criterion. Using the previously developed disks model (Yankelevsky and Adin, 1980), Yankelevsky (1983b) optimized the shape of a projectile penetrating into soil by minimizing the instantaneous resistance force. The optimal shape was found to be determined by a single parameter depending on the velocity and deceleration of the impactor and the properties of the medium.

As a direct criterion for optimization, the maximum DOP for a given impact velocity in the case of a SIS and the BLV for a SFT were used. Yankelevsky and Gluck (1980) obtained formulas for the DOP of an ogive-shaped projectile penetrating into soil and analyzed the influence of the impactor's shape parameters and the characteristics of the shield material on the criterion. Bondarchuk et al. (1982) used a simple LIM for shape optimization of 3-D impactors penetrating into SISs (soil and metal). Numerical calculations and experiments showed that 3-D impactors can offer advantages over bodies of revolution, when the DOP is taken as the criterion of optimization. Additional calculations associated with determining efficient 3-D penetrators can be found in the monographs by Vedernikov and Shchepanovsly (1995) and Vedernikov et al. (1995). Ostapenko et al. (1994) found numerically the optimum cross-section of a 3-D conical impactor with a maximum DOP for the class of three-terms LIMs with friction. Ostapenko and Yakunina (1999) used this criterion in their analytical investigation of a variational problem based on two-term LIMs with friction; they considered slender bodies with self-similar cross-sections.

Let us now discuss more comprehensively the optimization of impactors having the shape of bodies of revolution, using as the criterion the maximum DOP. We employ a two-term LIM and neglect incomplete immersion of the impactor in the shield at the initial stage of penetration (see *Section 2-9.3.2*). This problem merits attention since historically it was the first properly posed and correctly solved variational problem in projectile shape optimization (following Alekseev et al., 1987, we can designate this problem as the first problem of the optimal control theory) and since, as mentioned before (Ben-Dor et al., 2001a), investigation of this or similar problems is still fraught with difficulties. In his classic study, Isaac Newton (1687) considered motion of a body of revolution in a gaseous medium and assumed that the external rarefied medium interacts with the body through elastic collisions of its particles. Although Newton did not use modern mathematical theory (that was nonexistent at that time), his geometrical constructions yielded a correct answer. In particular, he found that the optimal body of revolution has flat bluntness at the leading edge. Newton also considered inelastic interaction of particles of the medium with a body. In the latter case, the magnitudes of the local and integral forces are smaller by a factor of two than in the elastic case, but the optimal shape of the body remains the same. The necessity to take into account the possibility of the existence of flat bluntness in the mathematical formulation of the problem renders this problem methodologically interesting and instructive. The study of the calculus of variations and of the optimal control theory is often accompanied by analysis of this problem (see, e.g., Alekseev et al., 1987; Edwards, 1997; Ivanov, 1998; Silva and Torres, 2004). The results obtained by Newton did not invoke particular interest among researchers in the field of aerodynamics because he used models for projectile-medium interaction that did not realize the desired degree of accuracy in practice. The situation changed drastically with the emergence of supersonic velocities and cosmic flights. Interestingly, Newton's model was used mainly not for calculating the drag experienced by a body in a free molecular flow regime at high altitudes but for aerodynamic calculations for flight in a dense atmosphere at supersonic and hypersonic velocities. In these calculations, the collisions of gas molecules with the flying body were considered completely inelastic, and the corresponding approach was referred to as "Newton's theory", "Newton's drag law" etc. (Hayes and Probstein, 1959; Chernyi, 1969). These developments stimulated numerous studies on the application of Newton's model to shape optimization of flying bodies. Many interesting results obtained in this field were surveyed in the monograph of Miele (1965). Naturally, the interest was regenerated in the problem of determining a body of revolution with minimum drag that had been solved by Newton at the end of the 17th Century. Eggers et al., 1957 (see also Miele, 1965) studied this problem and its modifications using the modern mathematical apparatus. They found an analytical solution of Newton's problem and determined the equation of the generator of the optimal shape in a closed form.

Let us return to optimization of impactors having the shape of bodies of revolution. Jones et al. (1998) considered such problem using the two-term Newton's LIM without friction determined by Eq. (2-9.31) and employed the "shape factor" as the optimization criterion. Later, Jones and Rule (2000) showed that the criterion of the maximum DOP implied the same variational problem. In the latter case, the optimization problem is identical to "Newton's optimization problem" in aerodynamics, which permits an analytical solution. In their solution, Jones et al. (1998) used two approximate methods. The first method is a perturbation method that assumes that the ratio R/L is a small parameter (R and L are the radius of a shank and impactor's length, respectively). The second method involves minimization of a target functional in 5-dimensional parameter space, where the chosen class of functions permits only generators that pass through the origin. Although the mathematical formulation of the problem does not permit the possibility of the existence of optimal shapes with flat bluntness, the "best" approximate solutions tend to describe impactors having exactly this shape (especially when the ratio R/L is not small). Based on their numerical findings, Jones et al. (1998) arrived at the correct conclusion that an "optimal penetrator" has a blunt tip. Certainly, this conclusion follows directly from the known analytical solution of the problem. In their further study, Jones and Rule (2000) investigated a new problem, namely, maximization of the DOP using the two-term LIM and taking friction into account. The mathematical formulation of the problem does not provide for the possibility of the existence of optimal shapes with flat bluntness, and the situation repeats itself. Numerical simulations performed by Jones and Rule (2000) showed that for some values of the parameters the predicted impactor tip is "as close to blunt-ended as possible". The latter effect can be explained by the fact that if friction is taken into account, optimal impactors also have flat bluntness (Ben-Dor et al., 2003a).

By emphasizing the necessity to account for the existence of flat bluntness in formulating these applied variational problems, we simply situation somewhat. The equation of the generator of a body of revolution can be written as $x=x(y)$, where the origin of the coordinate system is located at the leading edge, the x-axis is directed along the axis of symmetry of the body, and the y-axis is normal to the projectile's axis of symmetry. Then, in the formulation of the problem, one must include the condition $x'(y)\geq 0$ (Alekseev et al., 1987) that automatically implies the existence of shapes with flat bluntness and excludes from consideration saw-like generators (see *Figure 5-1*). Admitting the saw-like generators allows us to make a target function arbitrarily small, but this solution has no practical sense, since without the condition $x'(y) \geq 0$, our variational problem is not properly posed.

Despite the availability of well developed methods for investigating variational problems, practical solution of these problems requires some experience and intuition that allow one to obtain correct results by using reasonably less rigorous considerations. Note also that the investigation of practical problems is generally performed using the necessary conditions for the optimum.

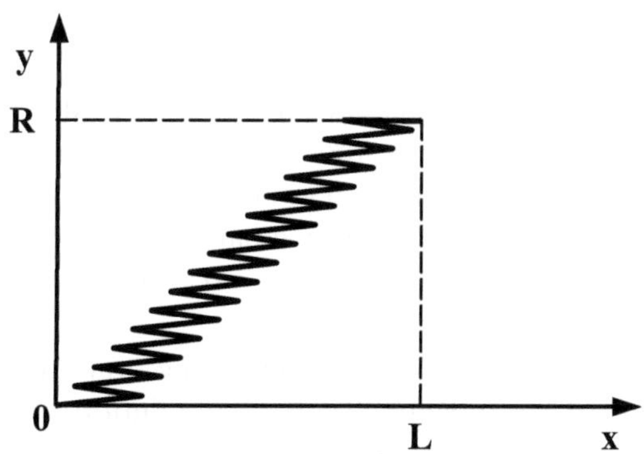

Figure 5-1. Saw-shaped generator of a body of revolution.

Variational problems associated with shape optimization of impactors have some peculiarities, which have also been encountered in application of LIMs in aerodynamics. Miele and Hull (Miele, 1965) investigated a problem of a minimum drag projectile, for which pressure was found using Newton's model and the shape of the shank or its surface area were known. They showed (with some additional restrictions) that for the optimal projectile $u = -\vec{v}^0 \cdot \vec{n}^0 = u_*$, where u_* is a constant that depends upon a coefficient of friction. Berdichevsky (1975) investigated a similar problem without friction but with some additional requirements for the shape of a projectile and demonstrated that certain versions of the optimization problem are well-posed while some others are ill-posed. The simplest version of the ill-posed problem is to find a minimum-drag conical projectile with a given shape of the shank, without additional restrictions. Clearly the drag can be made infinitely small by increasing the height of the cone. Dubinsky, 1980 (see also Bunimovich and Dubinsky, 1995) extended the results by Miele and Hull to LIMs. Yakunina (2000a,b) proposed procedures for constructing projectiles with minimum drag from conical and plane elements with constant $-\vec{v}^0 \cdot \vec{n}^0 = u_*$. Later, Yakunina (2001) used DOP as the criterion of optimization and two-term LIMs for the normal and tangent forces to extend this theory to 3-D impactors. Although at first sight this approach is quite neat and attractive, there are several factors that restrict its applicability. Mathematical formulation of the problem by Yakunina (2001) does not allow to introduce a restriction on the length of the projectile. The Euler-Lagrange equation yields a solution with $0 < u_* < 1$ only for particular values of the parameters of the model and impact velocity, while the physical meaning of the solutions with $u_* = 0$ and $u_* = 1$ is doubtful. Particularly disturbing is the fact that the solution is strongly affected by the value of friction coefficient μ_{fr}; this coefficient is often used in penetration models for calibration and its magnitude can be varied in a wide range. Experiments have shown that the friction coefficient μ_{fr} for penetration of a metal shield by a metal striker is negligibly small (Kraft, 1955; Recht, 1990), and at the same time the assumption $\mu_{fr} = 0$ renders the optimization problem degenerate.

Aptukov and Pozdeev (1982) considered the minimax problem for determining the shape of an impactor (body of revolution) that penetrates to the maximum depth under the most unfavorable distribution of the mechanical properties along the depth of a SIS with a given areal density. The two-term LIM without friction was used with a linear relationship between the parameters of the model.

Shape optimization of penetrating impactors were studied by Ben-Dor et al. (1997a,c; 1999d; 2000a; 2001a; 2002a,b,c; 2003a,b). In a number of sections of this monograph, we use some previously published studies, but we also present the results that are original and have not been published before.

2. BASICS OF THE CALCULUS OF VARIATIONS

In this section and the subsequent section, we present some analytical and numerical methods of the calculus of variations. This survey is by no means complete or rigorous. Our goal is to facilitate an understanding of the methods used in this monograph for solving particular engineering problems. Therefore, the material is presented in the form most suitable for achieving this goal.

The simplest problem of the calculus of variations for the function of one variable is presented in many textbooks and handbooks (see, e.g., Korn and Korn, 1968; Gelfand and Fomin, 1963; Ewing, 1985). This problem is to find, among a class of functions $y(x)$, the function (the *minimizer*) that minimizes the functional:

$$I[y(x)] = \int_a^b F(x, y(x), y'(x))\,dx\,, \tag{2.1}$$

where function $y(x)$ is defined for $a \le x \le b$, and function F has first and second continuous partial derivatives with respect to x, y and y'. The problem of maximization can be reduced to the problem of minimization by changing the sign of the functional $I[y(x)]$. The boundary values y_a and y_b (see *Figure 5-2*) where

$$y(a) = y_a\,,\; y(b) = y_b \tag{2.2}$$

can be given or left free.

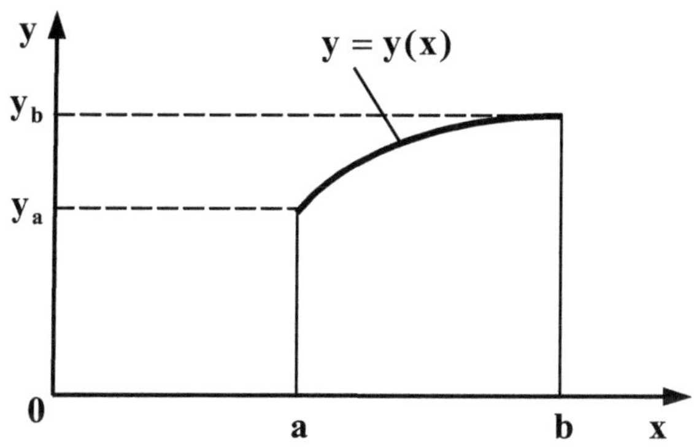

Figure 5-2. Notations.

The function $y(x)$ minimizes the functional determined by Eq. (2.1) if and only if it satisfies the *Euler-Lagrange differential equation:*

$$F_y - \frac{d}{dx} F_{y'} = 0, \tag{2.3}$$

where F_y and $F_{y'}$ are the partial derivatives of F with respect to its corresponding arguments. This equation can be rewritten as:

$$y''F_{y'y'} + y'F_{yy'} + F_{xy'} - F_y = 0 . \qquad (2.4)$$

The solution of this equation (the *extremal*) takes the form:

$$y = y(x;C,c) , \qquad (2.5)$$

where C and c are constants.

Several cases are known for which the first integral of the Euler-Lagrange equation can be determined. In particular, if F does not depend explicitly on x, i.e.,

$$F = F(y,y') , \qquad (2.6)$$

then the following expression is valid:

$$F - y'F_{y'} = C . \qquad (2.7)$$

The solution of the variational problem must satisfy the *Lagrange condition*:

$$F_{y'y'} \geq 0 . \qquad (2.8)$$

The *Erdmann-Weierstrass corner conditions* allow us to determine the locations of the points at which y' has a jump discontinuity (if these points exist). The following conditions must be satisfied at such points:

$$\Delta(F - y'F_{y'}) = 0, \quad \Delta(F_{y'}) = 0 , \qquad (2.9)$$

where $\Delta(...)$ denotes the difference between the corresponding values to the left and to the right of the point of discontinuity. The points where $F_{y'y'} \neq 0$ cannot be the corner points.

The *transversality conditions* are formulated as follows:

$$[F - y'F_{y'}]_{x=a} = 0 \text{ if } a \text{ is free,} \qquad (2.10)$$

$$[F - y'F_{y'}]_{x=b} = 0 \text{ if } b \text{ is free,} \qquad (2.11)$$

$$[F_{y'}]_{x=a} = 0 \text{ if } y_a \text{ is free,} \qquad (2.12)$$

$$[F_{y'}]_{x=b} = 0 \text{ if } y_b \text{ is free,} \tag{2.13}$$

where $[...]_{x=a}$ and $[...]_{x=b}$ mean that the expressions in the square brackets (values x, y and y') are calculated at $x = a$ and $x = b$, respectively.

Functionals of more general form than functionals determined by Eq. (2.1) are not uncommon in impact dynamics in applications concerned with shape optimization. Let us consider generalized functional of the following form (Brady, 1938; Bunimovich and Dubinsky, 1973; Cherkaev and Cherkaeva, 2003):

$$J[y(x)] = f(a, b, y_a, y_b, I_1, I_2, ..., I_N), \tag{2.14}$$

where f is some function,

$$I_i = I_i[y(x)] = \int_a^b F_i(x, y(x), y'(x))\,dx, \quad i = 1, 2, ..., N. \tag{2.15}$$

Investigation of such functionals can be reduced to the study of the functional in the form given by Eq. (2.1) with:

$$F(x, y(x), y'(x)) = \sum_{i=1}^{N} \hat{\lambda}_i F_i(x, y(x), y'(x)), \tag{2.16}$$

and with modified transversality conditions:

$$[F - y'F_{y'} - \partial f / \partial a]_{x=a} = 0 \text{ if } a \text{ is free,} \tag{2.17}$$

$$[F - y'F_{y'} + \partial f / \partial b]_{x=b} = 0 \text{ if } b \text{ is free,} \tag{2.18}$$

$$[F_{y'} - \partial f / \partial y_a]_{x=a} = 0 \text{ if } y_a \text{ is free,} \tag{2.19}$$

$$[F_{y'} + \partial f / \partial y_b]_{x=b} = 0 \text{ if } y_b \text{ is free.} \tag{2.20}$$

The constants $\hat{\lambda}_i$ ($i = 1, 2, ..., N$) are determined from the following equations:

$$\hat{\lambda}_i = \left.\frac{\partial f}{\partial I_i}\right|_{y=\tilde{y}(x;\hat{\lambda}_1,\hat{\lambda}_2,\ldots,\hat{\lambda}_N)}, \tag{2.21}$$

where $y = \tilde{y}(x;\hat{\lambda}_1,\hat{\lambda}_2,\ldots,\hat{\lambda}_N)$ is the solution of the optimization problem for the functional given by Eq. (2.16).

If some $\hat{\lambda}_{i_*}$ is positive (without the loss in generality, we can assume that $\hat{\lambda}_1 > 0$), the functional in Eq. (2.16) can be replaced by the functional:

$$F(x,y(x),y'(x)) = F_1(x,y(x),y'(x)) + \sum_{i=2}^{N} \lambda_i F_i(x,y(x),y'(x)), \tag{2.22}$$

and the solution of the optimization problem for this functional $y = \tilde{y}(x;\lambda_2,\ldots,\lambda_N)$ is used in the equations:

$$\lambda_i = \left.\left(\frac{\partial f}{\partial I_i} \Big/ \frac{\partial f}{\partial I_1}\right)\right|_{y=\tilde{y}(x;\lambda_2,\ldots,\lambda_N)}. \tag{2.23}$$

Note that well-known Bolza functional encountered in solving the variational problem with the same name is the particular case of the functional given by Eq. (2.14) with:

$$J[y(x)] = f_0(a,b,y_a,y_b) + I_1[y(x)] \tag{2.24}$$

where f_0 is some function.

3. NUMERICAL METHOD OF LOCAL VARIATIONS

In this section, we consider briefly a very convenient numerical method for solving variational problems for generalized functionals described in *Section 5-2*, namely, the *method of local variations* (Chernous'ko and Banichuk, 1973). As is the case for most existing numerical methods, it allows us to find only local extremum. Using the notations of the previous section, we have outlined the method applied to the functionals of the following form:

$$J[y(x)] = f(y_a, y_b, I_1, I_2, \ldots, I_N). \tag{3.1}$$

We assume that $y(x)$ varies in the range between y^{min} and y^{max}, when $a \leq x \leq b$. We can impose additional constraints on x, y and y' as well as on the integrals I_i. In particular, y_a and y_b can be given or left free.

To solve the minimization problem for the functional given by Eq. (3.1), the function $y(x)$ can be approximated as a piecewise linear function determined by the values $y^{(0)} = y_a, y^{(1)},..., y^{(j)},..., y^{(n-1)}, y^{(n)} = y_b$ in $n+1$ equally spaced mesh points of interpolation, respectively, $x^{(0)} = a,\ x^{(1)},..., x^{(j)},...,\ x^{(n-1)}, x^{(n)} = b$, where $y^{(j)} = y(x^{(j)}),\quad x^{(j)} = j\Delta x$, $\Delta x = (b-a)/n,\quad j = 1,2,...,n$. The unknown solution y varies in this range of arguments and assumes discrete values, i.e., y^{min}, $y^{min} + \Delta y$,..., $y^{min} + (m-1)\Delta y$, where $y^{min} + m\Delta y = y^{max}$ and $\Delta y = (y^{max} - y^{min})/m$ (see *Figure 5-3*). Then the integrals in Eq. (2.15) can be approximated as:

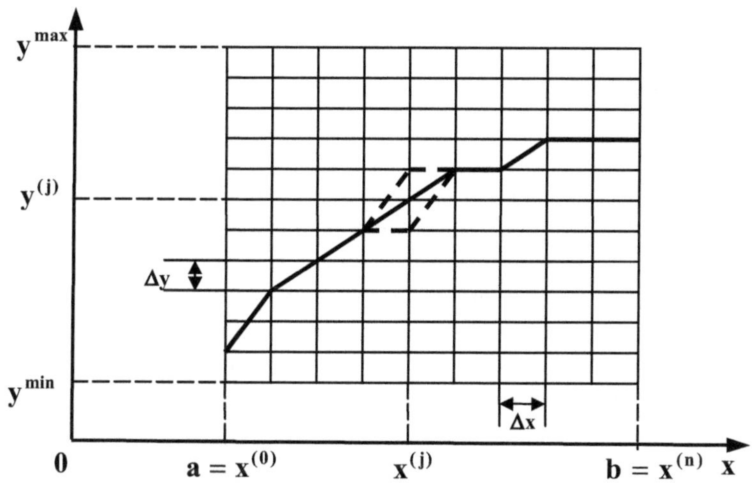

Figure 5-3. Discretization of the problem.

$$I_i\left(y^{(0)},..., y^{(n)}\right) = \sum_{i=1}^{N} I_i^{(j)}\left(y^{(j-1)}, y^{(j)}\right), \tag{3.2}$$

where

$$I_i^{(j)}\left(y^{(j-1)}, y^{(j)}\right) = \int_{x^{(j-1)}}^{x^{(j)}} F_i(x, y, y')\,dx$$
$$\approx \Delta x\, F_i\left(\frac{x^{(j-1)} + x^{(j)}}{2}, \frac{y^{(j-1)} + y^{(j)}}{2}, \frac{y^{(j)} - y^{(j-1)}}{\Delta x}\right). \tag{3.3}$$

Similar discretization must be performed for the expressions that describe the constraints.

The minimization problem is to find (with required accuracy) $n, y^{(0)},...,y^{(n-1)}$ that satisfy the constraints and provide the minimum value of the functional J determined by Eq. (3.1). In summary, the numerical minimization procedure comprises the following three steps.

Step 1. Choose the initial approximation, i. e., the initial values n and $y^{(0)},...,y^{(n)}$, taking into account the constraints, and calculate all $I_i^{(j)}$ using Eq. (3.3) and the current value of the criterion J using Eq. (3.2). Since the method allows us to determine only a local minimum, several calculations must be performed with various initial approximations.

Step 2. Perform sequentially the following calculations for $j = 0,...,n$. Change $y^{(j)}$ by the increments Δy and $(-\Delta y)$. If the constraints are satisfied, then a new value of the criterion $J = J^+$ ($J = J^-$) is calculated using Eqs. (3.1)-(3.3); otherwise set $J^+ = +\infty$ ($J^- = +\infty$). If $J^+ < min(J, J^-)$ then $y^{(j)} + \Delta y$ is set as a new value of the function at the point $x^{(j)}$ and $J = J^+$ is set as a new value of the criterion. If $J^- < min(J, J^+)$, then $y^{(j)} - \Delta y$ and $J = J^-$ are set as the new values of the function and of the criterion, respectively. If $J \le min(J^+, J^-)$, then J and $y^{(j)}$ retain their values. Note that only two integrals at most, $I_i^{(j)}$ and $I_i^{(j-1)}$, must be recalculated for correcting the criterion when the value $y^{(j)}$ is changed. If the value of the function is changed in the cycle of calculations for $j = 0,...,n$, the entire cycle is repeated. If the function and the criterion cease to change at *step 2,* then control is transferred to *step 3* of the procedure.

Step 3. If the function and the criterion were changed at *step 2* and the change of the criterion is within the allowable error limits, then the calculations are terminated. Otherwise, parameter n is increased by a factor of two, the values of the function y in new nodal points of the mesh are calculated using a linear approximation, and control is transferred to *step 2* of the procedure. The conditions of the convergence of the procedure are as follows: $\Delta x \to 0, \Delta y/(\Delta x)^{\zeta} \to 0$, where $\zeta \ge 2$.

Chapter 6

SHAPE OPTIMIZATION OF IMPACTORS PENETRATING INTO DUCTILE SHIELDS

1. TWO-TERM MODEL: NEWTON'S SOLUTION

Let us consider shape optimization of the impactor having the geometry of body of revolution penetrating into a SIS. We assume that the model given by Eq. (2-9.31) is valid and that friction between the impactor and the shield may be neglected. A dimensionless expression for the DOP can be obtained from Eqs. (2-9.30), using Eq. (2-9.25) for B_i, in the form:

$$\overline{H} = \frac{1}{\overline{J}_2} ln(1 + \lambda_2 \overline{J}_2), \tag{1.1}$$

where

$$\overline{x} = \frac{x}{L}, \overline{\Phi} = \frac{\Phi}{L}, \overline{\Phi}' = \frac{d\overline{\Phi}}{d\overline{x}}, \overline{r} = \frac{r}{L}, \tau = \frac{R}{L}, \tag{1.2}$$

$$\lambda_2 = \frac{a_2 v_{imp}^2}{a_0 \tau^2}, \quad \lambda_3 = \frac{2\pi a_2 L^2}{m}, \quad \overline{H} = \lambda_3 H$$

$$\overline{J}_2 = \frac{B_2}{\pi a_2 L^2} = \overline{J}_2[\overline{\Phi}(\overline{x})] = \overline{r}^2 + 2\int_0^1 \frac{\overline{\Phi}\overline{\Phi}'^3}{\overline{\Phi}'^2 + 1} d\overline{x} . \tag{1.4}$$

Since

$$\overline{H} = \int_0^{\lambda_2} \frac{d\zeta}{1+\zeta\overline{J}_2}, \quad \frac{d\overline{H}}{d\overline{J}_2} = -\int_0^{\lambda_2} \frac{\zeta d\zeta}{(1+\zeta\overline{J}_2)^2} < 0 \ , \tag{1.5}$$

it can be concluded that $\overline{H}$ is a decreasing function of $\overline{J}_2$ when λ_2 and λ_3 are fixed (this property for the case $r = 0$ was discussed by Jones and Rule, 2000). The latter conclusion has certain implications, as discussed below.

Let us consider the problem of determining the shape of the impactor having the maximum DOP determined by the functional $H[\Phi(x)]$ among the impactors with a given length L and shank radius R, where

$$\Phi(L) = R \tag{1.6}$$

and the parameters describing the properties of the shield, a_0 and a_2, the mass of impactor m and the impact velocity v_{imp} are given. Then the solution of the problem of minimization of the functional $\overline{J}_2[\overline{\Phi}(\overline{x})]$,

$$\overline{J}_2[\overline{\Phi}(\overline{x})] \to min \tag{1.7}$$

taking into account the constraint

$$\overline{\Phi}(1) = \tau \tag{1.8}$$

provides the solution of the above-described initial problem (after transforming the impactor's generator equation into the dimensional form).

The variational problem given by Eqs. (1.7) and (1.8) is the classical problem of Newton discussed in *Section 5-1*. Let us present here the solution of this problem in modern notations (Alekseev et al., 1987; Miele, 1965). The equation of the generator of the optimal impactor reads:

$$\overline{x}(t) = \overline{r}\varphi_1(t), \ \overline{\Phi}(t) = \overline{r}\varphi_2(t), \ t_1^* \le t \le 1, \tag{1.9}$$

where

$$\overline{r} = \frac{1}{\varphi_1(t_1^*)}, \tag{1.10}$$

$$\varphi_1(t)=\frac{3+4t^2+4t^4\ln t-7t^4}{16t^4},\ \varphi_2(t)=\frac{(t^2+1)^2}{4t^3}. \tag{1.11}$$

Parameter t_1^*, the derivative of the optimum impactor's generatrix at its end point, $x=L$, satisfies the following equation:

$$\varphi_2(t_1)-\tau\varphi_1(t_1)=0. \tag{1.12}$$

The solution of Eq. (1.12) is shown in *Figure 6-1*. The shape of the generatrix of the optimal impactor is shown in *Figure 6-2* for different values of τ. The optimum impactor has flat bluntness with a relative shank radius r/R that increases with τ.

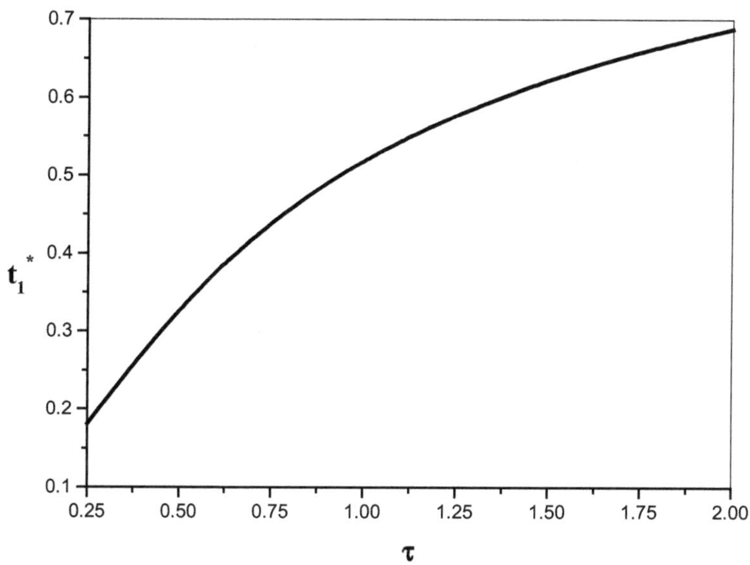

Figure 6-1. Dependence of t_1^* vs. τ.

The minimum value of the functional $\bar{J}_2$ is:

$$\bar{J}_2^{min}=\frac{\tau^2\varphi_3(t_1^*)}{8\varphi_2^2(t_1^*)},\quad \varphi_3(t)=\frac{2t^6+17t^4+10t^2-4t^4\ln t+3}{4t^4}. \tag{1.13}$$

As we shall see subsequently, many problems of shape optimization of penetrating impactors are reduced to Newton's problem.

It must be noted that if instead of the model determined by Eq. (2-9.31),

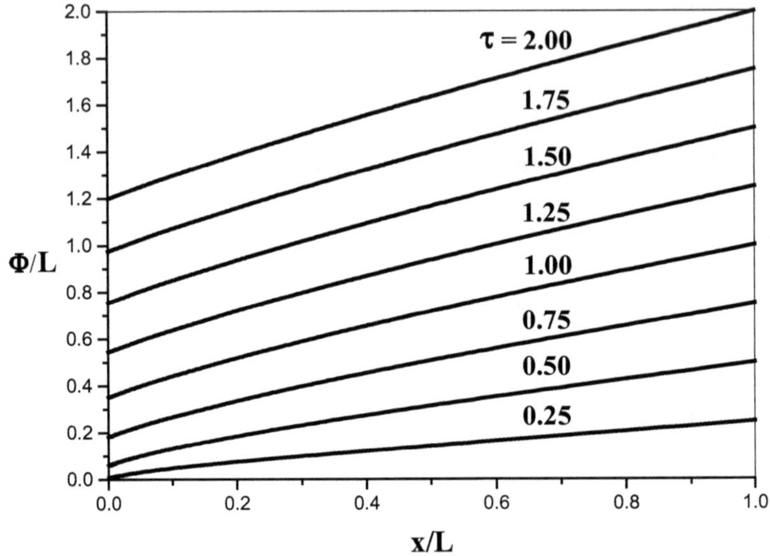

Figure 6-2. Shape of the generator of the optimal impactor.

we consider the model given by Eq. (2-9.27) with $A_0 = const$, then the optimization problem also reduces to minimization of some integral functional because in this case as well as in Eq. (2-9.27), $B_0 = const$ (see Eq. 2-9.26 for $\mu_{fr} = 0$). Eq. (2-9.17) implies that the classic variational problem for the integral functional is also obtained for 3-D impactors with a given area of the shank when the model with $A_0 = const$ and $A_1 = const$ is used. In this case also, $B_0 = const$.

2. TWO-TERM MODEL WITH FRICTION: NUMERICAL INVESTIGATION

Consider now the problem similar to the problem in *Section 6-1* taking into account the friction between the impactor and the shield. Here we use the results obtained by Ben-Dor et al. (2003a). Earlier, the same problem was considered by Jones and Rule (2000). The differences between the two approaches is discussed in *Section 5-1.*

A dimensionless expression for the DOP can be obtained from Eqs. (2-9.30) and (2-9.25) in the following form:

$$\overline{H} = \frac{1}{\overline{K}_2} ln\left(1 + \lambda_4 \frac{\overline{K}_2}{\overline{K}_0} \right), \tag{2.1}$$

where Eq. (1.2) remains valid, λ_3 is determined by Eq. (1.3),

$$\overline{K}_0 = \overline{r}^2 + 2\int_0^1 \overline{\Phi}(\overline{\Phi}' + \mu_{fr})d\overline{x}\,, \overline{K}_2 = \overline{r}^2 + 2\int_0^1 \frac{\overline{\Phi}\overline{\Phi}'^2(\overline{\Phi}' + \mu_{fr})}{\overline{\Phi}'^2 + 1} d\overline{x} \tag{2.2}$$

and

$$\lambda_4 = \frac{a_2 v_{imp}^2}{a_0}, \qquad \overline{K}_i = \frac{B_i}{\pi a_i L^2}, \quad i = 0,2\,. \tag{2.3}$$

The problem of optimization,

$$\overline{H} = \overline{H}(\overline{K}_0, \overline{K}_2) \to max, \tag{2.4}$$

is investigated taking into account the condition given by Eq. (1.8) and the following constraints:

$$\overline{\Phi}'(\overline{x}) \geq 0\,, \ \tau\overline{x} \leq \overline{\Phi}(\overline{x}) \leq \tau\,, \ 0 \leq \overline{x} \leq 1\,. \tag{2.5}$$

The constraint $\tau\overline{x} \leq \overline{\Phi}$ is introduced so as to eliminate from consideration sharp bodies with a needle-shaped part of the nose that are of no practical significance. This constraint implies that a sharp cone with a generatrix $\overline{\Phi} = \tau\overline{x}$ is an optimal impactor for relatively small λ_4. Indeed, expanding the logarithm in the expression for the DOP given by Eq. (2.1) in power series of $\phi\lambda_4$, where

$$\phi = \frac{\overline{K}_2}{\overline{K}_0}, \tag{2.6}$$

we obtain:

$$\overline{H} = \frac{1}{\overline{K}_2} \sum_{\nu=1}^{\infty} (-1)^{\nu-1} \frac{(\phi\lambda_4)^\nu}{\nu}\,. \tag{2.7}$$

Since

$$\overline{K}_0 - \overline{K}_2 = 2\int_0^1 \frac{\overline{\Phi}(\overline{\Phi}' + \mu_{fr})}{\overline{\Phi}'^2 + 1} d\overline{x} > 0, \tag{2.8}$$

and, consequently, $\phi < 1$, the series in Eq. (2.7) converge for $\lambda_4 \leq 1$. Then for small λ_4, Eq. (2.1) can be written as:

$$\overline{H} \approx \frac{\phi\lambda_4}{\overline{K}_2} = \frac{\lambda_4}{\overline{K}_0} = \frac{\lambda_4}{\tau^2 + 2\mu_{fr}\overline{K}_1}. \tag{2.9}$$

In Eq. (2.9), the following transformation of the integral $\overline{K}_0$ is used:

$$\overline{K}_0 = \overline{r}^2 + \int_0^1 \frac{d\overline{\Phi}^2}{d\overline{x}} d\overline{x} + 2\mu_{fr}\int_0^1 \overline{\Phi} d\overline{x} = \tau^2 + 2\mu_{fr}\overline{K}_1, \tag{2.10}$$

where

$$\overline{K}_1 = \int_0^1 \overline{\Phi} d\overline{x}. \tag{2.11}$$

Taking into account Eq. (2.5) we may conclude that in the limiting case under consideration the minimum $\overline{K}_1$ (maximum $\overline{H}$) is attained when $\overline{\Phi} = \tau\overline{x}$, i.e., for a cone-shaped impactor.

The results of numerical calculations were obtained using the method of local variations (see *Section 5-3*). The shape of the generatrix of the optimum impactor is shown in *Figure 6-3a-b*. For $\lambda_4 \leq \lambda_4^*$, where λ_4^* is a boundary value depending on τ and μ_{fr}, the optimal impactor is a sharp cone, while for $\lambda_4 > \lambda_4^*$, the optimal impactor has flat bluntness.

The maximum DOP, $\overline{H}_{max}$, as a function of the parameter λ_4 (for convenience, written in transformed variables) is shown in *Figure 6-4a-d*. Inspection of these figures reveals that the influence of friction coefficient on $\overline{H}_{max}$ is significantly enhanced with the increase of λ_4, i.e., with the increase of impact velocity for a given shield material.

Figure 6-5 shows that not only the shape of the optimum impactor is close to a blunt (in general case) cone but also that the maximum DOP of the optimal impactor $\overline{H}_{max}$ and the maximum DOP of the optimal truncated conical impactor $\overline{H}_{max\,cone}$ are also close. The difference in DOP between the optimal impactor and the optimal truncated conical impactor decreases with an increase of the friction coefficient. Therefore, the optimal truncated cone (a body having a rather simple shape) penetrates to the depth that is close to the optimal depth. Dimensionless radius of the bluntness of the optimal cone $r_{opt\,cone}/R$ as a function of λ_4 is shown for different τ and μ_{fr} in *Figure 6-6a-d.*

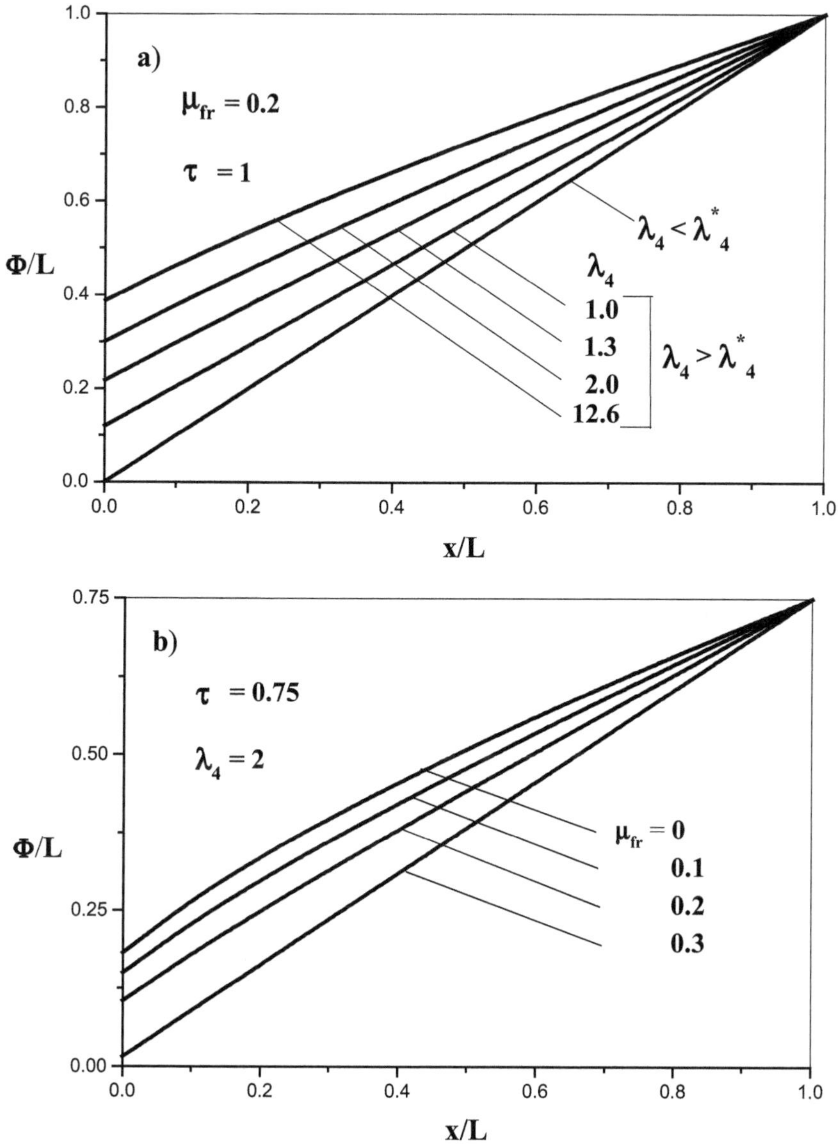

Figure 6-3. Shape of the generator of the optimal impactor.

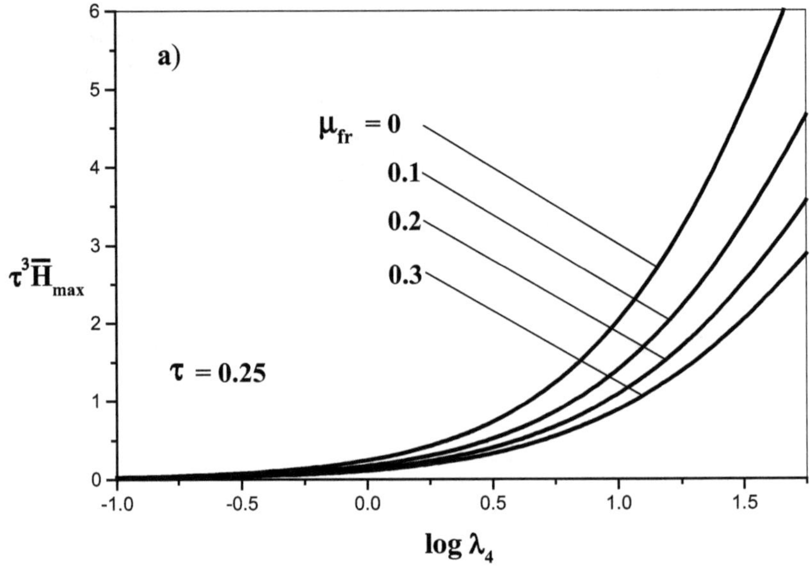
a)
μ_{fr} = 0
0.1
0.2
0.3
τ = 0.25
$\tau^3\overline{H}_{max}$
log λ_4
0
1
2
3
4
5
6
-1.0
-0.5
0.0
0.5
1.0
1.5

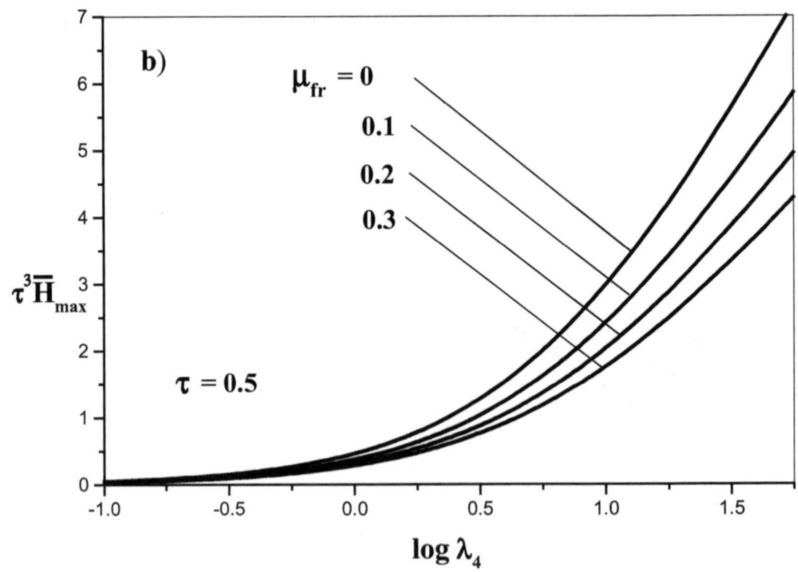
b)
μ_{fr} = 0
0.1
0.2
0.3
τ = 0.5
$\tau^3\overline{H}_{max}$
log λ_4
0
1
2
3
4
5
6
7
-1.0
-0.5
0.0
0.5
1.0
1.5

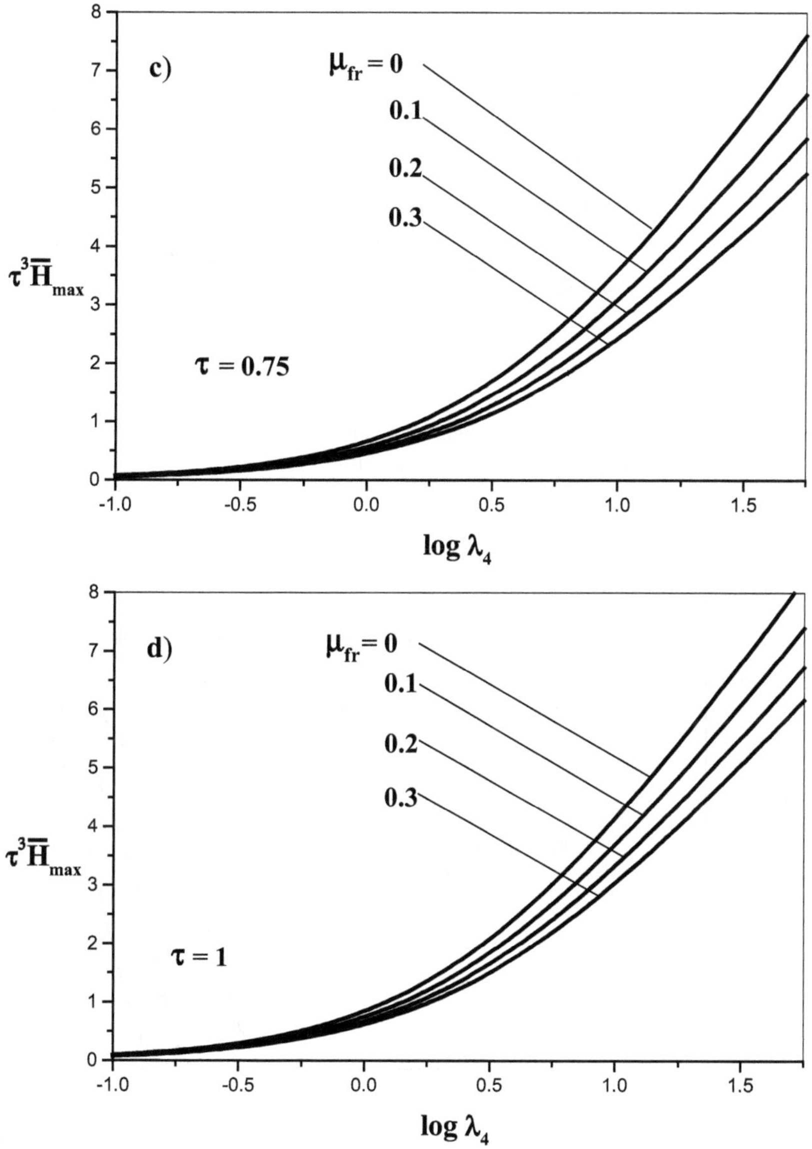

Figure 6-4. DOP of the optimal impactor, $\overline{H}_{max}$, as a function of parameter λ_4, for different τ and μ_{fr}.

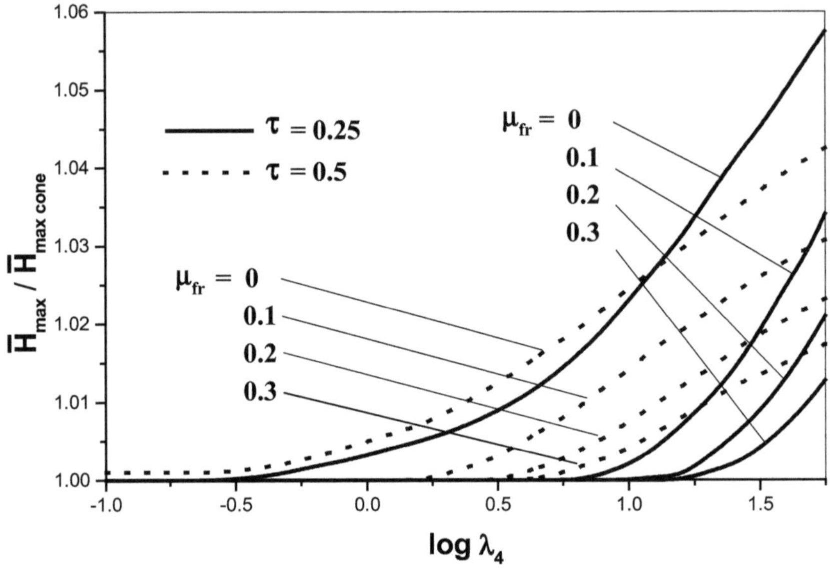

Figure 6-5. Comparison between the values of the DOP for the optimal impactor, $\overline{H}_{max}$, and the optimal truncated conical impactor, $\overline{H}_{max\ cone}$.

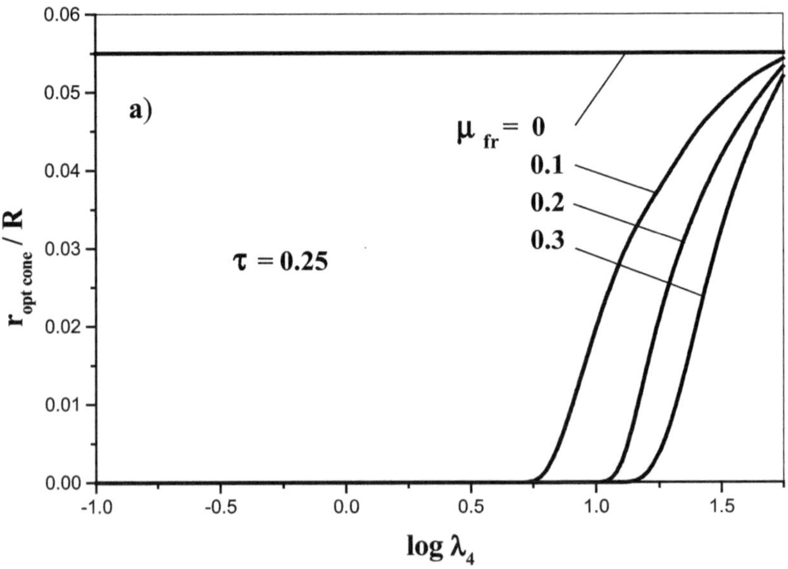

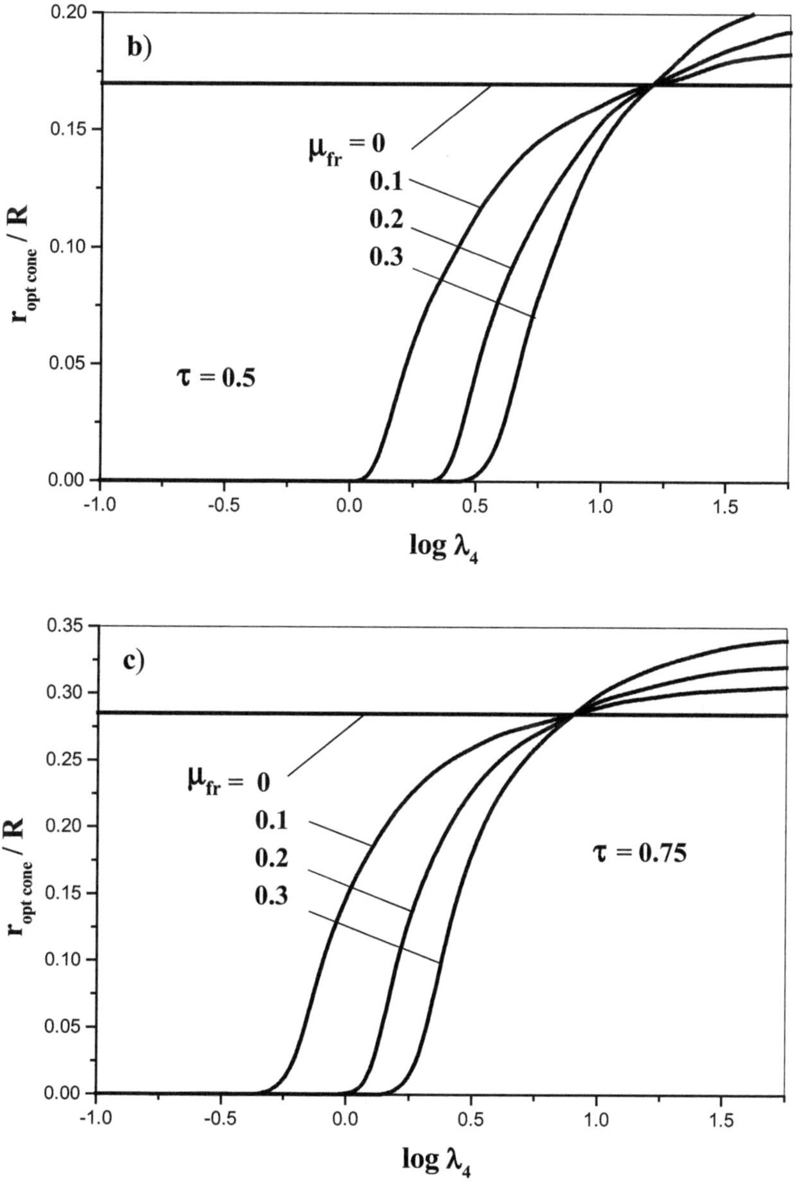
b)
$\mu_{fr} = 0$
0.1
0.2
0.3
$\tau = 0.5$
$r_{opt\ cone} / R$
$\log \lambda_4$
0.00
0.05
0.10
0.15
0.20
-1.0
-0.5
0.0
0.5
1.0
1.5
c)
$\mu_{fr} = 0$
0.1
0.2
0.3
$\tau = 0.75$
$r_{opt\ cone} / R$
$\log \lambda_4$
0.00
0.05
0.10
0.15
0.20
0.25
0.30
0.35
-1.0
-0.5
0.0
0.5
1.0
1.5

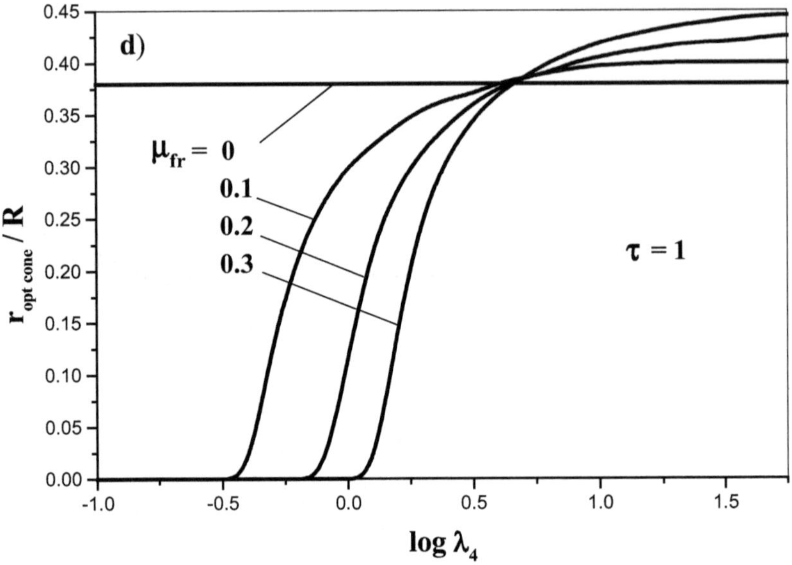

Figure 6-6. Radius of bluntness of the optimal truncated conical impactor $r_{opt\ cone}$ normalized by the radius of the shank of the impactor R , for different τ and μ_{fr} .

3. THREE-TERM MODEL: FORMULATION OF THE PROBLEM

To describe the impactor-shield interaction, we use the model that was proposed and studied by Warren and Forrestal (1998). This model is based on the SCEA in the form determined by Eq. (3-1.8), and it implies the LIM described by Eq. (2-9.22), $\Omega_n(u,v) = a_2 u^2 v^2 + a_1 uv + a_0$ with $\mu_{fr} = 0$ and

$$a_i = \tilde{a}_i \mu_7 \mu_3^i \, , i = 0,1,2, \, \mu_3 = \sqrt{\gamma_{sh} / \mu_7} \, , \tag{3.1}$$

where μ_7 is the uniaxial compressive strength of the material of the shield, γ_{sh} is density of the non-deformed material of the shield, and $\tilde{a}_i \, (i = 0,1,2)$ are the dimensionless coefficients determining the properties of the material of the shield. Values $\tilde{a}_i$ for aluminum are given in *Table 6-1* (Warren and Forrestal, 1998).

In this study, we undertook to investigate this problem analytically for the models 2-4 given in *Table 6-1*. Model 1 was studied in *Section 6-1*, and the results obtained for this model will be used only for the completeness of the presentation.

Table 6-1. Parameters of the models for an aluminum shield (Warren and Forrestal, 1998)

Model number	Characteristic of the model	$\widetilde{a}_0$	$\widetilde{a}_1$	$\widetilde{a}_2$
1	Incompressible without strain rate effects	4.8376	0.0	1.5
2	Incompressible with strain rate effects	5.5468	0.6255	1.4377
3	Compressible without strain rate effects	4.4534	0.4680	0.9926
4	Compressible with strain rate effects	5.0394	0.9830	0.9402

The expression for determining the DOP is given by Eq. (2-9.19):

$$H = mM(1,1,B_0,B_1,B_2;v_{imp}) = m\int_0^{v_{imp}} \frac{\zeta d\zeta}{B_2\zeta^2 + B_1\zeta + B_0}, \tag{3.2}$$

where B_0, B_1 and B_2 can be determined from transformed Eqs. (2-9.25) and (2-9.26):

$$\frac{B_0}{\pi a_0} = R^2, \tag{3.3}$$

$$\frac{B_i}{\pi a_i} = R^2 - 2\int_0^L \Phi\psi_i(\Phi')dx, \; i = 1,2, \tag{3.4}$$

$$\psi_1(z) = z\left(1 - \frac{z}{\sqrt{z^2+1}}\right), \; \psi_2(z) = \frac{z}{z^2+1}. \tag{3.5}$$

The integral in Eq. (3.2) can be evaluated through elementary functions as shown in *Appendix 2*.

Equations. (3.3) and (3.4) can be written as follows:

$$B_i = \pi\mu_7 R^2 \mu_3^i e_i, \tag{3.6}$$

where

$$e_0 = \widetilde{a}_0 > 0, \; e_i = \widetilde{a}_i(1 - \frac{2}{\tau^2}\bar{J}_i), \; i = 1,2, \; \tau = \frac{R}{L}, \tag{3.7}$$

$$\bar{J}_i = \int_0^1 \bar{\Phi}\psi_i(\bar{\Phi}')d\bar{x}, \quad i = 1,2, \tag{3.8}$$

$$\overline{x}=\frac{x}{L},\overline{\Phi}=\frac{\Phi}{L},\overline{\Phi}'=\frac{d\overline{\Phi}}{d\overline{x}}\ . \tag{3.9}$$

Using Eqs. (A2.15), (A2.16) and (3.6), we can transform the expression for the DOP (Eq. 3.2) as follows:

$$\begin{aligned}H &= mM(1,1,\pi\mu_7 R^2 e_0,\pi\mu_7 R^2 \mu_3 e_1,\pi\mu_7 R^2 \mu_3^2 e_2;v_{imp})\\ &=\frac{m}{\pi\mu_7 R^2}M(1,1,e_0,\mu_3 e_1,\mu_3^2 e_2;v_{imp})\\ &=\frac{m}{\pi\mu_7 R^2\mu_3^2}M(1,1,e_0,e_1,e_2;\mu_3 v_{imp})\\ &=\frac{m}{\pi R^2\gamma_{sh}}M(1,1,e_0,e_1,e_2;\mu_3 v_{imp}).\end{aligned} \tag{3.10}$$

Therefore, the dimensionless DOP can be written as:

$$\overline{H}=M(1,1,e_0,e_1,e_2;\overline{v}_{imp})=\int_0^{\overline{v}_{imp}}\frac{\zeta d\zeta}{e_2\zeta^2+e_1\zeta+e_0}, \tag{3.11}$$

where

$$\overline{v}_{imp}=\mu_3 v_{imp},\overline{H}=\lambda_5 H,\lambda_5=\frac{\pi\gamma_{sh}R^2}{m}\ . \tag{3.12}$$

Let us show that $e_1,e_2>0$ if $\widetilde{a}_1,\widetilde{a}_2>0$. The following estimations $(i=1,2)$

$$1-\frac{2}{\tau^2}\overline{J}_i\ge 1-\frac{2}{\tau^2}\int_0^1\overline{\Phi}\overline{\Phi}'d\overline{x}=1-\frac{2}{\tau^2}\cdot\frac{1}{2}\left[\tau^2-\left(\frac{r}{L}\right)^2\right]=\left(\frac{r}{R}\right)^2\ge 0 \tag{3.13}$$

imply that $e_1,e_2\ge 0$. Now let us prove that $e_1,e_2\ne 0$. Consider the equations $e_i=0\,(i=1,2)$, which can be written as:

$$\int_0^1\left[1-\frac{\overline{\Phi}'^i}{(\overline{\Phi}'^2+1)^{i/2}}\right]\overline{\Phi}\,\overline{\Phi}'d\overline{x}=\frac{\tau^2}{2} \tag{3.14}$$

or

$$\int_0^1 \overline{\Phi}\,\overline{\Phi}'d\overline{x} - \int_0^1 \frac{\overline{\Phi}\,\overline{\Phi}'^{i+1}}{(\overline{\Phi}'^2+1)^{i/2}}d\overline{x} = \frac{\tau^2}{2}. \qquad (3.15)$$

The first integral in the left-hand side of Eq. (3.15) is equal to $0.5[\tau^2-(r/L)^2]$, and therefore this equation yields:

$$\int_0^1 \frac{\overline{\Phi}\,\overline{\Phi}'^{i+1}}{(\overline{\Phi}'^2+1)^{i/2}}d\overline{x} = -\frac{1}{2}\left(\frac{r}{L}\right)^2. \qquad (3.16)$$

The integral in Eq. (3.16) is larger than or equal to zero. Therefore, the right-hand side of Eq. (3.16) is less than or equal to zero, and equality can be attained only if $r=0$ and $\overline{\Phi}'=0\,(\overline{\Phi}=const)$ for $0\le\overline{x}\le 1$. The contradiction of these conditions implies that $e_1,e_2\neq 0$.

The considered problem is to maximize the DOP for a SIS with the known impact velocity and parameters determining the mechanical properties of the shield. The mass, length and shank radius of the impactor are assumed to be given. The problem reduces to the optimization of the functional $\overline{H}$ in Eq. (3.11) with given parameters $\tilde{a}_0,\tilde{a}_1,\tilde{a}_2,\overline{v}_{imp},\tau$, where the solution must satisfy the following condition:

$$\overline{\Phi}(1)=\tau. \qquad (3.17)$$

Although the expression for the DOP $\overline{H}$ does not depend explicitly on a radius of a flat bluntness, the model also allows a truncated impactor. This is very important because a high-speed optimal impactor has bluntness, as is shown below.

4. THREE-TERM MODEL: INVESTIGATION OF THE VARIATIONAL PROBLEM

4.1 Outline of the method of solution

The optimization criterion, the functional $\overline{H}[\overline{\Phi}(\overline{x})]$, is determined by Eq. (3.11), and it depends on the functionals $\overline{J}_1$ and $\overline{J}_2$ through the parameters e_1 and e_2. Functionals $\overline{J}_1$ and $\overline{J}_2$, in turn, depend on the unknown solution $\overline{\Phi}(\overline{x})$. According to the method described in *Section 5-2*, this non-classical variational problem can be reduced to investigating the following functional:

$$\hat{J} = \int_0^1 \hat{F}(\overline{\Phi},\overline{\Phi}')\,d\overline{x}, \quad \hat{F}(\overline{\Phi},\overline{\Phi}') = \hat{\lambda}_1\overline{\Phi}\psi_1(\overline{\Phi}') + \hat{\lambda}_2\overline{\Phi}\psi_2(\overline{\Phi}'), \qquad (4.1)$$

where functions ψ_1 and ψ_2 are determined by Eq. (3.5) and $\hat{\lambda}_i\ (i = 1,2)$ are some unknown constants. Therefore, the solution of the variational problem for a functional given by Eq. (4.1) depends on these constants. These constants are determined by the following formulas:

$$\hat{\lambda}_i = \frac{\partial\overline{H}}{\partial\overline{J}_i} = \frac{\partial\overline{H}}{\partial e_i}\frac{\partial e_i}{\partial\overline{J}_i} = \frac{2\tilde{a}_i}{\tau^2}M(2,i+1,e_0,e_1,e_2;\overline{v}_{imp}), \qquad (4.2)$$

where the integrals $M(2,2,e_0,e_1,e_2;\overline{v}_{imp})$ and $M(2,3,e_0,e_1,e_2;\overline{v}_{imp})$ are calculated for the optimum solution of the variational problem. Since $\hat{\lambda}_2 > 0$, the criterion given by Eq. (4.1) can be replaced by the following functional:

$$\overline{J} = \frac{1}{\hat{\lambda}_2}\hat{J} = \int_0^1 F(\overline{\Phi},\overline{\Phi}')\,d\overline{x}, \qquad (4.3)$$

where

$$F(\overline{\Phi},\overline{\Phi}') = [\lambda\psi_1(\overline{\Phi}') + \psi_2(\overline{\Phi}')]\overline{\Phi} = \overline{\Phi}\psi(\overline{\Phi}'), \qquad (4.4)$$

$$\psi(z) = z\left[\lambda\left(1 - \frac{z}{\sqrt{z^2+1}}\right) + \frac{1}{z^2+1}\right], \qquad (4.5)$$

$$\lambda = \frac{\tilde{a}_1}{\tilde{a}_2}\frac{M(2,2,e_0,e_1,e_2;\overline{v}_{imp})}{M(2,3,e_0,e_1,e_2;\overline{v}_{imp})}, \quad \lambda = \frac{\hat{\lambda}_1}{\hat{\lambda}_2} \geq 0\ . \qquad (4.6)$$

Although the optimization criterion is represented in a form that does not depend explicitly on the dimensionless radius of the impactor's flat bluntness, $\overline{r} = r/L = \overline{\Phi}(0)$, the model also allows truncated impactors, i.e., $\overline{r}$ is considered as an unknown parameter to be determined.

We studied the variational problem for the functional $\overline{J}$ using the necessary conditions for the extrema. Generally, the variational problem is solved in the following manner. First, a solution, $\overline{\Phi} = \overline{\Phi}_{opt}(\lambda, \overline{x})$, is found for every parameter $\lambda > 0$. Then, a decreasing function $\overline{v}_{imp}(\lambda)$ is determined (for a given τ) such that if λ_* is a root of the equation $\overline{v}_{imp}(\lambda) = \overline{v}_{imp*}$, then $\overline{\Phi} = \overline{\Phi}_{opt}(\lambda_*, \overline{x})$ is the solution of the considered variational problem for a given impact velocity $\overline{v}_{imp*}$.

4.2 The necessary conditions

Since the integrand F in Eq. (4.3) does not depend on $\overline{x}$, the Euler-Lagrange equation has a first integral (see Eq. 5-2.7), $F - \overline{\Phi}' F_{\overline{\Phi}'} = C$, which can be written as follows:

$$\overline{\Phi} \varphi_0(\overline{\Phi}') = C, \tag{4.7}$$

where C is a constant and

$$\varphi_0(z) = \psi(z) - z\psi'(z) = z^2 \left(\frac{\lambda}{(z^2+1)^{3/2}} + \frac{2z}{(z^2+1)^2} \right). \tag{4.8}$$

A transversality condition written for the initial point $\overline{x} = 0$ (see Eq. 5-2.12), $(F_{\overline{\Phi}'})_{\overline{x}=0} = 0$, reads:

$$\overline{r} \varphi_1(t_0) = 0, \tag{4.9}$$

where

$$t_0 = \overline{\Phi}'(0), \tag{4.10}$$

$$\varphi_1(z) = \psi'(z) = \lambda \left[1 - \frac{z(z^2+2)}{(z^2+1)^{3/2}} \right] + \frac{1-z^2}{(z^2+1)^2} \tag{4.11}$$

A Legendre condition (see Eq. 5-2.8) written with the less than or equal to sign because the problem of maximization is considered, $F_{\overline{\Phi}'\overline{\Phi}'} \le 0$, yields:

$$\overline{\Phi} \varphi_2(\overline{\Phi}') \ge 0, \tag{4.12}$$

where

$$\varphi_2(z) = -\psi''(z) = \lambda\frac{2-z^2}{(z^2+1)^{5/2}} + \frac{2z(3-z^2)}{(z^2+1)^3}. \qquad (4.13)$$

4.3 Constructing the solution

4.3.1 Investigation of the equation for determining t_0^*

Since $\bar{r} \neq 0$ (see Eq. 4.7), Eq. (4.9) implies that t_0^* (optimal value t_0) must be determined as a root of the equation:

$$\varphi_1(t_0) = 0 . \qquad (4.14)$$

Let us now prove that the latter equation has a unique solution. The function $\varphi_1(z)$ is defined by Eq. (4.11), and it can be written as:

$$\varphi_1(z) = \lambda\varphi_1^*(z) + \varphi_1^{**}(z), \qquad (4.15)$$

where

$$\varphi_1^*(z) = 1 - \frac{z(z^2+2)}{(z^2+1)^{3/2}}, \quad \varphi_1^{**}(z) = \frac{1-z^2}{(z^2+1)^2}. \qquad (4.16)$$

Let us solve the inequality:

$$\varphi_1^*(z) < 0. \qquad (4.17)$$

After a change of variables $z^2 = Z$, inequality (4.17) reads:

$$\frac{\sqrt{Z}(Z+2)}{(Z+1)^{3/2}} > 1. \qquad (4.18)$$

After some algebra, we obtain:

$$Z^2 + Z - 1 > 0 \qquad (4.19)$$

Taking into account the condition $Z \geq 0$, the solution of the inequality (4.19) is $Z > 0.5(\sqrt{5}-1)$. Finally, the solution of the inequality given by Eq. (4.17) reads:

$$z > \tilde{t}, \quad \tilde{t} = \sqrt{0.5(\sqrt{5} - 1)} \approx 0.786. \tag{4.20}$$

Therefore, for $z > 1 > \tilde{t}$ the inequalities $\varphi_1^*(z) < 0$ and $\varphi_1^{**}(z) < 0$ are valid, and, consequently, $\varphi_1(z) < 0$. Thus, Eq. (4.14) cannot have a root that is larger or equal to 1.

Equations (4.11) and (4.13) imply that $\varphi_2(z) = -\varphi_1'(z) > 0$ for $0 < z < 1$, i. e., the function $\varphi_1(z)$ decreases in the range of z from 0 to 1. Since

$$\varphi_1(0) = \lambda + 1 > 0, \tag{4.21}$$

$$\varphi_1(1) = \lambda(1 - 3/2^{3/2}) \approx -0.061\lambda < 0, \tag{4.22}$$

there exists only one root. Calculating $\varphi_1(z)$ at some intermediate point:

$$\varphi_1(\tilde{t}) = 0.5(7 - 3\sqrt{5}) \approx 0.146 > 0 \tag{4.23}$$

we conclude that the unique root of Eq. (4.14), t_0^*, satisfies the following inequality:

$$\tilde{t} < t_0^* < 1. \tag{4.24}$$

4.3.2 Equation of the extremal

Using a parameter

$$t = \bar{\Phi}' \tag{4.25}$$

we may write the solution of Eq. (4.7) as:

$$\bar{\Phi} = \frac{C}{\varphi_0(t)}, \tag{4.26}$$

$$\bar{x} = C \int_{t_0^*}^{t} \varphi_3(z) dz, \tag{4.27}$$

where

$$\varphi_3(z) = \frac{1}{z}\frac{d}{dz}\left[\frac{1}{\varphi_0(z)}\right] = \frac{\psi''}{(\psi - p\psi')^2} = -\frac{\varphi_2(z)}{[\varphi_0(z)]^2} < 0, \qquad (4.28)$$

$$C > 0,\ 0 < t_1 < t \le t_0^*, \qquad (4.29)$$

and the value of parameter t, $t = t_1$, corresponds to the boundary point $\bar{x} = 1, \bar{\Phi} = \tau$.

Since

$$\frac{d^2\bar{\Phi}}{d\bar{x}^2} = \frac{dt}{d\bar{x}} = \left(\frac{d\bar{x}}{dt}\right)^{-1} = \frac{1}{C\varphi_3(t)} < 0\ , \qquad (4.30)$$

we conclude that Eqs. (4.26) and (4.27) determine a convex curve.

Clearly, a Legendre condition given by Eq. (4.12) is satisfied for every t, where $0 < t \le t_0 < 1$. Since $\varphi_2(t) \neq 0$, the corner points do not exist.

4.3.3 Investigation of the equation for determining t_1^*

Equations (4.26) and (4.27) written at the right boundary point read:

$$\tau = \frac{C}{\varphi_0(t_1)}, \qquad (4.31)$$

$$1 = C\int_{t_0^*}^{t_1} \varphi_3(z)dz\ . \qquad (4.32)$$

Equations. (4.31) and (4.32) yield an equation for t_1:

$$\Lambda(t_1) = 0, \qquad (4.33)$$

where

$$\Lambda(z) = \tau\int_{t_0^*}^{z}\frac{\psi''dt}{(\psi - t\psi')^2} - \frac{1}{\psi(z) - z\psi'(z)} = \tau\int_{t_0^*}^{z}\varphi_3(\zeta)d\zeta - \frac{1}{\varphi_0(z)}. \qquad (4.34)$$

Let us prove that Eq. (4.33) has a unique solution at the interval $(0, t_0)$. Taking into account that $\varphi_0, \varphi_2 > 0$ and:

$$\Lambda(t_0^*) = -\frac{1}{\psi(t_0^*) - t_0^*\psi'(t_0^*)} = -\frac{1}{\varphi_0(t_0^*)}, \tag{4.35}$$

$$\Lambda'(z) = \frac{\psi''(z)(\tau - z)}{[\psi(z) - z\psi'(z)]^2} = \frac{\varphi_2(z)(z-\tau)}{[\varphi_0(z)]^2}, \tag{4.36}$$

we can establish some properties of the function $\Lambda(t_1)$.

Let us investigate the behavior of the function $\Lambda(t_1)$ when $t_1 \to +0$. Equations (4.8) and (4.13) imply that $\varphi_0 \propto (t_1)^2, \varphi_2 \propto 1$ and $\varphi_0, \varphi_2 > 0$. Using Eq. (4.36), we conclude that $\Lambda' \propto (t_1)^{-4}$ and $\Lambda < 0$. Since

$$\Lambda(t_1) = \Lambda(t_0^*) + \int_{t_0^*}^{t_1} \Lambda'(t)dt, \quad \Lambda(t_0^*) < \infty, \; t_1 < t_0^*, \tag{4.37}$$

then

$$\lim_{t_1 \to +0} \Lambda(t_1) = +\infty. \tag{4.38}$$

If $t_0^* > \tau$ (see *Figure 6-7a*), then $\Lambda(t_1)$ decreases from $\Lambda(t_0^*) < 0$ to $\Lambda(\tau) < 0$, while t_1 decreases from t_0^* to τ; a further decrease of t_1 to $+0$ implies increase of $\Lambda(t_1)$ to $+\infty$. If $t_0^* < \tau$ (see *Figure 6-7b*), then $\Lambda(t_1)$ increases from $\Lambda(t_0^*) < 0$ to $+\infty$, while t_1 decreases from t_0^* to $+0$.

Consequently, Eq. (4.33) has the unique root $t_1 = t_1^* < t_0^*$.

4.3.4 Procedure for constructing a solution

When t_1^* has been determined, C and $\bar{r}$ can be found from Eqs. (4.31) and (4.26), respectively:

$$C = \tau\varphi_0(t_1^*), \tag{4.39}$$

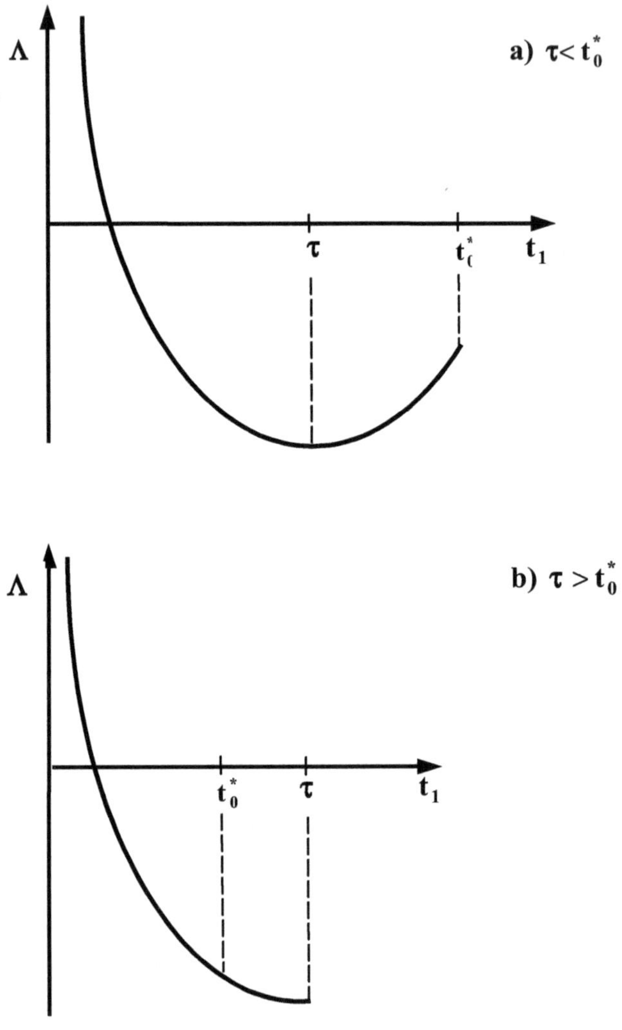

Figure 6-7. Behavior of the function $\Lambda(t_1)$.

$$\bar{r} = \frac{C}{\varphi_0(t_0^*)} = \frac{\tau\varphi_0(t_1^*)}{\varphi_0(t_0^*)} \,. \tag{4.40}$$

The above-described procedure allows us to determine the optimum generatrix of the impactor for a given τ and some $\lambda > 0$ and to calculate the parameters e_1 and e_2. Then, the first relationship in Eqs. (4.6) can be considered as the equation for determining $\overline{v}_{imp}$:

$$\Xi(\overline{v}_{imp}) = 0\,, \tag{4.41}$$

where

$$\Xi(W) = M(2{,}3{,}e_0{,}e_1{,}e_2;\overline{v}_{imp}) - cM(2{,}2{,}e_0{,}e_1{,}e_2;\overline{v}_{imp}) \\ = \int_0^W \frac{\zeta^2(\zeta - c)d\zeta}{(e_0 + e_1\zeta + e_2\zeta^2)^2}, \quad c = \frac{\widetilde{a}_1}{\widetilde{a}_2\lambda}. \tag{4.42}$$

Let us prove that Eq. (4.41) has only one root. To this end, we use the expression for the derivative $\Xi'(W)$:

$$\Xi'(W) = \frac{W^2(W - c)}{(e_0 + e_1 W + e_2 W^2)^2} \tag{4.43}$$

and the following property of the function $\Xi(W)$:

$$\lim_{W \to +\infty} \Xi(W) = +\infty\,. \tag{4.44}$$

The validity of Eq. (4.44) can be proved using Eqs. (A2.18)-(A2.22). Equations (4.42)-(4.44) show that the function $\Xi(W)$ decreases from $\Xi(0) = 0$ to $\Xi(c) < 0$ when W increases from 0 to c. A further increase of W to $+\infty$ results in an increase of the function $\Xi(W)$ from a negative value $\Xi(c)$ to $+\infty$. Consequently, Eq. (4.41) has only one root, $\overline{v}_{imp} = \overline{v}^*_{imp} > c$.

The optimum value of the dimensionless DOP, $\overline{H}^{max}$, can be determined from Eqs. (3.11) where e_i are found from Eq. (3.7) and the integrals $\overline{J}_i$ are calculated for the optimum function defined by Eqs. (4.26) and (4.27). The expression for the optimum values of integrals $\overline{J}_i$ reads:

$$\overline{J}_i^{opt} = \int_{t_0^*}^{t_1^*} \overline{\Phi}\psi_i(z)\frac{d\overline{\Phi}}{zdz}dz = -C^2\int_{t_0^*}^{t_1^*} \psi_i(z)\frac{\varphi_0'(z)}{z[\varphi_0(z)]^3}dz\,. \\ = C^2\int_{t_1^*}^{t_0^*} \psi_i(z)\frac{\varphi_2(z)}{[\varphi_0(z)]^3}dz\,. \tag{4.45}$$

The above-described procedure for constructing a solution is summarized in *Table 6-2*, in which the references to the relevant equations are given.

Table 6-2. Sequence of steps for constructing the optimum solution

Step	Determined parameter	Equation
1	t_0^*	(4.14)
2	t_1^*	(4.33)
3	C	(4.39)
4	$\overline{r}$	(4.40)
5	$\overline{v}_{imp}^*$	(4.41)
6	$\overline{J}_i^{opt}$	(4.45)

5. THREE-TERM MODEL: RESULTS AND DISCUSSION

Numerical calculations performed using the above-described optimization procedure show the following. The shape and the DOP of the optimum impactor are very close to the shape and the DOP of the optimum truncated cone, where the shape of the optimum cone is determined for the same $\overline{v}_{imp}$, τ and the same model is used for impactor-shield interaction. The difference in the two DOPs is less than 2% for every $1 \le \overline{v}_{imp} \le 4$, $0.25 \le \tau \le 1.5$ and for each of the considered models of the material of the shields. The ranges of variation of the bluntness radius when $\overline{v}_{imp}$ changes from 1 to 4 are shown in *Table 6-3*.

Table 6-3. Normalized (divided by the shank radius) radius of the bluntness of the optimum truncated cone

τ	Model 1	Model 2	Model 3	Model 4
0.25	0.06	0.07÷0.09	0.07÷0.09	0.08÷0.10
0.50	0.17	0.19÷0.20	0.19÷0.21	0.19÷0.23
1.00	0.38	0.40÷0.41	0.40÷0.42	0.41÷0.43
1.50	0.52	0.53÷0.55	0.53÷0.55	0.53÷0.56

Inspection of *Table 6-3* shows that the type of model and the magnitude of parameter $\overline{v}_{imp}$ have only a weak effect on the shape of the optimum truncated cone. Therefore, the difference between the curves in *Figure 6-8a-d* is determined mainly by the difference in the models used to calculate the DOP and not by the difference in the optimum impactor's shape for different models. Generally, the shape of the impactor can strongly affect the DOP. A comparison between the optimum impactor and the sharp cone is shown in *Figure 6-9*, where the DOP of the sharp cone and the DOP of the optimum impactor are the average values obtained for the considered models.

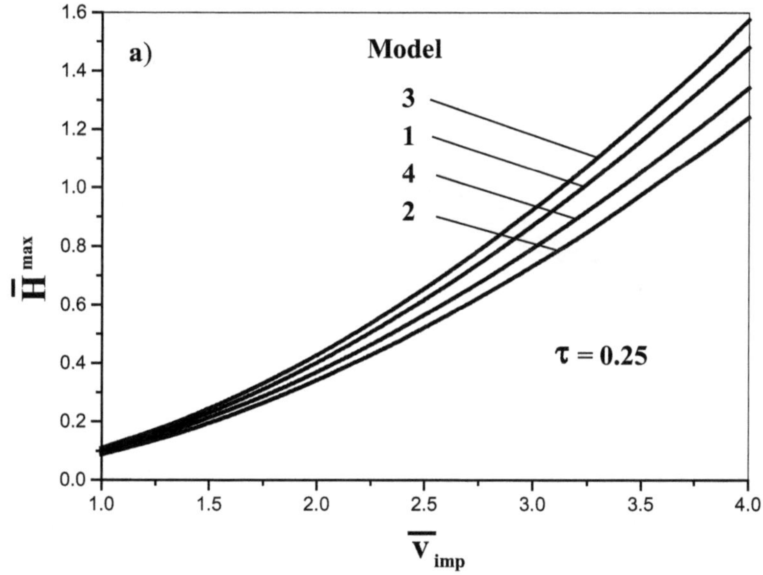

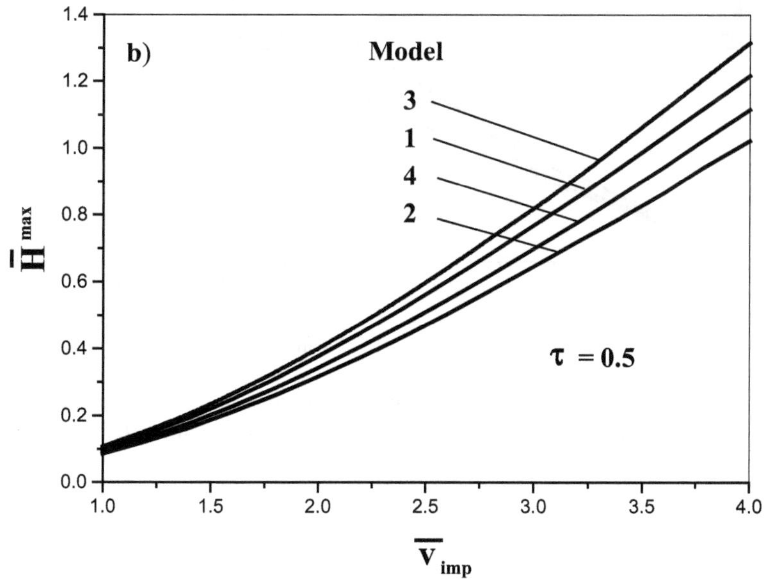

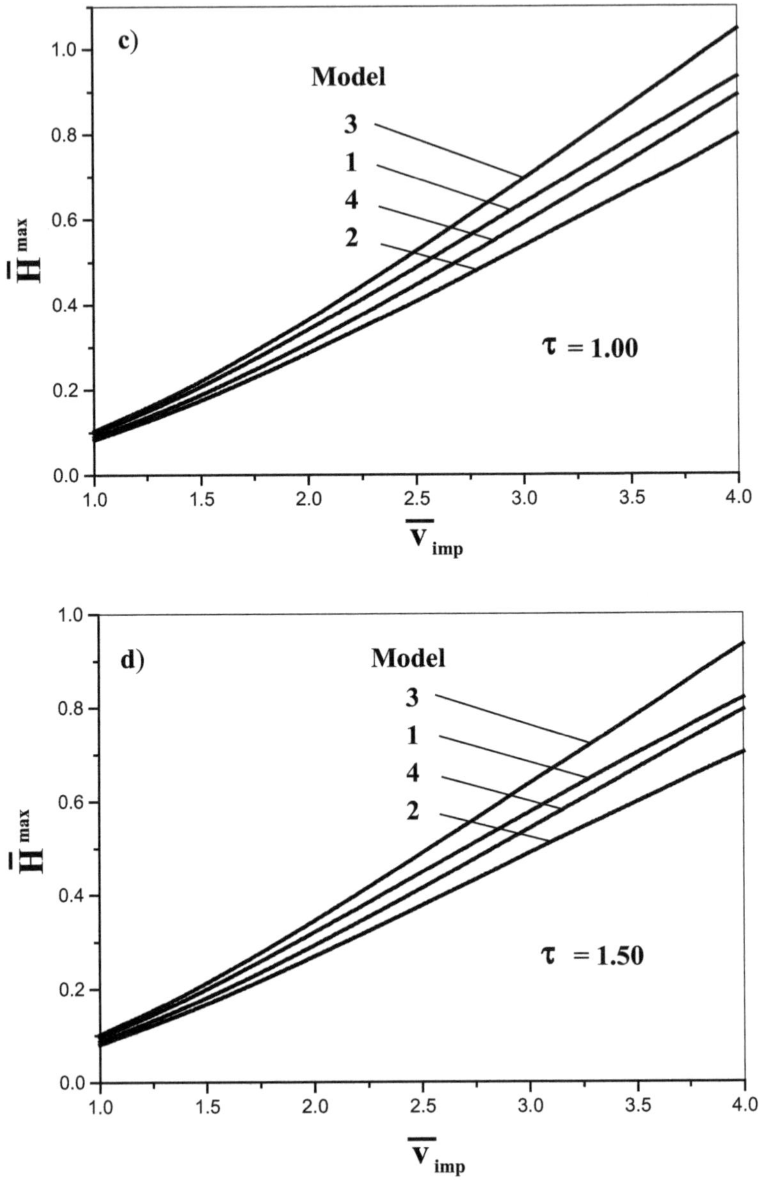

Figure 6-8. Dimensionless DOP for the optimum impactor, $\overline{H}^{max}$, vs. dimensionless impact velocity, $\overline{v}_{max}$, for different models of the material of the shield and different values of the normalized thickness of the impactor's nose, τ.

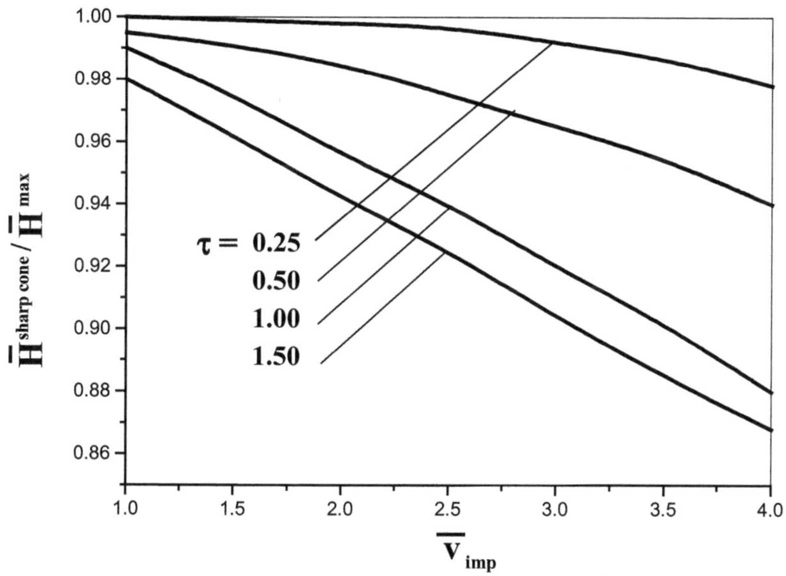

Figure 6-9. DOP of the impactor with a sharp conical nose normalized by the DOP of the optimum impactor vs. dimensionless impact velocity, $\overline{v}_{max}$, for different values of the normalized thickness of the impactor's nose, τ.

Chapter 7

SHAPE OPTIMIZATION OF IMPACTORS PENETRATING INTO CONCRETE SHIELDS

1. TWO-STEP MODELS OF PENETRATION

As a basic penetration model for shape optimization of impactors penetrating into concrete shields, we have adopted a slight modification of the three-term two-step model suggested by Forrestal and Tzou (1997). Here, we will not discuss the choice of the magnitude of parameter κ in Eq. (3-1.9), since it does not affect the optimal shape of the impactor. We will start by deriving a formula for calculating the DOP, which is valid for an arbitrary 3-D body, including a body with a flat bluntness. Clearly, the two-term two-step model of Forrestal et al. (1994), which will be also used, is a particular case of three-term two-step model.

Let us consider the penetration of a 3-D impactor with impact velocity v_{imp} into a semi-infinite concrete shield. We assume that the two-stage penetration model is valid; the drag force in the first stage of penetration is given by Eq. (3-1.9) and the impactor-shield interaction model for the second stage is generally described by Eq. (2-9.15). Then, the formula for the drag force may be written as:

$$D = \begin{cases} D_{\Diamond}(h) & if \quad 0 \le h \le \kappa R \\ D_0(v) & if \quad h \ge \kappa R \end{cases}, \tag{1.1}$$

where (see Eqs. 3-1.9 and 2-9.16):

$$D_{\lozenge}(h)=\tilde{c}h, \quad D_0(v)=B_2v^2+B_1v+B_0, \tag{1.2}$$

κ is some positive constant that does not affect the final result, and $\tilde{c}$ is a constant that is determined from the condition of continuity of the resistance force at $h=\kappa R$. Coefficients B_i are chosen from *Tables 2-6* and *2-7*; they depend on the particular penetration model adopted and the shape of the impactor.

The equation of motion of an impactor with mass m in the first stage of penetration may be written as:

$$mv\frac{dv}{dh}=-\tilde{c}h\,. \tag{1.3}$$

Now, let us consider the motion of the impactor between the initial $(h=0, v=v_{imp})$ and the final $(h=\kappa R, v=v_0)$ locations in the first stage of penetration. Equation (1.3) yields:

$$m\int_{v_{imp}}^{v_0} vdv=-\tilde{c}\int_0^{\kappa R} hdh\,. \tag{1.4}$$

Calculating the integrals, we obtain:

$$m(v_{imp}^2-v_0^2)=\tilde{c}(\kappa R)^2\,. \tag{1.5}$$

Similarly, integration of the equation of motion of the impactor in the second stage of penetration,

$$mv\frac{dv}{dh}=-(B_2v^2+B_1v+B_0)\,, \tag{1.6}$$

between the locations $h=\kappa R, v=v_0$ and $h=H,\ v=0$ (H is the DOP) yields:

$$m\int_{v_0}^{0}\frac{vdv}{B_2v^2+B_1v+B_0}=-\int_{\kappa R}^{H} dh\,. \tag{1.7}$$

Using the notations given in *Appendix 2,* we obtain in a general case:

$$H = \kappa R + mM(1,1,B_0,B_1,B_2;v_0). \tag{1.8}$$

The condition of continuity of the resistance force at $h = \kappa R$, $D_\Diamond(\kappa R) = D_0(v_0)$, yields the following relationship:

$$\tilde{c}(\kappa R) = B_2 v_0^2 + B_1 v_0 + B_0 . \tag{1.9}$$

Equations (1.5) and (1.9) allow us to exclude $\tilde{c}$ and to obtain the relationship between v_{imp} and v_0:

$$(B_2 + \lambda_6)v_0^2 + B_1 v_0 - (\lambda_6 v_{imp}^2 - B_0) = 0, \tag{1.10}$$

where

$$\lambda_6 = \frac{m}{\kappa R}. \tag{1.11}$$

If all $B_i > 0$ and

$$v_{imp} > \sqrt{B_0/\lambda_6}\,, \tag{1.12}$$

then Eq. (1.10) has the unique positive root:

$$v_0 = \frac{B_1}{2(B_2 + \lambda_6)}\left[\sqrt{1 + \frac{4(B_2 + \lambda_6)(\lambda_6 v_{imp}^2 - B_0)}{B_1^2}} - 1\right]. \tag{1.13}$$

If a model with

$$B_1 = 0 \tag{1.14}$$

is considered, then, instead of Eq. (1.13), the following equation

$$v_0 = \sqrt{\frac{\lambda_6 v_{imp}^2 - B_0}{B_2 + \lambda_6}} \tag{1.15}$$

must be used, and Eq. (1.8) can be written in the following form (see *Appendix 2*):

$$H = \kappa R + mM(1,1,B_0,0,B_2;v_*) = \kappa R + \frac{m}{2B_2} ln\left(1+\frac{B_2}{B_0}v_0^2\right). \tag{1.16}$$

2. SOLUTIONS FOR THE TWO-TERM MODEL

2.1 Bodies of revolution

Consider the problem of maximizing the DOP H for a semi-infinite concrete shield (Ben-Dor et al., 2003b), with known parameters determining its mechanical properties, namely, the parameters a_0, a_2 and κ in the two-term model without friction determined by Eq. (2-9.31). It is assumed that the mass of the impactor having the shape of a body of revolution m, length L, shank radius R and the impact velocity are given.

Equations (2-9.25) and (2-9.26) imply that the only parameter in Eqs. (1.15) and (1.16) that depends upon the shape of the impactor is B_2. The dependence of H on B_2 can be obtained by substituting v_0 from Eq. (1.15) into Eq. (1.16). To prove that H is a decreasing function of B_2, let us rewrite Eq. (1.16) as follows:

$$H = \kappa R + mB_0 \int_0^{v_0} \frac{\zeta d\zeta}{1+\frac{B_2}{B_0}\zeta^2}. \tag{2.1}$$

Differentiating Eq. (2.1) using the Leibnitz rule (Korn and Korn, 1968), we obtain:

$$\frac{1}{mB_0}\frac{dH}{dB_2} = \int_0^{v_0} \frac{d}{dB_2}\left(\frac{\zeta}{1+\frac{B_2}{B_0}\zeta^2}\right)d\zeta + \frac{dv_0}{dB_2}\left(\frac{\zeta}{1+\frac{B_2}{B_0}\zeta^2}\right)_{\zeta=v_0}$$
$$= -\left[\frac{1}{B_0}\int_0^{v_0}\frac{\zeta^3 d\zeta}{(1+\frac{B_2}{B_0}\zeta^2)^2} + \frac{1}{2}\sqrt{\frac{\lambda_6 v_{imp}^2 - B_0}{(B_2+\lambda_6)^3}}\left(\frac{v_0}{1+\frac{B_2}{B_0}v_0^2}\right)\right], \tag{2.2}$$

where the dependence of v_0 on B_2 is given by Eq. (1.15). Since $dH/dB_2 < 0$, H is a decreasing function of B_2, and the minimum B_2 provides the maximum H.

Substituting v_0 from Eq. (1.15) and B_0 and B_2 as indicated in *Table 2-7* into Eq. (1.16), and using Eq. (6-1.2) and the relationships:

$$\overline{H} = \frac{H}{0.5\kappa R}, \quad \lambda_7 = \frac{mv_{imp}^2}{\pi\kappa a_0 R^3} - 1, \quad \lambda_8 = \frac{m}{\pi\kappa a_2 R L^2}, \tag{2.3}$$

we may rewrite Eq. (1.16) as follows:

$$\overline{H} = Q(\overline{J}_2) = 2 + \frac{\lambda_8}{\overline{J}_2} \ln\left(1 + \frac{\lambda_7 \overline{J}_2}{\lambda_8 + \overline{J}_2}\right), \tag{2.4}$$

$$\overline{J}_2 = \overline{r}^2 + 2\int_0^1 \frac{\overline{\Phi}\,\overline{\Phi}'^3}{\overline{\Phi}'^2 + 1} d\overline{x} = \tau^2 - 2\overline{J}_3, \quad \overline{J}_3 = \int_0^1 \frac{\overline{\Phi}\,\overline{\Phi}'}{\overline{\Phi}'^2 + 1} d\overline{x}. \tag{2.5}$$

Since $B_2 = \pi a_2 L^2 \overline{J}_2$, using the dimensionless variables, the problem reduces to the variational problem given by Eqs. (6-1.7) and (6-1.8).

The solution of this problem was discussed in *Section 6-1*. In this section we determine the 'relative optimal' geometry of impactors for some simple classes of shapes and analyze how their efficiency compares with that of the "absolute optimum" impactor. Clearly, the minimization of $\overline{J}_2$ is equivalent to the maximization of $\overline{J}_3$. However, it is more convenient to consider the latter problem.

It must be noted that when a two-term penetration model is used shape optimization of impactors penetrating into ductile and into concrete shields reduces to the minimization of the same functional. Therefore, the results for the optimal shapes of various classes of projectiles that are obtained *Section 7-2* are also valid for ductile shields.

When Newton's model is used, the problem of finding a body of revolution having the minimum drag in a high speed gas flow reduces to the minimization of the same functional. Newton's model has been used in gasdynamics for many years for solving various variational problems. Therefore, it is quite probable that results similar to those derived for penetration problems in *Section 7-2.3* and *7-2.4* has been previously obtained in gasdynamics.

Finally it must be noted that some generalizations discussed at the end of *Section 6-1* are also valid in this case.

2.2 Truncated-conical impactors

Let us consider a class of impactors with a conical shape and analyze their efficiency as compared to that of the optimum impactor. The equation of the generatrix of such impactor may be written as:

$$\overline{\Phi}(\overline{x}) = \overline{r} + \tan\upsilon\,\overline{x}, \quad \overline{r} = \tau - \tan\upsilon, \tag{2.6}$$

where υ is an angle between the generatrix and the $\overline{x}$-axis. Substituting Eqs. (2.6) into the second equation in Eqs. (2.5), we obtain:

$$\overline{J}_3^{tr-cone} = Y(\upsilon) = \sin\upsilon(\tau\cos\upsilon - 0.5\sin\upsilon), \quad 0 \le \tan\upsilon \le \tau. \tag{2.7}$$

To determine the shape of the optimal conical impactor, let us find the maximum of function $Y(\upsilon)$. Since the first derivative is

$$Y' = -\cos^2\upsilon(\tau z^2 + z - \tau), \quad z = \tan\upsilon, \tag{2.8}$$

then $Y'(\upsilon) > 0$ for $0 \le z < z^{max}$, $Y'(\upsilon^{max}) = 0$, and $Y'(\upsilon) < 0$ for $z^{max} \le z < \tau$, where

$$z^{max} = \frac{\sqrt{4\tau^2 + 1} - 1}{2\tau}, \quad \upsilon^{max} = \tan^{-1}(z^{max}) = \frac{1}{2}\tan^{-1}(2\tau). \tag{2.9}$$

Consequently, $\upsilon = \upsilon^{max}$ is the location of the maximum of $Y(\upsilon)$. Then Eqs. (2.6) and (2.7) yield the following formulas for the radius of the bluntness of the optimal truncated conical impactor and the optimal value of the functional $\overline{J}_3$:

$$\frac{r^{opt\ tr-cone}}{R} = \frac{\overline{r}^{opt\ tr-cone}}{\tau} = 1 - \frac{2}{\sqrt{4\tau^2 + 1} + 1}, \quad \overline{J}_3^{opt\ tr-cone} = \frac{\sqrt{4\tau^2 + 1} - 1}{4} \tag{2.10}$$

For a sharp cone $\tan\upsilon = \tau$, and Eq. (2.7) yields the following formula for the value of the functional $\overline{J}_3$:

$$\overline{J}_3^{sh-cone} = \frac{\tau^2}{2(\tau^2 + 1)}. \tag{2.11}$$

Plots of $\overline{J}_3^{tr-cone}$ versus the normalized radius of bluntness

$$\widetilde{r} = \frac{r}{R} = \frac{\overline{r}}{\tau} \tag{2.12}$$

for different τ are shown in *Figure 7-1*. The variation of the function is quite small in the neighborhood of the maximum, specially for relatively small τ. This means that the radius of the flat bluntness can be changed in the vicinity of the optimal value (see *Figure 7-2*) without considerable loss in the value of the integral $\overline{J}_3$. Plots of $\overline{J}_3$ versus τ for an optimal truncated cone and a sharp cone are shown in *Figure 7-9*.

This problem was first investigated by Newton (1687). The solution of this problem in modern notation can be found in Edwards (1997).

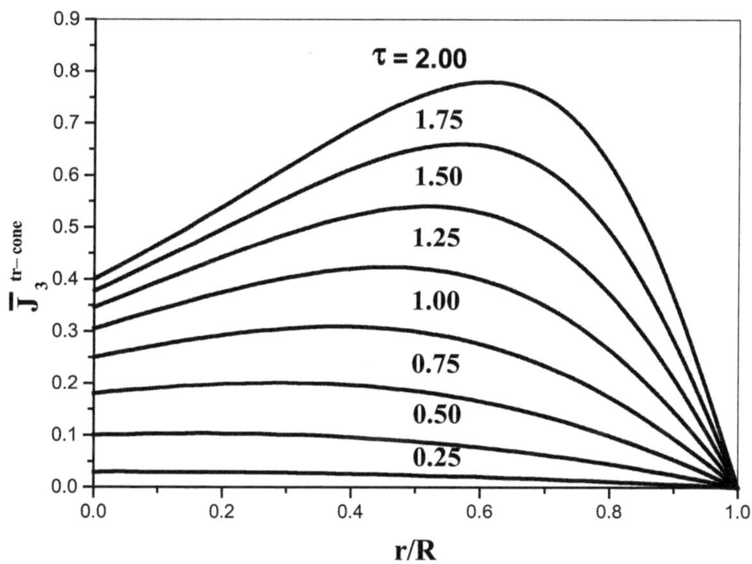

Figure 7-1. Functional $\overline{J}_3^{tr-cone}$ vs. normalized radius of the bluntness of a truncated-conical impactor, $\widetilde{r} = r/R$.

2.3 Truncated-ogive impactors

The equation of the generatrix of the impactor with a truncated-ogive nose (see *Figure 7-3*) in dimensionless variables may be written as:

$$\overline{\Phi}(\overline{x}) = \tau - \overline{q}_1 + \sqrt{\overline{q}_1^2 - (\overline{x} - 1)^2}, \quad 0 \le \overline{x} \le 1, \tag{2.13}$$

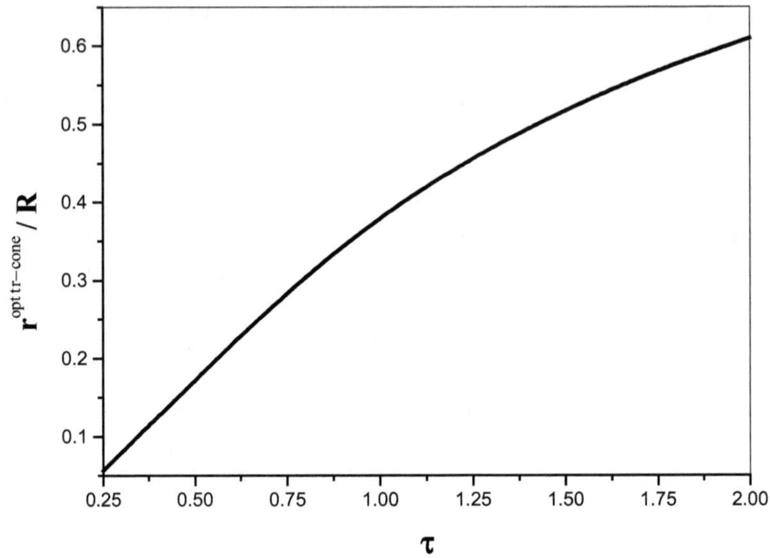

Figure 7-2. Normalized radius of the bluntness of the optimum truncated-conical impactor $r^{opt\ tr-cone}/R$ vs. normalized thickness τ.

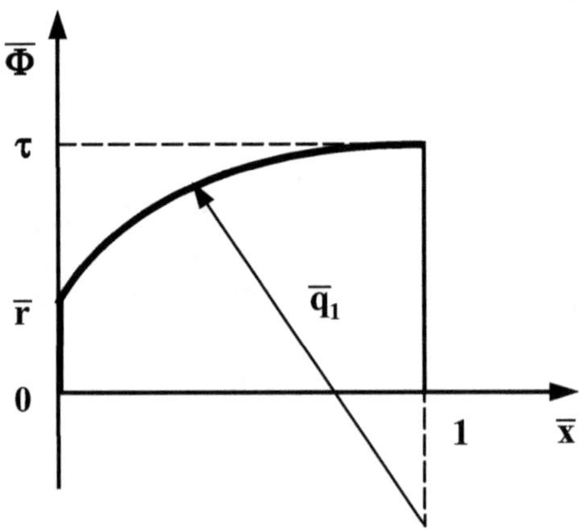

Figure 7-3. Shape of the nose of a truncated-ogive impactor.

where $\overline{q}_1 = q_1/L$ is the dimensionless ogive radius. This is an equation of the circle arc of the radius $\overline{q}_1$ with the center at the point $\overline{x} = 1, \overline{\Phi} = \tau - \overline{q}_1 \le 0$. We must now take into account the conditions that follow from the geometrical considerations:

$$\overline{q}_1 \geq \tau, \quad \overline{q}_1 \geq 1, \quad \overline{\Phi}(0) \geq 0 . \tag{2.14}$$

Solving the equation $\overline{\Phi}(0) = \overline{r}$, we can obtain the following relationship between $\overline{r}$ and $\overline{q}_1$:

$$\overline{q}_1 = \frac{q_1}{L} = \frac{(\tau - \overline{r})^2 + 1}{2(\tau - \overline{r})}, \tag{2.15}$$

and replace the constraints on $\overline{q}_1$ in Eq. (2.14) by the following constraints on $\overline{r}$:

$$max(0, \tau^2 - 1) \leq \overline{r}^2 \leq \tau^2 , \tag{2.16}$$

or by the constraints on $\widetilde{r} = r/R$:

$$max(0, 1 - \tau^{-2}) \leq \widetilde{r}^2 \leq 1 . \tag{2.17}$$

Substituting Eq. (2.13) into the expression for $\overline{J}_3$ in Eqs. (2.5), we can calculate the integral for a truncated-ogive nose:

$$\overline{J}_3^{tr-ogive} = \frac{1}{2} + \frac{\tau - \overline{q}_1}{3\overline{q}_1^2}[\overline{q}_1^3 - (\overline{q}_1^2 - 1)^{3/2}] - \frac{1}{4\overline{q}_1^2} . \tag{2.18}$$

In the case of an ogive-nose impactor without flat bluntness, $\overline{r} = 0$ and Eqs. (2.15) and (2.18) can be rewritten as:

$$\overline{q}_1 = \frac{\tau^2 + 1}{2\tau}, \quad \overline{J}_3^{ogive} = \frac{\tau^2(\tau^4 + 2\tau^2 + 3)}{6(\tau^2 + 1)^2}, \quad 0 < \tau \leq 1 . \tag{2.19}$$

Using the caliber-radius-head parameter

$$\psi = \frac{q_1}{2R} = \frac{\overline{q}_1}{2\tau} = \frac{\tau^2 + 1}{4\tau^2} = \frac{1}{4}\left(1 + \frac{1}{\tau^2}\right), \tag{2.20}$$

we may rewrite the second and third equations in Eqs. (2.19) as:

$$\overline{J}_3^{ogive} = \frac{24\psi^2 - 8\psi + 1}{48\psi^2(4\psi - 1)}, \quad \psi \geq \frac{1}{2} . \tag{2.21}$$

Plots of $\overline{J}_3^{tr-ogive}$ versus $\widetilde{r}$ for different τ are shown in *Figure 7-4*. If $\tau \le 1$, then $\widetilde{r}$ varies from 0 to 1; there is a maximum $\overline{J}_3^{tr-ogive}(\widetilde{r})$, but the values of the function are very close to this maximum value when $\widetilde{r}$ changes in the range from 0 to 0.5. If $\tau > 1$, then $\widetilde{r}$ varies in the range from $\widetilde{r} = \widetilde{r}_0$ to 1 where $\widetilde{r}_0 = \sqrt{1-\tau^{-2}}$. If $1 < \tau <\approx 1.15$, then the maximum $\overline{J}_3^{tr-ogive}(\widetilde{r})$ is attained for $\widetilde{r} > \widetilde{r}_0$. If $\tau >\approx 1.15$, then $\widetilde{r} = \widetilde{r}_0$ is the location of the maximum $\overline{J}_3^{tr-ogive}(\widetilde{r})$. A plot of the normalized radius of the bluntness of the optimum truncated-ogive impactor $r^{opt\ tr-ogive}/R$ versus the normalized thickness τ is shown in *Figure 7-5*. The plot of $\overline{J}_3$ versus τ for optimal truncated ogive is shown in *Figure 7-9*.

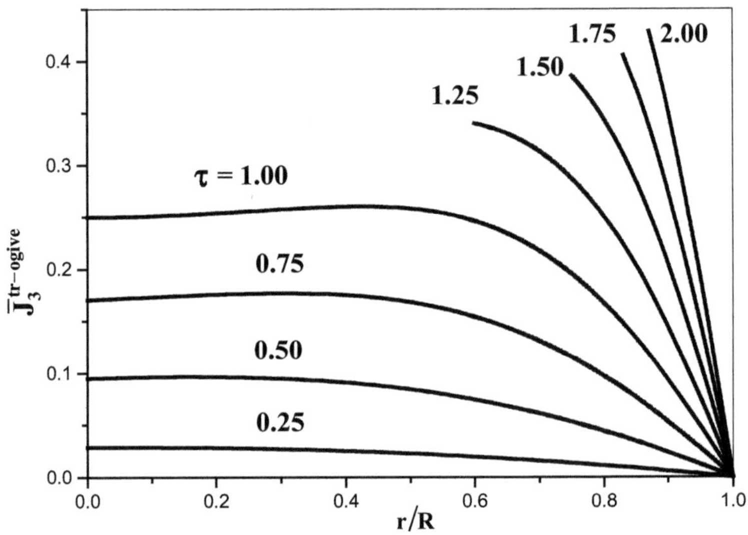

Figure 7-4. Functional $\overline{J}_3^{tr-ogive}$ vs. normalized radius of the bluntness of truncated-ogive impactor, $\widetilde{r} = r/R$.

2.4 Spherical-conical impactors

The equation of the generatrix of the impactor with a spherical-conical nose (see *Figure 7-6*) in dimensionless variables may be written as:

$$\overline{\Phi}(\overline{x}) = \begin{cases} \sqrt{(2\overline{q}_2 - \overline{x})\overline{x}} & \text{if} \quad 0 \le \overline{x} \le \overline{x}_0 \\ \overline{q}_2 \cos\upsilon + tg\,\upsilon(\overline{x} - \overline{x}_0) & \text{if} \quad \overline{x}_0 \le \overline{x} \le 1 \end{cases}, \tag{2.22}$$

where $\overline{q}_2$ is the radius of the spherical bluntness, υ is the angle between the generatrix of the cone and the $\overline{x}$-axis,

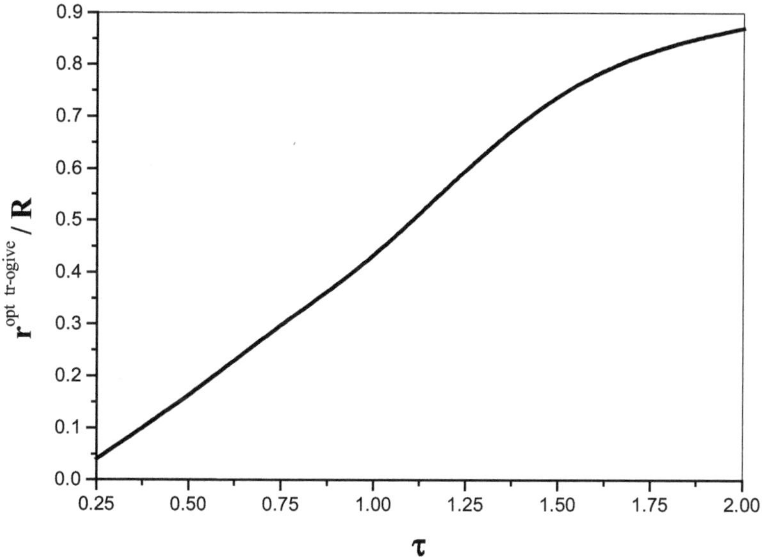

Figure 7-5. Normalized radius of the bluntness of the optimum truncated-ogive impactor $r^{opt\ tr-ogive}/R$ vs. normalized thickness τ.

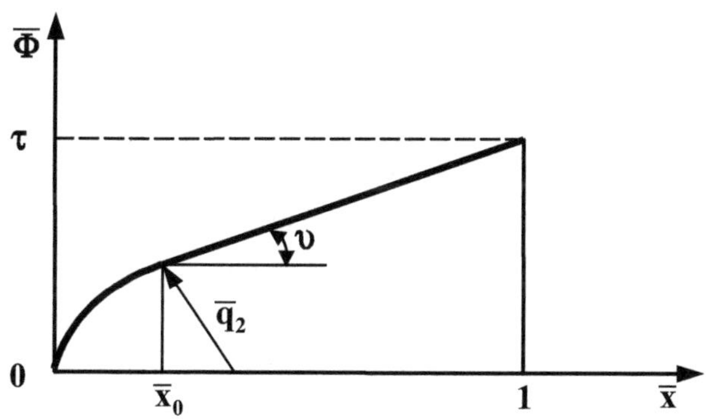

Figure 7-6. Shape of the nose of a spherical-conical impactor.

$$\bar{x}_0 = \frac{x_0}{L} = \tau \cos \upsilon - \sin \upsilon, \; \bar{q}_2 = \frac{q_2}{L} = \frac{\tau \cos \upsilon - \sin \upsilon}{1 - \sin \upsilon}. \quad (2.23)$$

The parameter $\bar{x}_0$ is the abscissa of the point where the circle arc smoothly transforms to the straight line:

$$\overline{\Phi}(\overline{x}_0-)=\overline{\Phi}(\overline{x}_0+),\quad \overline{\Phi}'(\overline{x}_0-)=\overline{\Phi}'(\overline{x}_0+). \tag{2.24}$$

The following conditions that follow from geometrical considerations must be satisfied:

$$0\le \overline{q}_2 \le min(1,\tau), \tag{2.25}$$

or

$$0\le \widetilde{q}_2 \le min(\tau^{-1},1), \tag{2.26}$$

where

$$\widetilde{q}_2=\frac{q_2}{R}=\frac{\overline{q}_2}{\tau}. \tag{2.27}$$

Substituting Eq. (2.22) into Eq. (2.5) we obtain after some algebra the expression for $\overline{J}_3$ in the case of spherical-conical impactor:

$$\overline{J}_3^{sph-cone}=0.25\overline{q}_2^2 cos^4 \upsilon+0.5 sin\upsilon(1-\overline{x}_0)[2\overline{q}_2^2 cos^2\upsilon+(1-\overline{x}_0)sin\upsilon]. \tag{2.28}$$

Equation (2.28) together with Eq. (2.23) determine $\overline{J}_3^{sph-cone}$ as a function of υ. It is more convenient to use the dependence $\overline{J}_3^{sph-cone}$ versus $\overline{q}_2$ or $\widetilde{q}_2$. Equation (2.23) implies the following formula for υ as a function of $\overline{q}_2$ and τ:

$$\upsilon=sin^{-1}\left(\frac{\tau\sqrt{\tau^2+1-2\overline{q}_2}-\overline{q}_2(1-\overline{q}_2)}{\tau^2+(1-\overline{q}_2)^2}\right). \tag{2.29}$$

Substituting $\overline{q}_2=\tau\widetilde{q}_2$ into Eq. (2.29), we can obtain the dependence υ versus $\widetilde{q}_2$ (for a given τ). The above-given formulas allow us to calculate $\overline{J}_3^{sph-cone}$ for a given $\widetilde{q}_2$. The procedure for this calculation comprises calculating υ and $\widetilde{q}_2$ using Eqs. (2.29) and (2.27), determining $\overline{x}_0$ using Eq. (2.23), and finding $\overline{J}_3^{sph-cone}$ from Eq. (2.28).

Plots of $\overline{J}_3^{sph-cone}$ versus $\widetilde{q}_2$ for different τ are shown in *Figure 7-7*. If $\tau \leq 1$, then $\widetilde{q}_2$ varies from 0 to 1; there is a maximum $\overline{J}_3^{sph-cone}(\widetilde{q}_2)$, but variation of the function is quite small in the neighborhood of the maximum. If $\tau > 1$, then $\widetilde{q}_2$ varies from 0 to $\widetilde{q}_2 = \tau^{-1}$. If $1 < \tau <\approx 1.2$, then the maximum $\overline{J}_3^{sph-cone}(\widetilde{q}_2)$ is attained for $\widetilde{q}_2 < \tau^{-1}$. If $\tau >\approx 1.2$, then $\widetilde{q}_2 = \tau^{-1}$ is the location of the maximum $\overline{J}_3^{sph-cone}(\widetilde{q}_2)$. The plot of the normalized radius of the spherical bluntness of the optimum spherical-conical impactor $\widetilde{q}_2^{sph-cone} = q_2^{sph-cone}/R$ versus the normalized thickness τ is shown in *Figure 7-8*. The plot of $\overline{J}_3$ versus τ for an optimal spherical-conical impactor is shown in *Figure 7-9*.

2.5 Comparison of different shapes

Plots of the maximum $\overline{J}_3$ versus τ are shown in *Figure 7-9* for the "absolute optimal" nose and the optimal noses among the considered nose shapes. The second equation in Eqs. (2.5) implies that the minimum value of $\overline{J}_3$ is 0, and this value is attained for the cylinder, $\overline{\Phi}(\overline{x}) = \tau$.

Remarkably, as is demonstrated in *Figure 7-9*, the functional $\overline{J}_3$ for the optimal truncated-conical nose assumes the values that are very close to the optimum for all τ. For $0 < \tau <\approx 0.5$, the difference in the values of $\overline{J}_3$ for all shapes is very small. For $\approx 0.5 < \tau <\approx 1.0$, the value of $\overline{J}_3$ for the optimum spherical-conical nose is close to the optimum. For a relatively large τ, the advantage of the absolute optimal nose and the optimal truncated-conical nose over other shapes becomes quite pronounced.

Generally, the difference in the magnitude of the functional $\overline{J}_3$ cannot be considered as a measure of the difference in the criterion of optimization. To compare the efficiencies of the "absolute optimal" impactor and the optimal truncated-conical impactor the following parameter:

$$\varepsilon_0 = \frac{H^{max} - H^{max\,tr-cone}}{H^{max}} 100\% \tag{2.30}$$

is used, where H^{max} and $H^{max\,tr-cone}$ are the DOP of the "absolute optimum" impactor and optimum truncated-conical impactor, respectively. *Figure 7-10a-d* shows that these impactors are practically equivalent. This property allows us to propose the following approximate analytical formula for the "absolute maximum" DOP, $\overline{H}^{max}$, namely,

$$\overline{H}^{max} \approx Q(\tau^2 - 2\overline{J}_3^{opt\,tr-cone}), \tag{2.31}$$

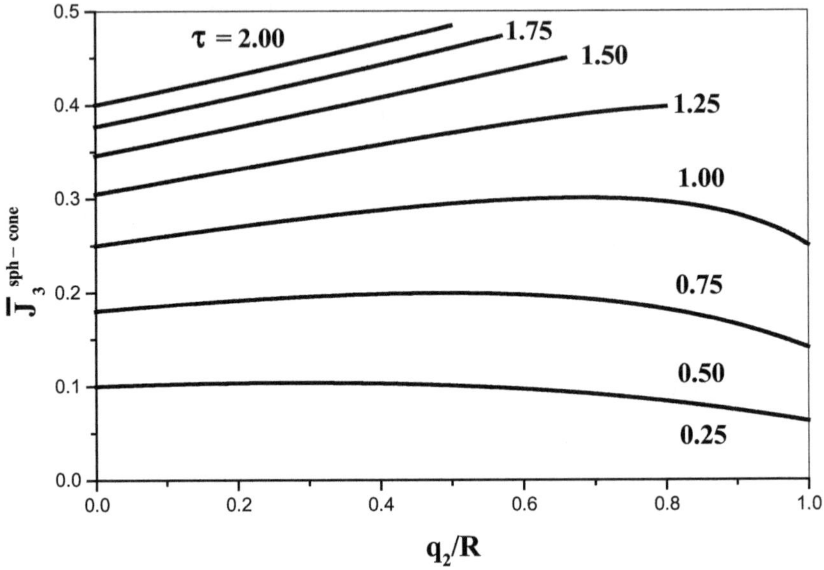

Figure 7-7. Functional $\overline{J}_3^{sph-cone}$ vs. normalized radius of the spherical bluntness of spherical-conical impactor, $\tilde{q}_2 = q_2/R$.

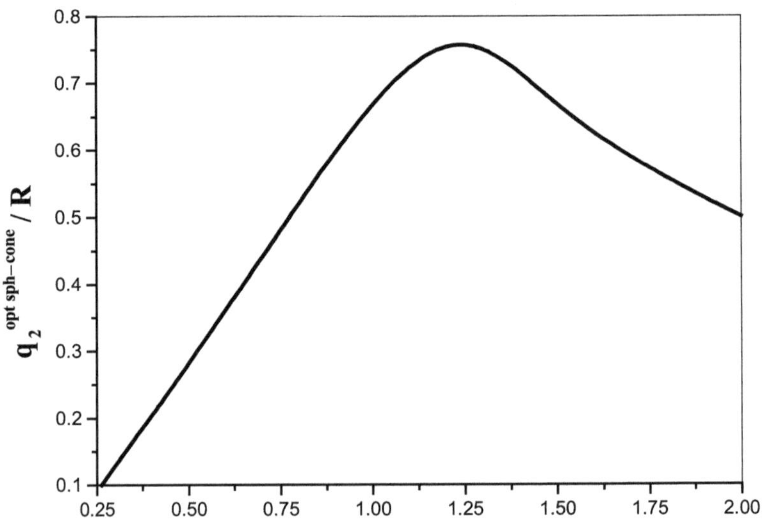

Figure 7-8. Normalized radius of the bluntness of the optimum spherical-conical impactor $q_2^{opt\ sph-cone}/R$, vs. normalized thickness τ.

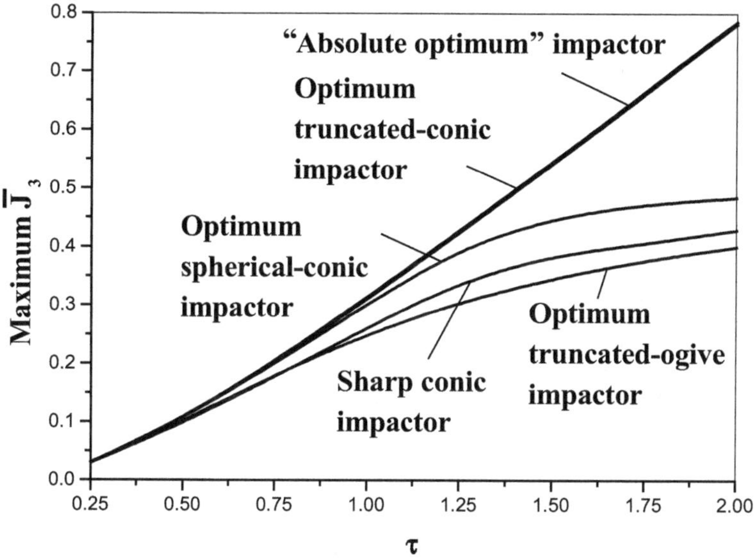

Figure 7-9. Comparison of the performance of impactors with different shapes of the nose.

where the function Q is determined by Eq. (2.4) and $\bar{J}_3^{opt\ tr-cone}$ is calculated using the second equation in Eqs. (2.10).

The "absolute optimum" impactor and the sharp conical impactor are compared in *Figure 7-11a-b* by using the following parameter:

$$\varepsilon_1 = \frac{H^{max} - H^{sharp\ cone}}{H^{max}} 100\% . \tag{2.32}$$

The pronounced advantage of the optimal shape over the sharp cone is clearly evident.

Thus, the optimal shape of a normally striking impactor penetrating at the maximum depth into a concrete semi-infinite shield is independent of the properties of the material of the shield in the framework of the employed two-term model of penetration. The optimum impactor has flat bluntness, and its shape and the DOP are very close to the shape and DOP of the optimal truncated-conical impactor. The typical shapes of the impactors in order of decreasing DOP are: optimal truncated-conical impactor, optimal spherical-conical impactor, sharp-conical impactor, and optimal truncated-ogive impactor. The difference in the efficiency between the optimal impactor or the optimal truncated-conical impactor and each of the remaining impactors in the list increases with an increase of the thickness of the impactor.

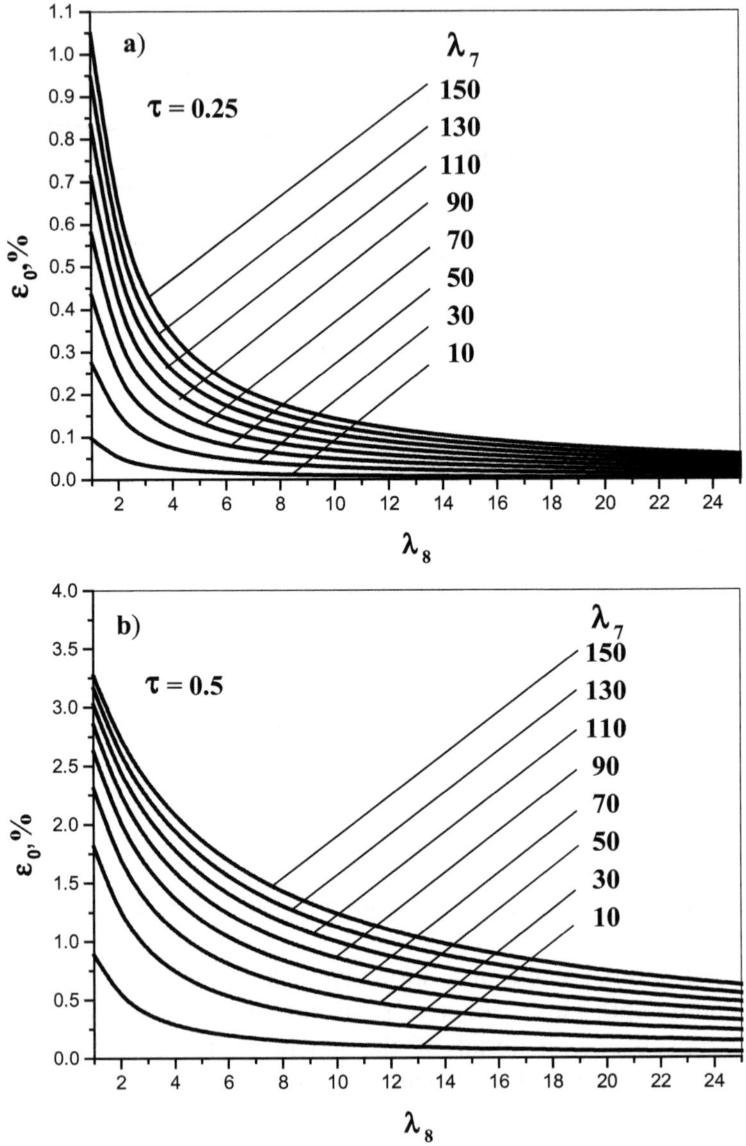
a)
τ = 0.25
λ7
150
130
110
90
70
50
30
10
ε0,%
λ8
b)
τ = 0.5
λ7
150
130
110
90
70
50
30
10
ε0,%
λ8

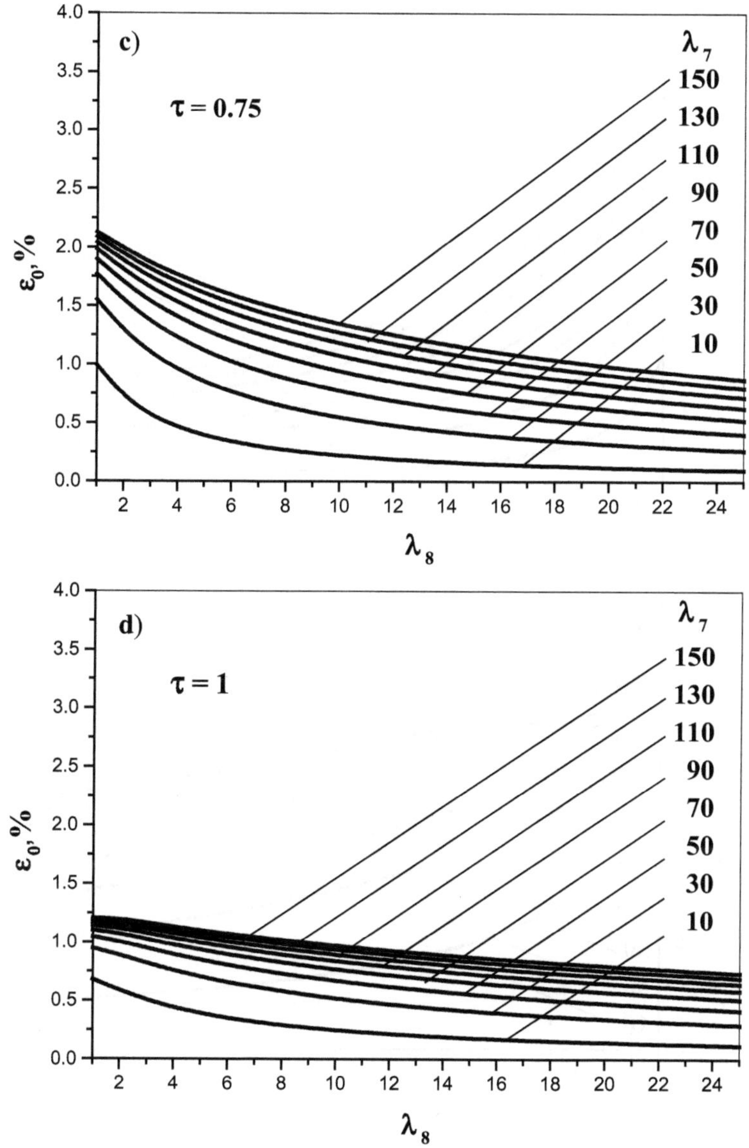

Figure 7-10. Comparison of the performance of the optimum truncated-conical impactor and the "absolute optimum" impactor.

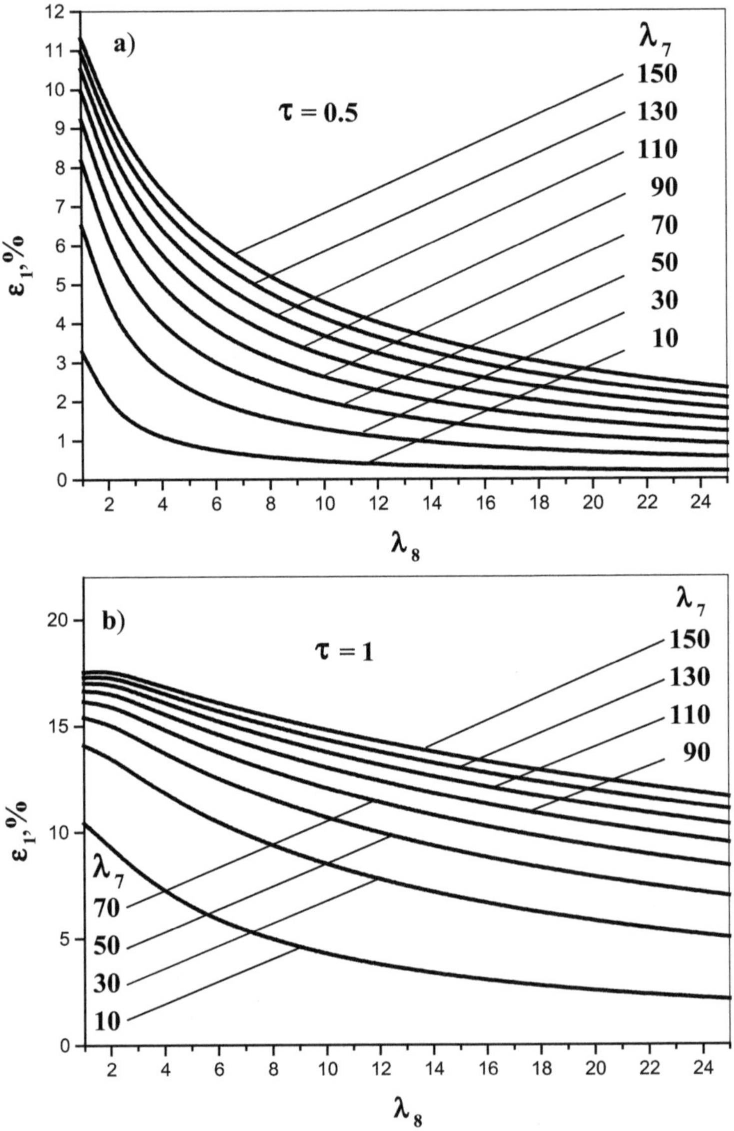

Figure 7-11. Comparison of the performance of the sharp conical impactor and the 'absolute optimum' impactor.

3. SOLUTION FOR THE THREE-TERM MODELS

3.1 Statement of the problem

In this section, we consider impactors having the shape of bodies revolution and employ the two-step three-term model that was described in *Section 7-1*. The references to the equations describing the model are given in *Table 7-1* . Parameters $\tilde{a}_0, \tilde{a}_1, \tilde{a}_2$ that determine a particular model are given in *Table 7-2*.

Table 7-1. References to the equations describing the two-step three-term model applied to a concrete shield

Determined parameter	Equation
H	(1.8)
v_0	(1.13)
λ_6	(1.11)
B_0, B_1, B_2	(6-3.3)-(6-3.5)
a_0, a_1, a_2	(6-3.1)

*Table 7-2.*Parameters of two-step three-term models for a concrete shield (Forrestal and Tzou, 1997).

Number of the model	Characteristic of the model	$\tilde{a}_0$	$\tilde{a}_1$	$\tilde{a}_2$
1	Incompressible, elastic-plastic	5.18	0.0	3.88
2	Incompressible, elastic-cracked-plastic	4.05	1.36	3.51
3	Compressible, elastic-plastic	4.50	0.75	1.29
4	Compressible, elastic-cracked-plastic	3.45	1.60	1.12

Using the dimensionless variables defined by Eq. (6-1.2), we may write the expression for the DOP as:

$$\overline{H} = \frac{H}{0.5\kappa R} = 2 + \lambda_9 M(1,1,e_0,e_1,e_2;\overline{v}_0), \tag{3.1}$$

where

$$\lambda_9 = \frac{2m}{\pi\kappa R^3 \gamma_{sh}}, \quad \lambda_{10} = \frac{\lambda_9 \overline{v}_{imp}^2}{2} - e_0 > 0, \quad \overline{v}_{imp} = \mu_3 v_{imp}, \tag{3.2}$$

$$\overline{v}_0 = \mu_3 v_0 = \frac{e_1}{2e_2 + \lambda_9}\left[\sqrt{1 + \frac{2\lambda_{10}(2e_2 + \lambda_9)}{e_1^2}} - 1\right], \tag{3.3}$$

and e_i, $\bar{J}_i$ and ψ_i are determined by Eqs. (6-3.7), (6-3.8) and (6-3.5), respectively.

The problem under consideration is to maximize the DOP for an impactor with known impact velocity and known parameters determining the mechanical properties of the shield. It is assumed that the mass and length of the impactor and the shank radius of impactor's nose are given. The problem then reduces to optimization of the functional $\overline{H}$ in Eq. (3.1) with given parameters $\tilde{a}_0, \tilde{a}_1, \tilde{a}_2, \lambda_9, \lambda_{10}, \tau$, provided that the solution satisfies the condition given by Eq. (6-3.17).

3.2 Investigation of the variational problem

Analysis of the variational problem is similar to that performed for the one-step three-term model (see *Section 6-4*). Therefore, in this section we omitted obvious explanations and proofs.

The variational problem under consideration may be reduced to investigating the functional given by Eq. (6-4.1), where functions ψ_1 and ψ_2 are determined by Eq. (6-3.5), and $\hat{\lambda}_i\,(i=1,2)$ are some unknown constants determined by means of the following formula:

$$\hat{\lambda}_i = \frac{\partial \overline{H}}{\partial \bar{J}_i} = \left(\frac{\partial \overline{H}}{\partial e_i} + \frac{\partial \overline{H}}{\partial \bar{v}_0} \frac{\partial \bar{v}_0}{\partial e_i} \right) \frac{\partial e_i}{\partial \bar{J}_i}, \quad i=1,2. \tag{3.4}$$

Let us calculate the derivatives in Eq. (3.4). Using the definition of the function M (see *Appendix 2*) and Eq. (6-3.7), we find that:

$$\frac{\partial \overline{H}}{\partial e_i} = -M(2,i+1,e_0,e_1,e_2;\bar{v}_0), \frac{\partial e_i}{\partial \bar{J}_i} = -\frac{2\tilde{a}_i}{\tau^2}, \frac{\partial \overline{H}}{\partial \bar{v}_0} = \frac{\bar{v}_0}{\hat{D}(\bar{v}_0)}, \tag{3.5}$$

where

$$\hat{D}(v) = e_2 v^2 + e_1 v + e_0. \tag{3.6}$$

To simplify the calculations of the derivatives $\partial \bar{v}_0 / \partial e_i$, we note that $\bar{v}_0$ (Eq. 3.3) is the positive root of the following equation:

$$(2e_2 + \lambda_9)\bar{v}_0^2 + 2e_1\bar{v}_0 - 2\lambda_{10} = 0. \tag{3.7}$$

Differentiating this equating with respect to e_1 and e_2, we obtain:

$$2(2e_2+\lambda_9)\overline{v}_0\frac{\partial\overline{v}_0}{\partial e_1}+2\overline{v}_0+2e_1\frac{\partial\overline{v}_0}{\partial e_1}=0, \tag{3.8}$$

$$2\overline{v}_0^2+2(2e_2+\lambda_9)\overline{v}_0\frac{\partial\overline{v}_0}{\partial e_2}+2e_1\frac{\partial\overline{v}_0}{\partial e_2}=0. \tag{3.9}$$

Equations (3.8) and (3.9) yield:

$$\frac{\partial\overline{v}_0}{\partial e_i}=-\frac{\overline{v}_0^i}{g_0(v_0)}, \tag{3.10}$$

where

$$g_0(\zeta)=(2e_2+\lambda_9)\zeta+e_1. \tag{3.11}$$

Therefore,

$$\hat{\lambda}_i=\frac{2\tilde{a}_i}{\tau^2}\left\{M(2,i+1,e_0,e_1,e_2;\overline{v}_0)+\frac{\overline{v}_0^{i+1}}{\hat{D}(\overline{v}_0)g_0(\overline{v}_0)}\right\}, \tag{3.12}$$

where $\overline{v}_0$ is determined by Eq. (3.3) and the integrals $\overline{J}_1[\overline{\Phi}(\overline{x})]$ and $\overline{J}_2[\overline{\Phi}(\overline{x})]$ in expressions for e_1 and e_2 are calculated for the optimum solution of the variational problem. Since $\hat{\lambda}_2>0$, the criterion given by Eq. (6-4.1) can be replaced by the functional in Eq. (6-4.3), where functions F and ψ are defined by Eqs. (6-4.4) and (6-4.5), correspondingly, and

$$\lambda=\frac{\hat{\lambda}_1}{\hat{\lambda}_2}=\frac{\tilde{a}_1}{\tilde{a}_2}\frac{\hat{D}(\overline{v}_0)g_0(\overline{v}_0)M(2,2,e_0,e_1,e_2;\overline{v}_0)+\overline{v}_0^2}{\hat{D}(\overline{v}_0)g_0(\overline{v}_0)M(2,3,e_0,e_1,e_2;\overline{v}_0)+\overline{v}_0^3}>0. \tag{3.13}$$

The following procedure may be used for solving the variational problem (for a given τ). First, a solution, $\overline{\Phi}=\overline{\Phi}_{opt}(\lambda,\overline{x})$, is found for every parameter $\lambda>0$. Then, a decreasing function $\overline{v}_0(\lambda)$ is determined, since Eq. (3.13) allows us, for every $\lambda=\lambda_*$, to find $\overline{v}_0=\overline{v}_{0*}$. Eliminating λ_{10} from Eq. (3.7) and using Eq. (3.2), we obtain the relationship between $\overline{v}_0$ and $\overline{v}_{imp}$

$$\overline{v}_{imp}=\sqrt{\overline{v}_0^2+\frac{2\hat{D}(\overline{v}_0)}{\lambda_9}} \tag{3.14}$$

which allows us to determine $\bar{v}_{imp*}$ for every $\bar{v}_{0*}$, i. e., we determine the function $\bar{v}_{imp}(\lambda)$ and the inverse function $\lambda(\bar{v}_{imp})$. Using the latter relationship, we can express the solution of the problem in terms of $\bar{v}_{imp}$.

The procedure for the derivation of the equation for the extremal is similar to that discussed in *Section 6-4*. Therefore, we use below the same notations taking into account that the equations for λ are different.

The equation for the extremal is given by Eqs. (6-4.26) and (6-4.27) where $0<t_1^*<t\le t_0^*$; parameters t_0^* and t_1^* are determined from Eqs. (6-4.14) and (6-4.31), correspondingly. The constant C and the dimensionless radius of the bluntness are determined from Eqs. (6-4.39) and (6-4.40), respectively.

We have thus devised a procedure that allows us, for a given τ and some $\lambda>0$, to determine the optimum generatrix of the impactor and to calculate the parameters e_1 and e_2. Then, Eq. (3.13) can be considered as the equation with respect to $\bar{v}_0$:

$$\Xi(\bar{v}_0)=0\,, \tag{3.15}$$

where

$$\Xi(\bar{v}_0)=\Xi_1(\bar{v}_0)+\Xi_2(\bar{v}_0)\,, \tag{3.16}$$

$$\Xi_1(\bar{v}_0)=M(2,3,e_0,e_1,e_2;\bar{v}_{imp})-cM(2,2,e_0,e_1,e_2;\bar{v}_{imp}) \\ =\int_0^{\bar{v}_0}\frac{\zeta^2(\zeta-c)d\zeta}{(e_0+e_1\zeta+e_2\zeta^2)^2}, \tag{3.17}$$

$$\Xi_2(\bar{v}_0)=\frac{\bar{v}_0^2(\bar{v}_0-c)}{\hat{D}(\bar{v}_0)g_0(\bar{v}_0)},\quad c=\frac{\tilde{a}_1}{\tilde{a}_2\lambda}\,. \tag{3.18}$$

Let us now prove that Eq. (3.15) has only one root. Eqs. (3.17) and (3.18) imply that $\Xi_1(\bar{v}_0)<0$, $\Xi_2(\bar{v}_0)<0$ and, consequently, $\Xi(\bar{v}_0)<0$ when $0\le\bar{v}_0\le c$. Therefore, Eq. (3.15) cannot have roots in this range. To investigate the behavior of the function $\Xi(\bar{v}_0)$ for $\bar{v}_0>c$, let us calculate the derivative $\Xi'(\bar{v}_0)$:

$$\Xi'(\bar{v}_0)=\bar{v}_0\frac{\lambda_9\bar{v}_0^2(\bar{v}_0-c)g_0(\bar{v}_0)+\hat{D}(\bar{v}_0)g_1(\bar{v}_0)}{[\hat{D}(\bar{v}_0)g_0(\bar{v}_0)]^2}, \tag{3.19}$$

where

$$g_1(\bar{v}_0) = 2(2e_2 + \lambda_9)\bar{v}_0(\bar{v}_0 - \frac{1}{2}c) + 3e_1(\bar{v}_0 - \frac{2}{3}c). \tag{3.20}$$

Equations (3.19) and (3.20) imply that $\Xi'(\bar{v}_0) > 0$ when $\bar{v}_0 > c$, i. e., $\Xi(\bar{v}_0)$ increases from the negative value $\Xi(c)$ when $\bar{v}_0$ increases from $\bar{v}_0 = c$.

Let us now estimate $\Xi(\bar{v}_0)$ when $\bar{v}_0 \to +\infty$. Equations (A2.19)-(A2.22) imply that:

$$\lim_{\bar{v}_0 \to +\infty} \Xi_1(\bar{v}_0) = +\infty . \tag{3.21}$$

Taking into account Eqs. (3.9) and (3.11), we obtain:

$$\lim_{\bar{v}_0 \to +\infty} \Xi_2(\bar{v}_0) = \lim_{\bar{v}_0 \to +\infty} \frac{\bar{v}_0^2(\bar{v}_0 - c)}{(e_2\bar{v}_0^2 + e_1\bar{v}_0 + e_0)[(2e_2 + \lambda_9)\bar{v}_0 + e_1]} = \frac{1}{e_2(2e_2 + \lambda_9)}. \tag{3.22}$$

Therefore,

$$\lim_{\bar{v}_0 \to +\infty} \Xi(\bar{v}_0) = +\infty . \tag{3.23}$$

Since $\Xi(\bar{v}_0)$ increases from $\Xi(c) < 0$ to $\Xi(+\infty) = +\infty$, Eq. (3.15) has the unique root $\bar{v}_0 = \bar{v}_{0*} > c$. The corresponding value of the impact velocity, $\bar{v}_{imp*}$, can be obtained by substituting $\bar{v}_0 = \bar{v}_{0*}$ into Eq. (3.14).

The optimum value of the dimensionless DOP, $\overline{H}^{max}$, can be determined from Eqs. (3.1), where e_i are found from Eq. (6-3.7) and the integrals $\bar{J}_i$ are calculated for the optimum function. The expression for the optimum values of the integrals $\bar{J}_i$, $\bar{J}_i^{opt}$ are given by Eq. (6-4.45).

The above described procedure for finding a solution is summarized in *Table 7-3*, in which references to the relevant equations are given.

Table 7-3 The sequence of steps for finding the optimum solution.

Step	Determined parameter	Equation
1	t_0^*	(6-4.14)
2	t_1^*	(6-4.33)
3	C	(6-4.39)
4	$\bar{r}$	(6-4.40)

Step	Determined parameter	Equation
5	v_{0*}	(3.15)
6	v_{imp*}	(3.14)
7	$\bar{J}_i^{opt}$	(6-4.45)

3.3 Numerical results and discussion

Values of the dimensionless DOP of the optimum impactor for different $\bar{v}_{imp}$, τ, λ_9 and different models are presented in *Table 7-4*.

Table 7-4. Dimensionless DOP for the optimum impactor.

$\bar{v}_{imp}$	$\lambda_9 = 12$				$\lambda_9 = 20$				$\lambda_9 = 28$			
Model →	1	2	3	4	1	2	3	4	1	2	3	4
						$\tau = 0.25$						
1.0	2.0	2.4	2.2	2.6	2.8	3.4	3.2	4.0	3.6	4.2	4.0	4.8
1.5	3.6	4.0	3.8	4.6	5.2	6.0	5.8	6.8	6.8	8.0	7.6	9.2
2.0	5.4	6.0	6.0	7.0	8.2	9.4	9.2	11.0	11.0	12.8	12.6	15.0
2.5	7.6	8.4	8.6	10.0	11.8	13.4	13.6	16.0	16.2	18.4	18.6	22.0
3.0	10.0	11.2	11.6	14.4	16.0	18.0	18.8	21.8	22.0	24.6	25.8	30.2
3.5	12.8	14.0	15.0	17.4	20.6	22.6	24.4	28.2	28.4	31.2	33.6	39.0
						$\tau = 0.50$						
1.0	2.2	2.4	2.2	2.6	2.8	3.2	3.0	3.6	3.6	4.0	4.0	4.6
1.5	3.4	3.6	3.8	4.2	4.8	5.4	5.4	6.4	6.4	7.2	7.2	8.4
2.0	4.8	5.4	5.4	6.4	7.4	8.0	8.6	9.8	9.8	10.8	11.6	13.4
2.5	6.4	6.8	7.8	8.6	10.0	10.8	12.2	13.8	13.6	14.6	16.8	19.0
3.0	8.2	8.6	10.2	11.2	12.8	13.6	16.2	18.2	17.4	18.6	22.4	25.0
3.5	9.8	10.2	12.6	14.0	15.6	16.4	20.4	22.6	21.4	22.6	28.2	31.4
						$\tau = 0.75$						
1.0	2.2	2.4	2.2	2.6	2.8	3.0	3.0	3.4	3.4	3.8	3.8	4.4
1.5	3.2	3.4	3.6	4.0	4.6	5.0	5.2	6.0	6.0	6.6	7.0	8.0
2.0	4.4	4.8	5.2	3.8	6.6	7.2	8.2	9.0	8.8	9.6	11.0	12.2
2.5	5.8	6.0	7.2	7.8	8.8	9.2	11.2	12.4	11.8	12.4	15.4	17.0
3.0	7.0	7.2	9.2	10.0	10.8	11.4	15.4	16.0	14.8	15.4	20.0	22.0
3.5	8.2	8.4	11.2	12.2	12.8	13.2	18.0	19.6	17.6	18.2	24.8	27.0

$\overline{v}_{imp}$	$\lambda_9 = 12$				$\lambda_9 = 20$				$\lambda_9 = 28$			
Model →	1	2	3	4	1	2	3	4	1	2	3	4
						τ = 1.00						
1.0	2.2	2.2	2.2	2.4	2.8	3.0	3.0	3.4	3.4	3.8	3.8	4.4
1.5	3.2	3.4	3.6	3.8	4.4	4.8	5.2	5.8	5.8	6.2	6.8	7.6
2.0	4.2	4.4	5.0	5.6	6.2	6.6	7.8	8.6	8.2	8.8	10.4	11.6
2.5	5.2	5.4	6.8	7.4	8.0	8.4	10.6	11.6	10.8	11.2	14.4	15.8
3.0	6.2	6.4	8.6	9.2	9.8	10.0	13.6	14.6	13.2	13.6	18.4	20.2
3.5	7.2	7.4	10.2	11.0	11.4	11.6	16.4	17.8	15.4	15.8	22.6	24.6

The shape and the DOP of the optimum impactor are very close to the shape and the DOP of the optimum truncated cone, respevtively, if the shape of the optimum cone is determined for the same $\overline{v}_{imp}$, τ, λ_9 and the same model is used. We calculated the performance parameter:

$$\delta_1 = \frac{H^{max\ sharp\ cone}}{H^{max}}, \tag{3.24}$$

where H^{max} is the DOP of the optimum impactor and $H^{max\ sharp\ cone}$ is the DOP of the impactor that is the best among the impactors with truncated conical noses. We found that δ_1 varies from 0.96 to 1.00 when $1 \le \overline{v}_{imp} \le 3.5$, $12 \le \lambda_9 \le 32$, $0.25 \le \tau \le 1$, and it increases with increasing τ, e.g., $\delta_1 = 0.99 \div 1.00$ for $\tau = 1$.

Generally, the impactor's shape can strongly affect the DOP; the comparison between the DOP of the optimum impactor and the DOP of the sharp cone, $H^{sharp\ cone}$, is shown in *Figure 7-12* for different τ and $\overline{v}_{imp}$.

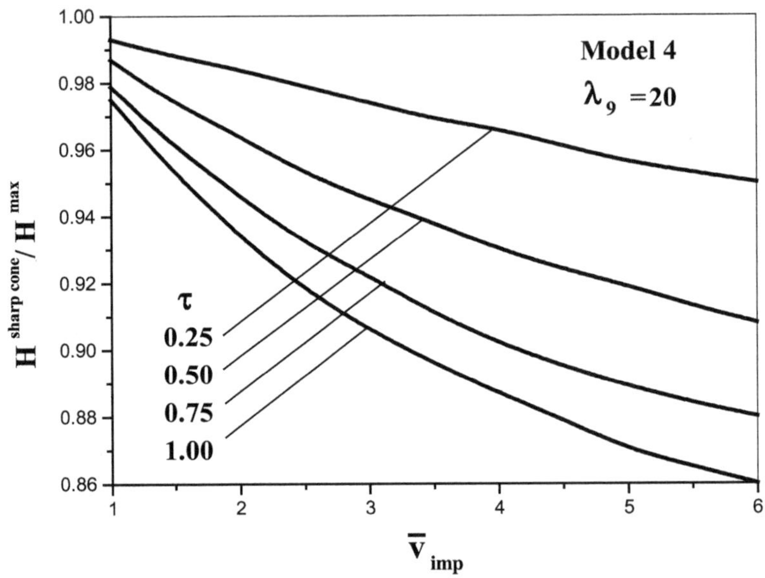

Figure 7-12. DOP of the sharp cone impactor, $H^{sharp\ cone}$, normalized by the DOP of the optimum impactor, H^{max}, as a function of the dimensionless impact velocity, $\bar{v}_{imp}$, for different $\tau = R/L$.

The estimates given in *Figure 7-12* for $\lambda_9 = 20$ are valid for other values of λ_9, with an accuracy of about 1%.

Since the optimum impactors are close to the truncated cones, the radius of the bluntness of the optimum cone can be used as the general characteristic parameter of the shape of the optimum impactor. This parameter is used in *Figure 7-13a-d* where $r^{opt\ tr-cone}/R$ is the radius of the bluntness of the optimum cone divided by the shank radius; parameter λ_9 practically does not influences $r^{opt\ tr-cone}/R$.

Therefore, the shape of the theoretically optimum impactor is close to that of a truncated cone and its radius of a flat bluntness depends mainly on the given normalized thickness of the impactor.

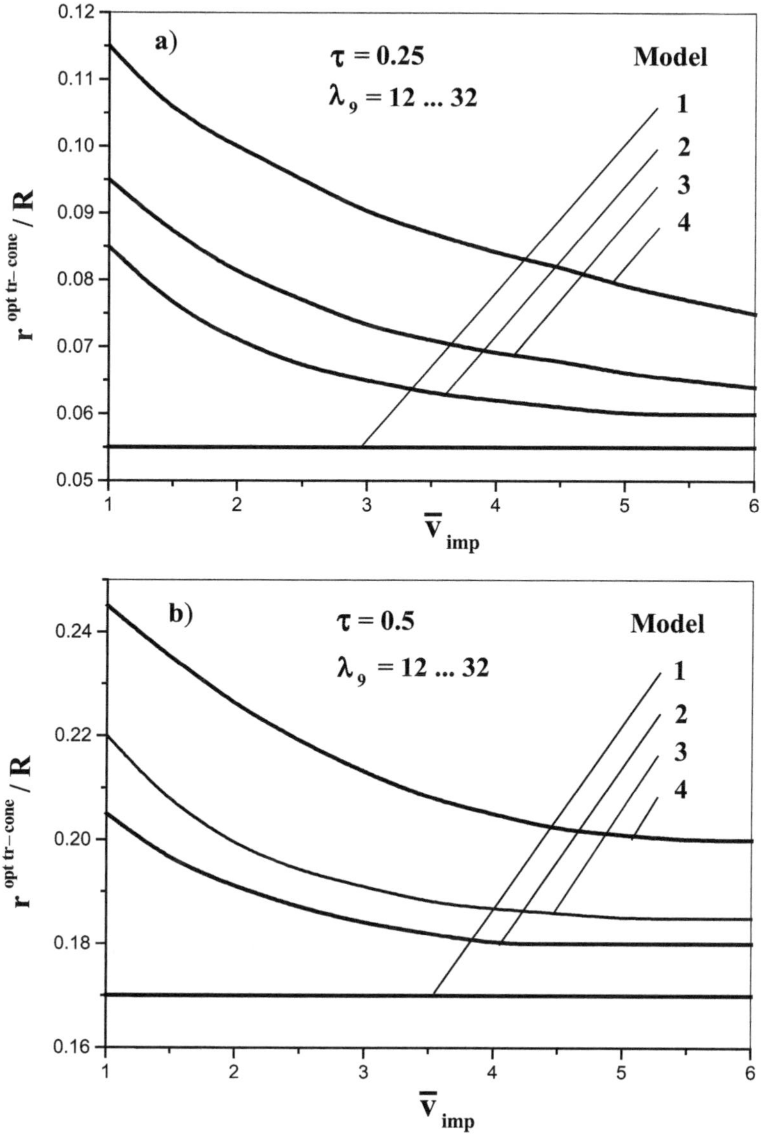
a)
τ = 0.25
λ9 = 12 ... 32
Model
1
2
3
4
0.12
0.11
0.10
0.09
0.08
0.07
0.06
0.05
r opt tr−cone / R
1
2
3
4
5
6
v imp
b)
τ = 0.5
λ9 = 12 ... 32
Model
1
2
3
4
0.24
0.22
0.20
0.18
0.16
r opt tr−cone / R
1
2
3
4
5
6
v imp

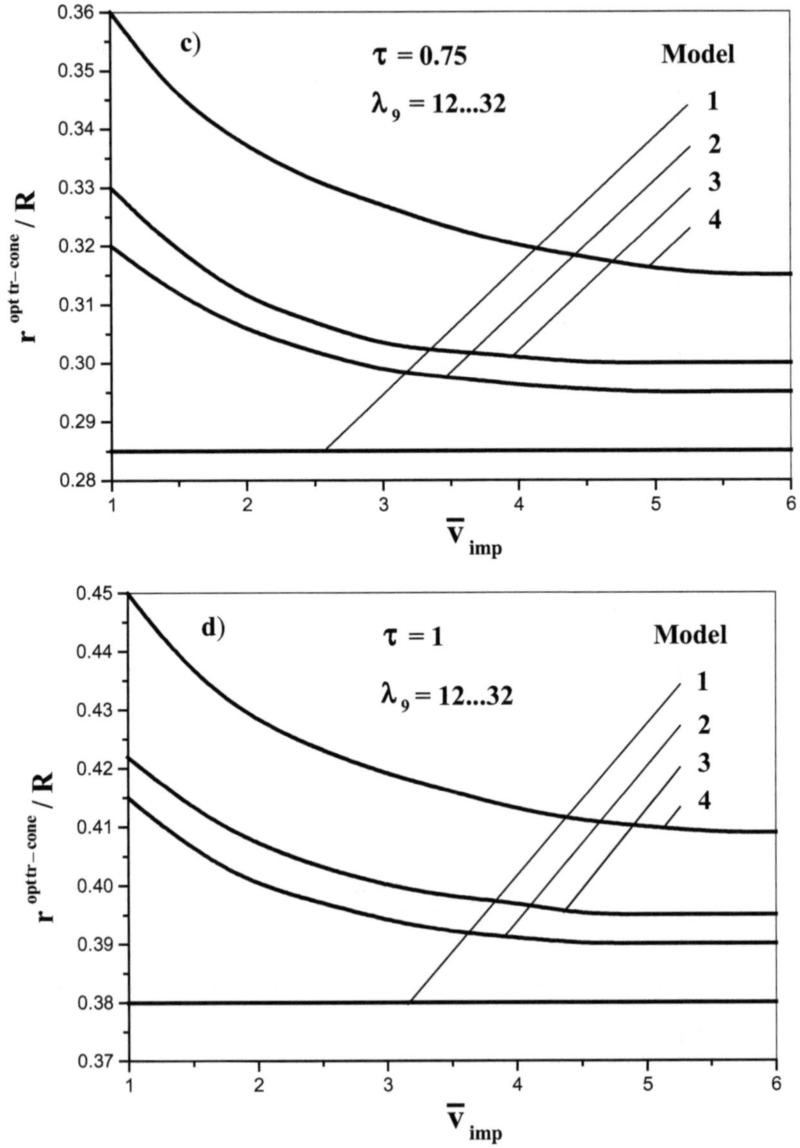

Figure 7-13. Normalized radius of the bluntness for the optimum truncated cone-shaped impactor, $r^{opt\ tr-cone}/R$, as a function of the dimensionless impact velocity, $\overline{v}_{imp}$, for different models of the shield material

Chapter 8

OPTIMUM SHAPE OF IMPACTORS AGAINST FIBRE-REINFORCED PLASTIC LAMINATES

1. GENERAL PENETRATION MODEL FOR THREE-DIMENSIONAL IMPACTORS

1.1 Basic model

The main results for ballistic impact dynamics of composites have been summarized in a number of reviews and monographs, e.g., Abrate (1994, 1998), Reid and Zhou (2000), Naik and Shrirao (2004), Kasano (1999). In the study of shape optimization of impactors penetrating into fibre-reinforced plastic (FRP) laminates, we are primarily interested in those models that allow us to describe the effect of the impactor's shape (over a wide range of shape variation) on the ballistic properties of the impactor (BLV, DOP). Analysis has shown that there are no models (even for projectiles having the shape of bodies of revolution) that completely meet our requirements. The most appropriate model for our purposes is that suggested by Wen (2000, 2001) (see also Reid and Wen, 2000 and some preceding studies by Wen et al., 1998; Reddy et al., 1998; Reid et al., 1999). Subsequently, Wen (2002a, 2002b) generalized his model to metal, concrete and soil shields, as well as to eroding penetrators.

Although the model suggested by Wen (2000, 2001) is not directly applicable to arbitrary impactors having the shape of bodies of revolution, it is suitable for our purposes for two reasons. First, it facilitates analytical representation for conical impactors. Therefore, in the framework of the LIA this model can be generalized to projectiles with more complicated shapes, including 3-D projectiles (see *Section 2-6.2.2*). Second, since Wen (2000, 2001) demonstrated good performance of this model for impactors with flat and semi-spherical bluntness, the generalized model may hopefully be applicable to projectiles having a nose deviating strongly from the conical shape.

Therefore, as a basic model, we use the model suggested by Wen (2000, 2001), where it is assumed that the local pressure σ_n applied normally to the surface of the projectile by an FRP laminate material during penetration by an impactor can be represented as:

$$\sigma_n = \mu_6 + \sqrt{\gamma_{sh}\mu_6}\,\beta_{imp} v_{imp}, \quad \mu_{fr} = 0 , \tag{1.1}$$

where γ_{sh} and μ_6 are the density and the quasi-static linear elastic limit of the material of the shield, respectively, v_{imp} is the impact (initial) velocity of the impactor, μ_{fr} is a friction coefficient between the impactor and the shield, and β_{imp} is a constant, which is determined empirically. Formulas for calculating β_{imp} as a function of geometrical parameters of the impactor are given in Wen (2000, 2001) for several typical shapes of bodies of revolution; in particular, $\beta_{imp} = 2\sin\upsilon$ for a conical-nosed impactor, where υ is the half angle of the apex of the cone. Let us now generalize this model for a projectile having the shape of an arbitrary 3-D body. To this end, we will use the method described in *Section 2-6.2.2.*

Let us rewrite the above-mentioned expression for β_{imp} for a conical impactor in the following form:

$$\beta_{imp} = 2(-\vec{v}^0 \cdot \vec{n}^0) = 2\cos\hat{\upsilon}, \tag{1.2}$$

where, as before, $\hat{\upsilon}$ is the angle between the vector $\vec{n}^0$ and the vector $(-\vec{v}^0)$, $\vec{n}^0$ is the inner normal vector at a given location at the impactor's surface, $\vec{x}^0 = -\vec{v}^0$, $\vec{v}^0$ is the unit vector of the velocity of the impactor, $\vec{x}^0$ is the unit vector of the corresponding axis; $\hat{\upsilon}$ may be defined as $\hat{\upsilon} = \pi/2 - \upsilon$ for a conical impactor. Thus, the model given by Eq. (1.1) can be written for a conical impactor in the form given by Eqs. (2-1.1), (2-3.1) and (2-3.2) with:

$$\Omega_n(u) = \mu_6 + 2\sqrt{\gamma_{sh}\mu_6}\, v_{imp} u, \quad u = -\vec{v}^0 \cdot \vec{n}^0 . \tag{1.3}$$

Assuming that the same LIM describes the local interactions between the shield and the impactor, independent of the shape of the impactor, we use

Eq. (1.3) for impactors of arbitrary shapes. Clearly, this generalized model coincides with the model suggested by Wen (2000, 2001) for flat-faced (cylindrical) projectiles and sharp cone-shaped projectiles. Therefore, the formulas for the BLV and DOP that yield this generalized model are similar to the formulas from Wen (2000, 2001) obtained by other methods for impactors having these shapes.

The model given by Eq. (1.3) is similar to the model considered in *Section 2-7* for $\alpha = 0$. However, in contrast to above-mentioned model, function Ω_n in Eq. (1.3) depends on v_{imp}. Therefore, we do not use all the formulas from *Section 2-7* directly.

Equation (2-3.3) implies the following expression for the drag force:

$$D(h, v_{imp}) = [\mu_6 + 2\sqrt{\gamma_{sh}\mu_6}\, v_{imp}]\,\delta(h)\sigma(0) + \int_{\theta(h)}^{\Theta(h)} dx \int_0^{2\pi} [\mu_6 + 2\sqrt{\gamma_{sh}\mu_6}\, v_{imp} u] u_1 d\vartheta, \tag{1.4}$$

where u and $u_1 = U\ (\mu_{fr} = 0)$ are determined by Eqs. (2-2.11) and (2-2.12), respectively.

The solution of the equation of motion of the impactor implies an equation similar to Eqs. (2-7.6) and (2-7.7):

$$v^2 = v_{imp}^2 - \frac{2}{m}\int_0^h D(\tilde{h}, v_{imp})\, d\tilde{h}. \tag{1.5}$$

1.2 Shield with a finite thickness

In the case of a shield with a finite thickness (SFT), the formula for the BLV v_{bl} is obtained by substituting $h = b + L, v = 0, v_{imp} = v_{bl}$ into Eq. (1.5):

$$v_{bl}^2 - \frac{2}{m}\int_0^{b+L} D(h, v_{bl})\, dh = 0. \tag{1.6}$$

Using Eq. (1.4) for $D(h, v_{bl})$, we may rewrite Eq. (1.6) as:

$$v_{bl}^2 - \frac{2}{m}\left\{ [\mu_6 + 2\sqrt{\gamma_{sh}\mu_6}\, v_{bl}]\, \sigma(0) \int_0^{b+L} \delta(h)dh \right.$$
$$\left. + \mu_6 \int_0^{b+L} dh \int_{\theta(h)}^{\Theta(h)} dx \int_0^{2\pi} u_1 d\vartheta + 2\sqrt{\gamma_{sh}\mu_6}\, v_{bl} \int_0^{b+L} dh \int_{\theta(h)}^{\Theta(h)} dx \int_0^{2\pi} uu_1 d\vartheta \right\} = 0 . \tag{1.7}$$

Applying Eq. (2-2.7) for the functions

$$\Xi(x) = \int_0^{2\pi} u_1(x,\vartheta)d\vartheta \tag{1.8}$$

and

$$\Xi(x) = \int_0^{2\pi} u(x,\vartheta)u_1(x,\vartheta)d\vartheta \tag{1.9}$$

we obtain, respectively:

$$\int_0^{b+L} dh \int_{\theta(h)}^{\Theta(h)} dx \int_0^{2\pi} u_1 dx = b\int_0^{L} dx \int_0^{2\pi} u_1 d\vartheta$$
$$= \frac{b}{2}\int_0^{2\pi} d\vartheta \int_0^{L} \frac{\partial \Phi^2}{\partial x} dx = b\,[\sigma(L) - \sigma(0)] , \tag{1.10}$$

and

$$\int_0^{b+L} dh \int_{\theta(h)}^{\Theta(h)} dx \int_0^{2\pi} uu_1 d\vartheta = bJ_0 , \tag{1.11}$$

where

$$J_0 = \int_0^{L} dx \int_0^{2\pi} uu_1\, d\vartheta . \tag{1.12}$$

Now using Eqs. (1.10), (1.11) and (2-7.38), we can write Eq. (1.7) for v_{bl} as follows:

$$v_{bl}^2 - \frac{4b\sqrt{\gamma_{sh}\mu_6}\,[\sigma(0)+J_0]}{m} v_{bl} - \frac{2b\mu_6\sigma(L)}{m} = 0 . \tag{1.13}$$

Eliminating m from Eq. (1.13), using Eqs. (2-4.1), (2-4.2) and dimensionless variables

$$\bar{x} = x/L,\ \bar{\Phi} = \Phi/L , \tag{1.14}$$

we may rewrite Eq. (1.13) in the following form:

$$\bar{v}_{bl}^2 - 2\lambda_{11}K_2\bar{v}_{bl} - \lambda_{11}K_1 = 0 , \tag{1.15}$$

where

$$\bar{v}_{bl} = \sqrt{\frac{\gamma_{sh}}{\mu_6}}\, v_{bl} , \qquad \lambda_{11} = \frac{2\gamma_{sh}b}{\gamma_{imp}L} , \tag{1.16}$$

$$K_1 = \frac{\bar{\sigma}(1)}{\bar{V}_{imp}} ,\quad K_2 = \frac{\bar{\sigma}(0)+\bar{J}_0}{\bar{V}_{imp}} , \tag{1.17}$$

$$\begin{gathered} \bar{V}_{imp} = \frac{V_{imp}}{L^3} = \frac{m}{\gamma_{imp}L^3} = \int_0^1 \bar{\sigma}(\bar{x})d\bar{x} + \bar{L}_0\,\bar{\sigma}(1), \\ \bar{\sigma}(\bar{x}) = \frac{1}{2}\int_0^{2\pi} \bar{\Phi}^2(\bar{x},\vartheta)d\vartheta,\ \ \bar{\Phi}_{\bar{x}} = \frac{\partial\bar{\Phi}}{\partial\bar{x}} ,\ \ \bar{\Phi}_\vartheta = \frac{\partial\bar{\Phi}}{\partial\vartheta} , \end{gathered} \tag{1.18}$$

$$\bar{J}_0 = \frac{J_0}{L^2} = \int_0^1 d\bar{x} \int_0^{2\pi} \frac{\bar{\Phi}^2\bar{\Phi}_{\bar{x}}^2}{\sqrt{\bar{\Phi}^2(\bar{\Phi}_{\bar{x}}^2+1)+\bar{\Phi}_\vartheta^2}} d\vartheta . \tag{1.19}$$

Solution of Eq. (1.15) yields a formula for the BLV:

$$\bar{v}_{bl} = \varphi(\lambda_{11}K_1, \lambda_{11}K_2) , \tag{1.20}$$

where

$$\varphi(z_1,z_2) = z_2 + \sqrt{z_2^2 + z_1} = z_2(1+\sqrt{1+z_1/z_2^2}) . \tag{1.21}$$

1.3 Semi-infinite shield

In the case of a semi-infinite shield (SIS), substituting $h = H, v = 0$ into Eq. (1.5), we obtain the correlation between the impact velocity, v_{imp} , and the DOP, H:

$$v_{imp}^2 - \frac{2}{m}\int_0^H D(h, v_{imp})dh = 0 . \tag{1.22}$$

Substituting $D(h, v_{bl})$ from Eq. (1.4) into Eq. (1.22), we obtain:

$$v_{imp}^2 - \frac{2}{m}\left\{ [\mu_6 + 2\sqrt{\gamma_{sh}\mu_6}\, v_{imp}]\,\sigma(0)H \right. \\ \left. + \mu_6 \int_0^H dh \int_0^{\Theta(h)} dx \int_0^{2\pi} u_1 d\vartheta + 2\sqrt{\gamma_{sh}\mu_6}\, v_{imp} \int_0^H dh \int_0^{\Theta(h)} dx \int_0^{2\pi} u u_1 d\vartheta \right\} = 0. \tag{1.23}$$

Applying Eq. (2-2.3) for the functions $\Xi(x)$ given by Eqs. (1.8) and (1.9), we obtain the following equation that relates the impact velocity to the DOP:

$$\overline{v}_{imp}^2 - 2\lambda_{12}\hat{K}_2\overline{v}_{imp} - \lambda_{12}\hat{K}_1 = 0 , \tag{1.24}$$

where

$$\overline{v}_{imp} = \sqrt{\frac{\gamma_{sh}}{\mu_6}} v_{imp} , \quad \lambda_{12} = \frac{2\gamma_{sh}}{\gamma_{imp}} , \tag{1.25}$$

$$\hat{K}_1 = \frac{\overline{\sigma}(0)\overline{H} + \hat{\overline{J}}_1}{\overline{V}}, \quad \hat{K}_2 = \frac{\overline{\sigma}(0)\overline{H} + \hat{\overline{J}}_0}{\overline{V}}, \quad \overline{H} = \frac{H}{L}, \tag{1.26}$$

$$\hat{\overline{J}}_0 = \int_0^{X_\Diamond(\overline{H})} (\overline{H} - \overline{x})d\overline{x} \int_0^{2\pi} \frac{\overline{\Phi}^2\overline{\Phi}_{\overline{x}}^2}{\sqrt{\overline{\Phi}^2(\overline{\Phi}_{\overline{x}}^2 + 1) + \overline{\Phi}_\vartheta^2}} d\vartheta, \tag{1.27}$$

$$\hat{\overline{J}}_1 = \int_0^{X_\Diamond(\overline{H})} (\overline{H} - \overline{x})d\overline{x} \int_0^{2\pi} \overline{\Phi}\overline{\Phi}_{\overline{x}}\, d\vartheta, \tag{1.28}$$

and function $X_\Diamond(z)$ is determined by Eq. (2-2.10).

Equation (1.24) allows us to express $\overline{v}_{imp}$ as a function of the DOP in the form:

$$\overline{v}_{imp} = \varphi(\lambda_{12}\hat{K}_1, \lambda_{12}\hat{K}_2), \tag{1.29}$$

where function φ is determined by Eq. (1.21).

When the impactor's shape is known, $\hat{K}_1$ and $\hat{K}_2$ are increasing functions of $\overline{H}$. In turn, φ is an increasing function of each of its arguments. Therefore, φ is an increasing function of $\overline{H}$, and there exists an inverse increasing function $\overline{H} = \psi(\overline{v}_{imp})$. Let us show that the expression for this function can be obtained in analytical form for the stage of motion of the impactor when $\overline{H} \geq 1$.

Let $\overline{v}_{imp\,0}$ be the impact velocity corresponding to the value of the DOP $\overline{H} = 1$. This impact velocity $\overline{v}_{imp\,0}$ can be obtained using Eq. (1.29) together with Eqs. (1.26)-(1.28) or $\overline{H} = 1$ and $X_\Diamond = 1$. If $\overline{v}_{imp} \geq \overline{v}_{imp\,0}$ ($\overline{H} \geq 1$), then $X_\Diamond = 1$, and Eqs. (1.27) and (1.28) yield:

$$\hat{\overline{J}}_0 = \overline{J}_0\overline{H} - \overline{J}_2, \; \overline{J}_2 = \int_0^1 \overline{x}d\overline{x} \int_0^{2\pi} \frac{\overline{\Phi}^2\overline{\Phi}_{\overline{x}}^2}{\sqrt{\overline{\Phi}^2(\overline{\Phi}_{\overline{x}}^2 + 1) + \overline{\Phi}_\vartheta^2}} d\vartheta, \tag{1.30}$$

$$\begin{aligned}\hat{\overline{J}}_1 &= \frac{1}{2}\int_0^{2\pi} d\vartheta \int_0^1 (\overline{H} - \overline{x})\frac{\partial\overline{\Phi}^2}{\partial\overline{x}} d\overline{x} = \frac{1}{2}\int_0^{2\pi} d\vartheta \left[(\overline{H} - \overline{x})\overline{\Phi}^2\right]_{\overline{x}=0}^{\overline{x}=1} \\ &+ \frac{1}{2}\int_0^1 d\overline{x} \int_0^{2\pi} \overline{\Phi}^2 d\theta = (\overline{H} - \overline{L}_0 - 1)\overline{\sigma}(1) - \overline{H}\,\overline{\sigma}(0) + \overline{V}_{imp},\end{aligned} \tag{1.31}$$

where integration by parts is used in Eq. (1.31). Then

$$\hat{K}_1 = \frac{\overline{\sigma}(1)(\overline{H} - \overline{L}_0 - 1) + \overline{V}_{imp}}{\overline{V}_{imp}}, \hat{K}_2 = \frac{[\overline{\sigma}(0) + \overline{J}_0]\overline{H} - \overline{J}_2}{\overline{V}_{imp}} \tag{1.32}$$

and Eq. (1.24) when considered as an equation with respect to $\overline{H}$ yields:

$$\overline{H} = \frac{\overline{v}_{imp}^2\overline{V}_{imp} / \lambda_{12} + 2\overline{v}_{imp}\overline{J}_2 + \overline{\sigma}(1)(\overline{L}_0 + 1) - \overline{V}_{imp}}{2\overline{v}_{imp}[\overline{\sigma}(0) + \overline{J}_0] + \overline{\sigma}(1)}, \quad \overline{v}_{imp} \geq \overline{v}_{imp\,0}, \tag{1.33}$$

where $\overline{v}_{imp\,0}$ can be determined from Eq. (1.29) by substituting $\hat{K}_1$ and $\hat{K}_2$ from Eqs. (1.32).

2. PENETRATION MODEL FOR THREE-DIMENSIONAL CONICAL IMPACTORS

For a 3-D conical impactor, function $\overline{\Phi}(\overline{x},\vartheta)$ is generally of the following type:

$$\overline{\Phi}(\overline{x},\vartheta) = (\overline{k}_0 + k\overline{x})\overline{\eta}(\vartheta), \tag{2.1}$$

where $\overline{k}_0$ and k are dimensionless parameters,

$$k = 1 - \overline{k}_0, \quad 0 \le \overline{k}_0 \le 1, \tag{2.2}$$

and dimensionless function $\overline{\eta}(\vartheta)$ determines the shape of the impactor at the cross-section $\overline{x} = 1$.

2.1 Shield with a finite thickness

Substitution of Eqs. (2.1) and (2.2) into Eqs. (1.17)–(1.19) in the case of a finite shield yields:

$$K_1 = k_1, \quad K_2 = k_1\{\overline{k}_0^2 + (1-\overline{k}_0)^2(1+\overline{k}_0)\overline{I}[\overline{\eta}(\vartheta),\overline{k}_0]\}, \tag{2.3}$$

where

$$k_1 = 3/(3\overline{L}_0 + k_0^2 + k_0 + 1), \tag{2.4}$$

$$\overline{I}_0 = \int_0^{2\pi} \frac{\overline{\eta}^4 d\vartheta}{\sqrt{(1-k_0^2)\overline{\eta}^4 + \overline{\eta}^2 + \overline{\eta}_\vartheta^2}}, \quad \overline{I}_1 = \int_0^{2\pi} \overline{\eta}^2 d\vartheta, \tag{2.5}$$

$$\overline{I} = \frac{\overline{I}_0}{\overline{I}_1}, \quad \overline{\eta}_\vartheta = \frac{d\overline{\eta}}{d\vartheta}, \tag{2.6}$$

and Eq. (1.20) for the BLV remains as before, with K_1 and K_2 given by Eqs. (2.3).

In the particular case of a *sharp 3-D cone*:

$$\bar{k}_0 = 0 \tag{2.7}$$

and Eq. (1.20) for the BLV is valid with:

$$K_1 = k_1, \quad K_2 = k_1 \bar{I}[\bar{\eta}(\vartheta), 0], \tag{2.8}$$

$$k_1 = 3/(3\bar{L}_0 + 1)\ . \tag{2.9}$$

2.2 Semi-infinite shield

In the case of a SIS, expressions for $\hat{K}_1$ and $\hat{K}_2$ can be obtained by substituting Eqs. (2.1) and (2.2) into Eqs. (1.26)-(1.28):

$$\hat{K}_1 = k_1 [\bar{k}_0^2 \bar{H} + (1/3)(1 - \bar{k}_0) Z(\bar{H})]\ , \tag{2.10}$$

$$\hat{K}_2 = k_1 \{\bar{k}_0^2 \bar{H} + (1/3)(1 - \bar{k}_0)^2 Z(\bar{H}) \bar{I}[\bar{\eta}(\vartheta), \bar{k}_0]\}\ , \tag{2.11}$$

where

$$Z(\bar{H}) = \begin{cases} \bar{H}^2 [3\bar{k}_0 + (1 - \bar{k}_0)\bar{H}] & if\ \bar{H} \le 1 \\ 3(\bar{k}_0 + 1)\bar{H} - \bar{k}_0 - 2 & if\ \bar{H} \ge 1 \end{cases}, \tag{2.12}$$

k_1 is determined by Eq. (2.4), and Eq. (1.29) remains as before with $\hat{K}_1$ and $\hat{K}_2$ given by Eqs. (2.10) and (2.11).

In the case of a sharp 3-D conical impactor when Eq. (2.7) is valid, the expressions for the parameters and functions given by Eqs. (2.10)-(2.12) are as follows:

$$\hat{K}_1 = (k_1/3) Z(\bar{H}), \hat{K}_2 = k_1 (1/3) Z(\bar{H}) \bar{I}[\bar{\eta}(\vartheta), 0]\ , \tag{2.13}$$

$$Z(\bar{H}) = \begin{cases} \bar{H}^3 & if\ \bar{H} \le 1 \\ 3\bar{H} - 2 & if\ \bar{H} \ge 1 \end{cases}, \tag{2.14}$$

where k_1 is determined by Eq. (2.9). In this case the DOP, $\bar{H}$, as a function of the impact velocity, $\bar{v}_{imp}$, can be represented in analytical form. Solving Eq. (1.29) with $\hat{K}_1$ and $\hat{K}_2$ given by Eq. (2.13), we obtain:

$$\overline{H}=\begin{cases}\sqrt[3]{z} & if\ \overline{v}_{imp}\le\overline{v}_{imp\,0}\\ (z+2)/3 & if\ \overline{v}_{imp}\ge\overline{v}_{imp\,0}\end{cases}, \tag{2.15}$$

where

$$z=\frac{3\overline{v}_{imp}^{2}}{k_1\lambda_{12}\{2\overline{I}[\overline{\eta}(\vartheta),\overline{k}_0]\overline{v}_{imp}+1\}},\ \overline{v}_{imp\,0}=\varphi\left(\frac{k_1\lambda_{12}}{3},\frac{k_1\lambda_{12}}{3}\overline{I}[\overline{\eta}(\vartheta),\overline{k}_0]\right) \tag{2.16}$$

3. PENETRATION MODEL FOR BODIES OF REVOLUTION

3.1 Shield with a finite thickness

Substituting $\overline{\Phi}=\overline{\Phi}(\overline{x})$ into Eqs. (1.18) and (1.19), we obtain:

$$\overline{\sigma}(0)=\pi\overline{r}^2,\ \ \overline{\sigma}(1)=\pi\overline{R}^2, \tag{3.1}$$

$$\overline{V}_{imp}=\frac{V_{imp}}{L^3}=\frac{m}{\gamma_{imp}L^3}=\pi\left[\int_0^1\overline{\Phi}^2(\overline{x})d\overline{x}+\overline{L}_0\,\overline{R}^2\right], \tag{3.2}$$

$$\overline{J}_0=2\pi G_0, \tag{3.3}$$

$$G_0[\overline{\Phi}(\overline{x})]=\int_0^1\frac{\overline{\Phi}\,\overline{\Phi}'^2}{\sqrt{\overline{\Phi}'^2+1}}d\overline{x},\ \overline{\Phi}'=\frac{d\overline{\Phi}}{d\overline{x}}. \tag{3.4}$$

Then Eqs. (1.20) and (1.21) yield the following formula for the BLV:

$$\overline{v}_{bl}=\lambda_{13}(G+\sqrt{G^2+\lambda_{14}}), \tag{3.5}$$

where

$$\lambda_{13}=\frac{4\pi\gamma_{sh}bL^2}{m},\ \ \lambda_{14}=\frac{mR^2}{8\pi\gamma_{sh}bL^4}, \tag{3.6}$$

$$G = 0.5\bar{r}^2 + G_0 = 0.5\bar{r}^2 + \int_0^1 \frac{\bar{\Phi}\bar{\Phi}'^2}{\sqrt{\bar{\Phi}'^2 + 1}} d\bar{x}. \tag{3.7}$$

Equation (1.16) relating the BLV and the dimensionless BLV remains valid. The mass of the impactor can be expressed through the parameters that determine its shape and impactor's density by using Eq. (3.2), if required.

Let us now consider impactors with different shapes.

The generatrix of the simplest *ogive-shaped impactor without bluntness* is determined by Eq. (7-2.13), with $\bar{q}_1$ given by Eq. (7-2.19). Substituting these equations into Eq. (3.7), we obtain:

$$G = \frac{1}{4\tau}\left[\frac{3\tau^4 + 2\tau + 3}{3(\tau^2 + 1)} - \frac{1 - \tau^4}{2\tau} \sin^{-1}\left(\frac{2\tau}{\tau^2 + 1}\right)\right]. \tag{3.8}$$

For a hemisphere that is a special case of the ogive with $\tau = 1$:

$$G = 1/3. \tag{3.9}$$

In the case of a *cylindrical impactor*:

$$\bar{\Phi}(\bar{x}) = \tau, \bar{r} = \tau. \tag{3.10}$$

Substituting Eq. (3.10) into Eq. (3.7), we obtain:

$$G = 0.5\tau^2. \tag{3.11}$$

Then Eqs. (3.5) can be rewritten as follows:

$$\bar{v}_{bl} = \varphi(\lambda_{15}, \lambda_{15}) = \lambda_{15} + \sqrt{\lambda_{15}(\lambda_{15} + 1)}, \tag{3.12}$$

where

$$\lambda_{15} = \frac{2\pi\gamma_{sh} bR^2}{m}. \tag{3.13}$$

The generatrix of a *conical impactor* is determined by Eq. (7-2.6), and the expression for G in Eq. (3.5) reads:

$$G = 0.5[\tau^2 - \tan\upsilon(1 - \sin\upsilon)(2\tau - \tan\upsilon)] . \qquad (3.14)$$

For a *sharp cone*, $\tan\upsilon = \tau$ and Eq. (3.14) yields:

$$G = 0.5\tau^3 \Big/ \sqrt{\tau^2 + 1} . \qquad (3.15)$$

3.2 Semi-infinite shield

In a case of a SIS, Eqs. (1.24)-(1.26), (1.29) and (1.33) remain valid; and expressions for $\overline{\sigma}(0), \overline{\sigma}(1)$ and $\overline{V}_{imp}$ can be simplified by using Eqs. (3.1) and (3.2). The formula for the integrals in Eqs. (1.27) and (1.28) reads:

$$\hat{J}_i = 2\pi \int_0^{X_0(\overline{H})} (\overline{H} - \overline{x})\overline{\Phi}\overline{\Phi}' \left(\frac{\overline{\Phi}'}{\sqrt{\overline{\Phi}'^2 + 1}} \right)^{1-i} d\overline{x} , \; i = 0,1 . \qquad (3.16)$$

3.3 Comparison with the experimental data

The proposed model coincides with the model developed by Wen (2000, 2001) for sharp cones and flat-faced (cylindrical) projectiles. In the latter study, it was shown that for these shapes of the nose this model is in good agreement with the experimental results of Reid et al. (1995), Reid et al. (1999), Wen et al. (1998), Mines et al. (1999), Zhu et al. (1992), and Reddy et al. (1998).

The results presented in Wen (2000, 2001) also demonstrate good agreement between the predictions of the model and experimental data obtained by Kumar and Bhat (1998) for ogival-nosed projectiles and the experimental results of Mines et al., (1999), Wen et al. (1998), Reddy et al. (1998), and Reid et al. (1999) for hemispherical nosed missiles. Reyes-Villanueva and Cantwell (2004) showed good predictive properties of Wen's model applied to the penetration of fibre-metal-laminate-skinned aluminum foam sandwich structures by a hemispherical nosed impactor. Ulven et al. (2003) studied experimentally the penetration of vacuum assisted resin transfer molding processed carbon/epoxy laminates by impactors of different shapes (hemispherical, conical, flat, and fragment simulating) and compared their results with the predictions obtained using Wen's model. Despite large quantitative discrepancies between the theoretical and experimental results, the model correctly describes the tendencies in variation of the BLV during variation of the impactor's shape.

Figure 8-1a-b shows the comparison between the predictions obtained using Eqs. (1.16), (3.6), (3.5) and (3.9) and the experimental data of Mines et al. (1999) for the perforation of E-glass/polyester laminates struck transversely by hemispherical-ended impactors when the generalized model and Wen's model do not coincide. Following Wen (2000), where these data are also analyzed, we used in the theoretical calculations $\gamma_{sh} = 1650\, kg/m^3$, $\mu_6 = 225\, MPa$. Inspection of *Figure 8-1a-b* shows that the theoretically predicted ballistic limits are also in a good agreement with the experimental data. Thus, there is reason to believe that the proposed model can indeed be used for determining the optimal nose geometry of an impactor.

4. SOME OPTIMAL PROPERTIES OF THREE-DIMENSIONAL CONICAL IMPACTORS

Here, we do not formulate or solve the shape optimization problem for 3-D impactors, although we do describe some results obtained in this field (Ben-Dor et al., 2002c). The emphasis is placed on the analysis of some peculiarities of these problems, and at the same time we elucidate some considerations presented in *Section 5-1*.

Consider a variational problem of finding for a SFT the cross section of the impactor [the function $\overline{\eta}(\vartheta)$] that provides the minimum BLV $\overline{v}_{bl}$ for given $\lambda_{11}, \overline{L}_0, k_0$. In this case, Eqs. (2.3) and (2.4) imply that K_1 is known and K_2 is an increasing function of the functional $\overline{I}$. Eqs. (1.20) and (1.21) imply that $\overline{v}_{bl}$ is an increasing function of $\overline{I}$, and the problem is reduced to the minimization of the functional $\overline{I} = \overline{I}[\overline{\eta}(\vartheta), \overline{k}_0]$; the change of λ_{11} and/or $\overline{L}_0$ does not affect the optimal shape of the impactor.

Consider now for a SIS, the problem of finding the cross section of the impactor that provides the minimum $\overline{v}_{imp}$ for given $\lambda_{12}, \overline{L}_0, \overline{k}_0$ and the DOP $\overline{H}$. Equations (2.10), (2.11), (1.29) and (1.21) show that this problem reduces to the minimization of the same functional $\overline{I} = \overline{I}[\overline{\eta}(\vartheta), \overline{k}_0]$, and the change of λ_{11} and/or $\overline{L}_0$, $\overline{H}$ does not affect the optimal contour of the impactor.

Consider finally another variational problem for SIS: to determine the cross section of the impactor that provides the maximum DOP, $\overline{H}$, for given $\lambda_{12}, \overline{L}_0, \overline{k}_0$ and $\overline{v}_{imp}$. Equation (1.29) can be considered as the definition of the dependence between $\overline{H}$ and $\overline{I}$. Differentiating Eq. (1.29) over $\overline{I}$, taking into account Eqs. (2.10)-(2.12) and (1.21) and differentiability of the function $Z(\overline{H})$ for $\overline{H} > 0$, we obtain:

$$0 = \sum_{i=1}^{2} \frac{\partial \varphi}{\partial z_i} \cdot \frac{dz_i}{d\hat{K}_i} \cdot \frac{d\hat{K}_i}{dI}, \tag{4.1}$$

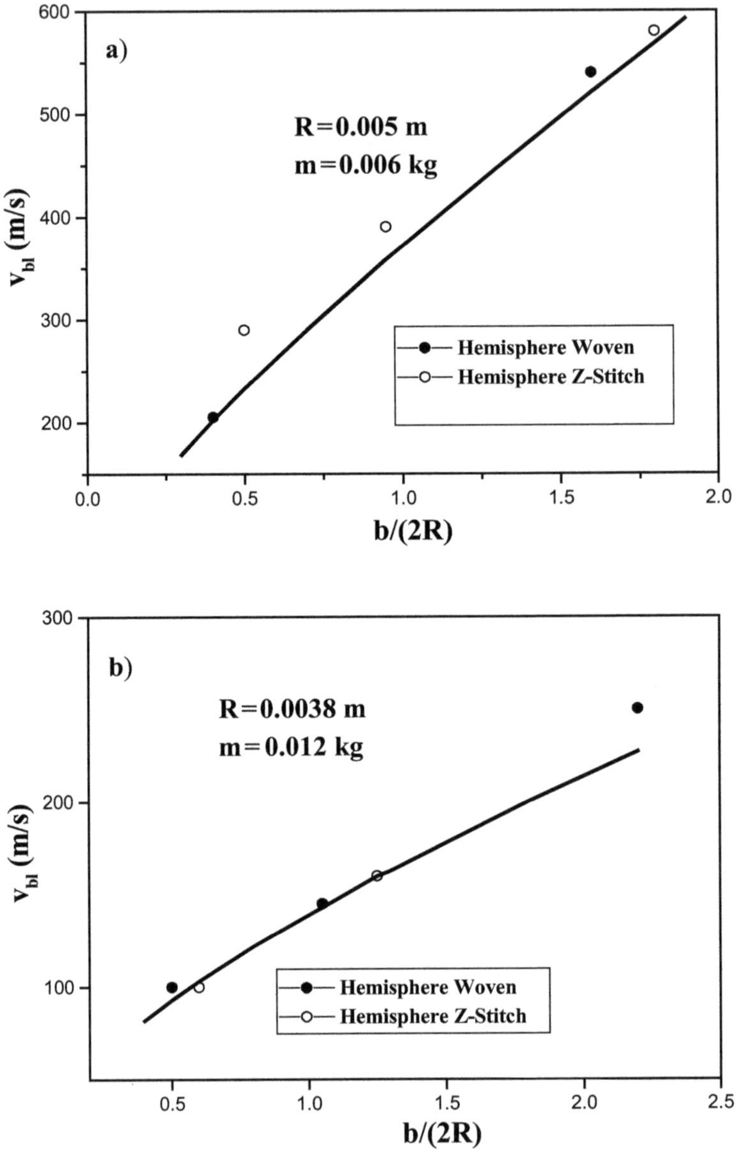

Figure 8-1. Comparison of the theoretical predictions (solid lines) with experimental data for perforation of glass reinforced plastic laminates struck by projectiles having a hemispherical shape (Mines et al., 1999).

where

$$z_i = \lambda_{12}\hat{K}_i, \frac{dz_i}{d\hat{K}_i} = \lambda_{12}, \; i = 1,2, \tag{4.2}$$

$$\frac{\partial \varphi}{\partial z_1} = \frac{1}{2\sqrt{z_1 + z_2^2}}, \; \frac{\partial \varphi}{\partial z_2} = 1 + \frac{z_2}{\sqrt{z_1 + z_2^2}}, \tag{4.3}$$

$$\frac{d\hat{K}_1}{d\bar{I}} = \bar{k}_1 \left[\bar{k}_0^2 \frac{d\bar{H}}{d\bar{I}} + \frac{1-\bar{k}_0}{3} \frac{dZ}{d\bar{H}} \cdot \frac{d\bar{H}}{d\bar{I}} \right], \tag{4.4}$$

$$\frac{d\hat{K}_2}{d\bar{I}} = \bar{k}_1 \left\{ \bar{k}_0^2 \frac{d\bar{H}}{d\bar{I}} + \frac{(1-\bar{k}_0)^2}{3} \left[\frac{dZ}{d\bar{H}} \cdot \frac{d\bar{H}}{d\bar{I}} \bar{I} + Z \right] \right\}, \tag{4.5}$$

$$\frac{dZ}{d\bar{H}} = \begin{cases} 3\bar{H}[2\bar{k}_0 + (1-\bar{k}_0)\bar{H}] & if \; \bar{H} \le 1 \\ 3(\bar{k}_0 + 1) & if \; \bar{H} \ge 1 \end{cases}. \tag{4.6}$$

Substituting Eqs. (4.2)-(4.6) into Eq. (4.1), after simple transformations, we obtain:

$$\frac{d\bar{H}}{d\bar{I}} = - \frac{Z(1-\bar{k}_0)^2 \frac{\partial \varphi}{\partial z_2}}{\sum_{i=1}^{2} \frac{\partial \varphi}{\partial z_i} [3\bar{k}_0^2 + (1-\bar{k}_0)^i \frac{dZ}{d\bar{H}} \bar{I}^{i-1}]}. \tag{4.7}$$

Since

$$\frac{\partial \varphi}{\partial z_1} > 0, \; \frac{\partial \varphi}{\partial z_2} > 0, \; \frac{dZ}{d\bar{H}} > 0, \; \bar{I} > 0, \; Z > 0, \; 0 \le \bar{k}_0 \le 1, \tag{4.8}$$

Eq. (4.7) implies that $d\bar{H} / d\bar{I} < 0$. Therefore the maximum DOP is attained for given $\lambda_{12}, \bar{L}_0, \bar{k}_0$ and $\bar{v}_{imp}$ when the value of $\bar{I}$ is minimal. Hence all three above-considered problems reduce to minimization of the same functional, where the optimal solution is independent of the properties of the shield.

Let us show that the proposed model predicts an advantage of 3-D conical impactors over conical impactors having the shape of a body of revolution. Consider a star-shaped impactor with the cross-section consisting of N identical segments (see *Figure 8-2a*, $N=4$). The equation of the straight-line boundary of a half-segment for $0 \le \vartheta \le \vartheta_0$ in polar coordinates $\hat{\rho}, \vartheta$ (in cross-section $\bar{x}=1$) is $\hat{\rho}=\bar{\eta}(\vartheta)$, where

$$\bar{\eta}(\vartheta)=\frac{\bar{R}_1\bar{R}_2 \sin\vartheta_0}{\bar{R}_2 \sin\vartheta_0-\bar{R}_1 \sin(\vartheta-\vartheta_0)}, \tag{4.9}$$

$$\vartheta_0=\frac{\pi}{N}, \bar{R}_1=\frac{R_1}{L}, \bar{R}_2=\frac{R_2}{L}, \tag{4.10}$$

and all the notations are shown in *Figure 8-2b*.

Substituting Eq. (4.9) into Eqs. (2.5) and (2.6), we obtain the following expression for the integral $\bar{I}$ for a star-shaped impactor:

$$\bar{I}_{star}=\frac{\bar{R}_1\bar{R}_2 \sin\vartheta_0}{\sqrt{\bar{R}_1^2+\bar{R}_2^2+(1-\bar{k}_0)^2\bar{R}_1^2\bar{R}_2^2 \sin^2\vartheta_0-2\bar{R}_1\bar{R}_2\cos\vartheta_0}}. \tag{4.11}$$

In the case of a conical impactor with radius of the base R_0, the value of the integral $\bar{I}$ can be obtained by substituting $\bar{\eta}(\vartheta)=\bar{R}_0$ $(\bar{R}_0=R_0/L)$ into Eqs. (2.5) and (2.6):

$$\bar{I}_{cone}=\frac{\bar{R}_0}{\sqrt{(1-\bar{k}_0)^2\bar{R}_0^2+1}}=\frac{1}{\sqrt{(1-\bar{k}_0)^2+(1/\bar{R}_0)^2}}. \tag{4.12}$$

Clearly, Eq. (4.12) implies that among the cones with $\bar{R}_1 \le \bar{R}_0 \le \bar{R}_2$ the minimum $\bar{I}_{cone}$ is attained if $\bar{R}_0=\bar{R}_1$. This cone is selected for comparison with a star-shaped impactor. After some algebra we obtain:

$$\varepsilon=\left(\frac{\bar{I}_{cone}}{\bar{I}_{star}}\right)^2=1+\frac{(\bar{R}_1-\bar{R}_2\cos\vartheta_0)^2}{\bar{R}_2^2\sin^2\vartheta_0[1-\bar{k}_0)^2\bar{R}_1^2+1]}, \tag{4.13}$$

and

$$\frac{d\varepsilon}{d\vartheta_0}=\frac{2\cos\vartheta_0(1-\phi)(\phi\cos^2\vartheta_0-1)}{\sin^3\vartheta_0[1-\bar{k}_0)^2\bar{R}_1^2+1]}, \quad \phi=\frac{\bar{R}_1}{\bar{R}_2\cos\vartheta_0}, \tag{4.14}$$

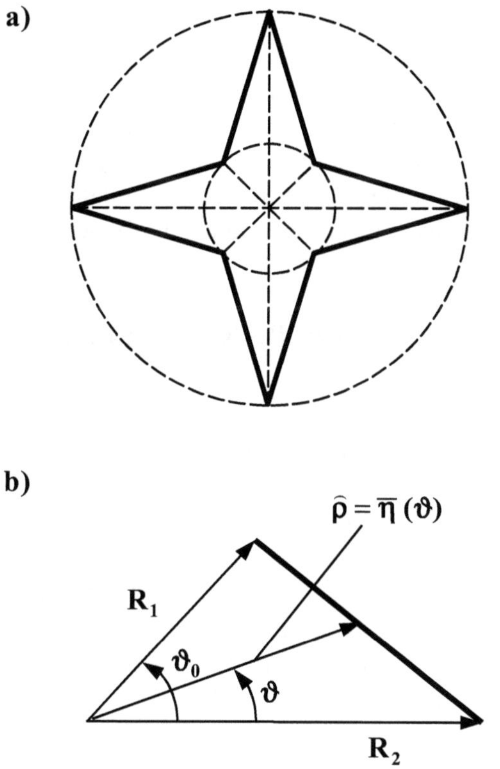

Figure 8-2. Cross-section of the star-shaped impactor: (a) general view for $N = 4$ and (b) half-segment.

where geometric considerations imply that $\phi < 1$. In differentiation with respect to ϑ_0, it is assumed that $N \geq 3$ can take real values.

Equations (4.13) and (4.14) show that:

$$\varepsilon > 1, \quad \varepsilon \xrightarrow[\vartheta_0 \to 0]{} \infty, \quad \frac{d\varepsilon}{d\vartheta_0} > 0 . \tag{4.15}$$

Hence, always $\bar{I}_{star} < \bar{I}_{cone}$. Moreover, $\bar{I}_{star} / \bar{I}_{cone}$ monotonically decreases with increasing number of segments N and becomes arbitrary close to zero. Since $\bar{I}_{cone}$ does not depend on N, this means that $\bar{I}_{star}$ can be made as small as desired by increasing N.

Therefore, increasing the number of segments in a star-shaped impactor causes the magnitude of the functional $\bar{I}$ approach arbitrarily close to the optimal value, which in this case equals zero. From a methodological point of view effects of this kind were discussed in *Section 5-1*.

It may be concluded that this model predicts that the use of impactors with complicated 3-D shapes allows us to increase considerably the performance of the impactor. Although it is possible to solve a number of different shape optimization problems, one has to keep in mind that there are no experimental data available that allow us to validate the applicability of this model to impactors having a complicated 3-D shape.

5. OPTIMIZATION OF PENETRATING BODIES OF REVOLUTION

5.1 The statement of the problem

Consider (Ben-Dor et al., 2002a) the problem of minimizing the BLV v_{bl} for a SFT when the thickness of the shield and the parameters determining its mechanical properties are known. The mass of the impactor, the length and shank radius are assumed to be given. It is easy to see from Eq. (3.5) that the BLV is an increasing function of G. Therefore, the problem can be reduced to the optimization of the functional:

$$G[\bar{\Phi}(\bar{x})] = 0.5\bar{r}^2 + \int_0^1 F(\bar{\Phi}, \bar{\Phi}')\,d\bar{x}, \tag{5.1}$$

where

$$F = \bar{\Phi}\,\varphi_0(\bar{\Phi}'),\ \varphi_0(z) = \frac{z^2}{\sqrt{z^2+1}} \tag{5.2}$$

and, as before, $\bar{\Phi} = \Phi/L, \bar{x} = x/L, \bar{\Phi}' = d\bar{\Phi}/d\bar{x}$.

The solution of the variational problem, $\bar{\Phi}(\bar{x})$, must satisfy the boundary condition:

$$\bar{\Phi}(1) = \tau \tag{5.3}$$

while the value $\bar{r} = \bar{\Phi}(0)$ is unknown.

It should be noted that a similar problem for a SIS is more complicated and can be solved only numerically (Ben-Dor et al., 2002b).

5.2 Investigation of the variational problem

We study the variational problem using the necessary conditions for the extremum (see *Section 5-2*).

Since F does not depend on $\overline{x}$, the Euler-Lagrange equation has a first integral:

$$\overline{\Phi} = C\varphi_1(\overline{\Phi}'), \quad \varphi_1(z) = \frac{(z^2+1)^{3/2}}{z^2}, \tag{5.4}$$

where C is a constant.

The transversality condition of the type given by Eq. (5-2.12) reads:

$$\overline{r}\varphi_2(t_0) = 0, \ \varphi_2(z) = \frac{z(z^2+2)}{(z^2+1)^{3/2}} - 1, \tag{5.5}$$

where t_0 is the derivative of the optimal generatrix at the initial point,

$$t_0 = \overline{\Phi}'(0). \tag{5.6}$$

Analysis of the Erdmann-Weierstrass corner conditions:

$$\Delta\left(\frac{\overline{\Phi}'^2}{(\overline{\Phi}'^2+1)^{3/2}}\right) = 0, \ \Delta\left(\frac{\overline{\Phi}'(\overline{\Phi}'^2+2)}{(\overline{\Phi}'^2+1)^{3/2}}\right) = 0 \tag{5.7}$$

shows that corner points do not exist.

The Legendre condition yields:

$$\frac{\overline{\Phi}(2-\overline{\Phi}'^2)}{(\overline{\Phi}'^2+1)^{5/2}} \geq 0 \tag{5.8}$$

Since $\overline{\Phi} \neq 0$ (see Eq. 5.4), t_0 is determined by equation $\varphi_2(t_0) = 0$, which can be written as follows:

$$\sqrt{T}(T+2) = (T+1)^{3/2}, T = t_0^2 \tag{5.9}$$

Equation (5.9) yields a quadratic equation $T^2 + T - 1 = 0$. Solving this equation allows us to determine t_0:

$$t_0 = t_0^*,\ t_0^* = \sqrt{0.5(\sqrt{5} - 1)} \approx 0.786. \qquad (5.10)$$

The constant C can expressed through $\overline{r}$ by considering Eq. (5.4) at the initial point $\overline{x} = 0$ where $\overline{\Phi} = \overline{r}$ and $\overline{\Phi}' = t_0$:

$$C = c_* \overline{r}, \quad c_* = 1/\varphi_1(t_0^*) = (t_0^*)^5 \approx 0.300, \qquad (5.11)$$

and Eq. (5.4) can be rewritten as follows:

$$\overline{\Phi} = c_* \overline{r}\, \varphi_1(\overline{\Phi}'). \qquad (5.12)$$

The solution of the ordinary differential equation (5.12) can be represented in a parametric form choosing as a parameter:

$$t = \overline{\Phi}'. \qquad (5.13)$$

Equation (5.12) can be rewritten as:

$$\overline{\Phi} = c_* \overline{r}\, \varphi_1(t), \qquad (5.14)$$

where $\overline{x}$ is considered a function of t and can be determined using the expression $d\overline{x} = d\overline{\Phi}/t$. Therefore,

$$\overline{x} = c_* \overline{r}\, \varphi_3(t), \qquad (5.15)$$

where

$$\varphi_3(z) = \int_{t_0^*}^{z} \frac{1}{\zeta} \frac{d\varphi_1(\zeta)}{d\zeta} d\zeta = \int_{t_0^*}^{z} \frac{(\zeta^2 - 2)\sqrt{\zeta^2 + 1}}{\zeta^4} d\zeta \qquad (5.16)$$
$$= \varphi_4(z) - \varphi_4(t_0^*),$$

$$\varphi_4(z) = \frac{2 - z^2}{3z^3} \sqrt{z^2 + 1} + \ln(z + \sqrt{z^2 + 1}). \qquad (5.17)$$

Taking into account Eqs. (5.14) and (5.15), Eq. (5.3) yields two correlations for determining $\bar{r}$ and the derivative of the optimal generatrix at the endpoint t_1:

$$\tau = c_* \bar{r}\, \varphi_1(t_1)\,, \tag{5.18}$$

$$1 = c_* \bar{r}\, \varphi_3(t_1)\,. \tag{5.19}$$

Eliminating $\bar{r}$ from Eqs. (5.18) and (5.19), we obtain the equation for determining t_1:

$$\Lambda(t_1) = 0\,, \tag{5.20}$$

where

$$\begin{aligned}\Lambda(z) &= \tau\varphi_3(z) - \varphi_1(z) = \tau[\varphi_4(z) - \varphi_4(t_0^*)] - \varphi_1(z) = \\ &= \frac{2\tau - z(3z^2 + \tau z + 3)}{z^3} \cdot \frac{\sqrt{z^2+1}}{3} + \tau[\ln(z + \sqrt{z^2+1}) - \varphi_4(t_0^*)].\end{aligned} \tag{5.21}$$

Let us calculate the derivative:

$$\varphi_1'(z) = \frac{(z^2-2)\sqrt{z^2+1}}{z^3}\,. \tag{5.22}$$

Equations (5.16) and (5.22) show that the moving a point along the curve determined by Eqs. (5.14), (5.15) from the initial point to the end point is associated with the decrease of the parameter t from $t = t_0^*$ to some value $t = t_1$, where t_1 is a root of Eq. (5.20). Hence this curve is convex. Let us prove that Eq. (5.20) has only one solution in the interval $(0, t_0^*)$.

Equation (5.21) implies that

$$\lim_{t_1 \to +0} \Lambda(t_1) = +\infty\,, \ \Lambda(t_0^*) = -\varphi_1(t_0^*) < 0 \tag{5.23}$$

and, consequently, there exists at least one root of Eq. (5.20) in the interval $(0, t_0^*)$. It can be easily shown that this equation has only one root. Using Eq. (5.16), we can write the derivative $\Lambda'(t_1)$ as:

$$\Lambda'(t_1) = \frac{\varphi_1'(t_1)}{t_1}(\tau - t_1)\,. \tag{5.24}$$

Equation (5.22) implies that $\varphi_1'(t_1) < 0$ for $0 < t_1 \le t_0^*$. Consequently, as show Eqs. (5.23) and (5.24), two versions of the behavior of the function are $\Lambda(t_1)$ possible (see *Figure 3-7a-b*), depending on whether $t_0^* \ge \tau$ or $t_0^* \le \tau$. Clearly, Eq. (5.20) has the unique solution $t_1 = t_1^*$ in both cases. Since $0 < t_1^* < t < t_0^* < 1$, the Legendre condition (Eq. 5.8) is satisfied.

After t_1^* has been determined, the parameter $\bar{r}$ can be found from Eqs. (5.18) or Eq. (5.19):

$$\bar{r} = \frac{1}{c_* \varphi_3(t_1^*)} \quad \text{or} \quad \bar{r} = \frac{\tau}{c_* \varphi_1(t_1^*)}\,. \tag{5.25}$$

Let us calculate the minimum of the optimized functional, i.e., the value of the functional G when the function $\overline{\Phi}(\bar{x})$ is the solution of the considered variational problem. We denote the minimum value of this functional by G_{min}. It is convenient to calculate the integral in Eq. (5.2) using the variable t given by Eq. (5.13) instead of $\bar{x}$. Since

$$d\bar{x} = \frac{d\overline{\Phi}}{t} = \frac{1}{t} \cdot \frac{d\overline{\Phi}}{dt} dt = \frac{c_* \bar{r} \varphi_1'(t)}{t} dt = c_* \bar{r} \varphi_3'(t) dt\,, \tag{5.26}$$

then

$$G_{min} = \bar{r}^2 (0.5 + c_*^2 \widetilde{G}_{min})\,, \tag{5.27}$$

where $\bar{r}$ is determined by Eq. (5.25) and

$$\begin{aligned} \widetilde{G}_{min} &= \int_{t_0^*}^{t_1^*} \varphi_0(\zeta)\varphi_1(\zeta)\varphi_3'(\zeta)\,d\zeta \\ &= \int_{t_0^*}^{t_1^*} \frac{(\zeta^2 - 2)(\zeta^2 + 1)^{3/2}}{\zeta^4} d\zeta = \varphi_5(t_1^*) - \varphi_5(t_0^*)\,, \end{aligned} \tag{5.28}$$

$$\varphi_5(z) = \frac{(3z^4 + 10z^2 + 4)\sqrt{z^2 + 1}}{6z^3} + \frac{1}{2} ln(\sqrt{z^2 + 1} - z)\,. \tag{5.29}$$

Figure 8-3 shows the shape of the generatrix of the optimum impactor's nose for different τ. The optimum projectile has plane bluntness with a radius that increases as τ increases (see *Figure 8-4*).

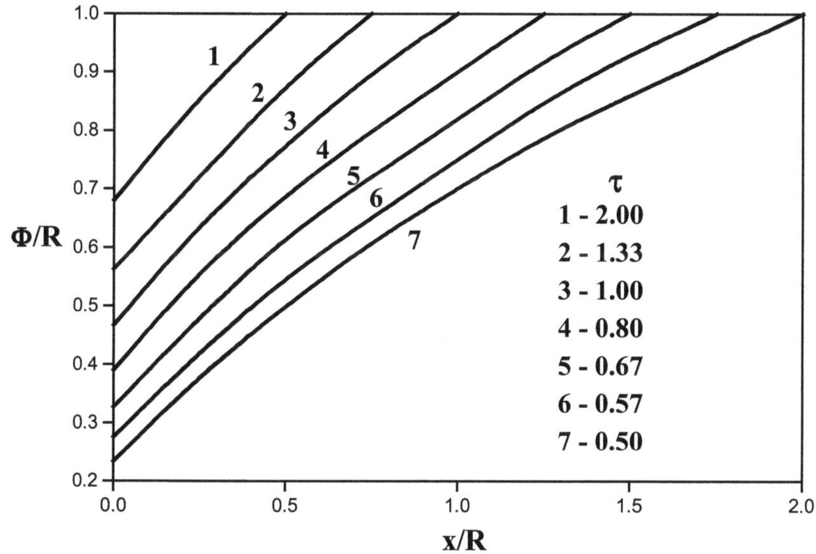

Figure 8-3. Shape of the generatrix of the optimal impactor.

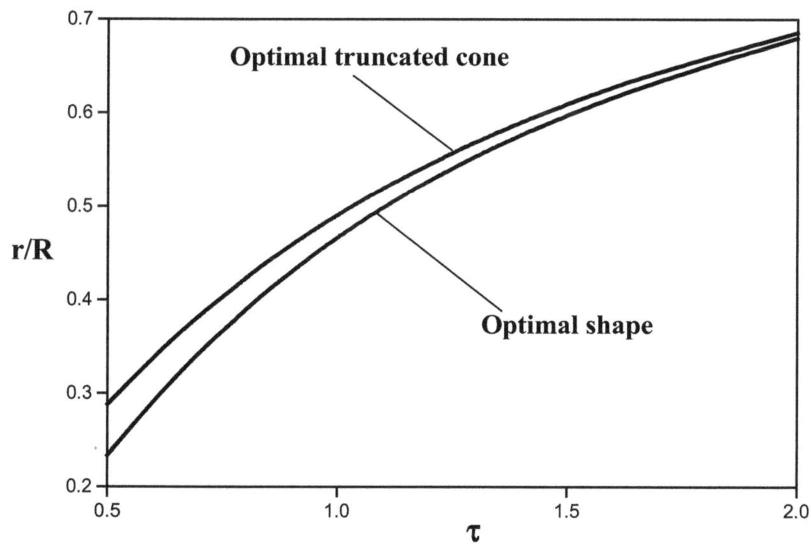

Figure 8-4. Radius of bluntness of the optimum impactor and the optimum truncated conical impactor.

6. COMPARISON BETWEEN DIFFERENT SHAPES

6.1 The worst value for the criterion

Let us first find the shape of the nose having the maximum G. Interchanging the dependent and independent variables in Eq. (5.1) and denoting $\bar{x}' = d\bar{x}(\bar{\Phi})/d\bar{\Phi}$, we obtain:

$$G = 0.5\bar{r}^2 + \int_{\bar{r}}^{\tau} \frac{\bar{\Phi} \cdot (\bar{x}')^{-2}}{\sqrt{(\bar{x}')^{-2} + 1}} \bar{x}' d\bar{\Phi} = 0.5\bar{r}^2 + \int_{\bar{r}}^{\tau} \frac{\bar{\Phi}\, d\bar{\Phi}}{\sqrt{\bar{x}'^2 + 1}}. \tag{6.1}$$

Since the equation of the generatrix for $0 \le \bar{\Phi} \le \bar{r}$ is $\bar{x}(\bar{\Phi}) = 0$, we can write:

$$\bar{r}^2 = 2\int_0^{\bar{r}} \frac{\bar{\Phi}\, d\bar{\Phi}}{\sqrt{\bar{x}'^2 + 1}}. \tag{6.2}$$

Substituting $\bar{r}^2$ from Eq. (6.2) to Eq. (6.1) yields:

$$G = \int_0^{\tau} \frac{\bar{\Phi}\, d\bar{\Phi}}{\sqrt{\bar{x}'^2 + 1}}. \tag{6.3}$$

Since G attains its maximum value when $\bar{x}(\bar{\Phi}) = 0$ for $0 \le \bar{\Phi} \le \tau$, Eq. (6.3) implies:

$$G \le \int_0^{\tau} \bar{\Phi}\, d\bar{\Phi} = \frac{\tau^2}{2}. \tag{6.4}$$

Therefore, a flat-faced cylindrical projectile has the maximum BLV for impactor noses that are the bodies of revolution and the upper bound for the functional G given by Eq. (6.4) is attained. The lower bound for this functional can be found using Eq. (6.4)

6.2 Optimum truncated-conical impactor

Consider the problem of shape optimization of a projectile in the class of cones with plane bluntness. The equation of the generatrix and the formula for calculating G are given by Eqs. (7-2.6) and (3.14), respectively. Clearly, minimization of G is equivalent to the maximization of the function

$$f(\upsilon)=\tau^2-2G=\tan\upsilon(2\tau-\tan\upsilon)(1-\sin\upsilon), \quad (6.5)$$

taking into account the constraints arising due to geometrical considerations:

$$0\leq\upsilon\leq\tan^{-1}(\tau)=\upsilon_0 . \quad (6.6)$$

Since $f(\upsilon_0)>f(0)$ where

$$f(0)=0,\, f(\upsilon_0)=\tau^2\left(1-\frac{\tau}{\sqrt{\tau^2+1}}\right)>0, \quad (6.7)$$

the point $\upsilon=0$ can not be a point of the maximum $f(\upsilon)$. On the other hand,

$$f'(\upsilon_0)=-\tau/\sqrt{\tau^2+1}<0, \quad (6.8)$$

where

$$f'(\upsilon)=\frac{2(1-\sin\upsilon)(\tau-\tan\upsilon)}{\cos^2\upsilon}-\sin\upsilon(2\tau-\tan\upsilon). \quad (6.9)$$

Equation (6.8) implies that there exists some $\tilde{\upsilon}\,(0<\tilde{\upsilon}\leq\upsilon_0)$ for which $f(\tilde{\upsilon})>f(\upsilon_0)$. Therefore, we have proved that the maximum value of the function f and the minimum value of the functional G are attained inside the interval $(0,\upsilon_0)$, i.e., the optimum projectile among the conical bodies has a finite bluntness $r<R$. The dependence of the normalized radius of the bluntness of the optimum cone vs. τ is shown in *Figure 8-4*.

6.3 Results of numerical simulation

Comparison of the values of G for the optimum shapes, optimum truncated cone, ogive, and sharp cone is shown in *Figure 8-5*. Equations (3.8) and (3.15) are used in calculations for ogives and sharp cones. *Figure 8-5* shows that the difference in G between the optimum impactor and the blunt cone impactor is relatively small for all practically relevant values of τ, whereas the advantage of the optimum impactor over the sharp cone and ogive impactors is significant for all τ.

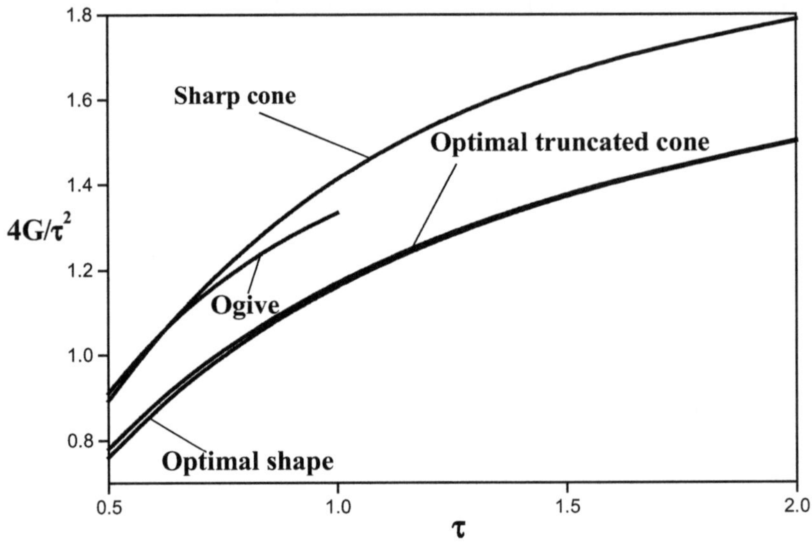

Figure 8-5. Comparison of the performance of impactors with different shapes of the nose.

Chapter 9

AREA RULES FOR PENETRATING IMPACTORS

1. INTRODUCTION

Area rules, in the commonly accepted sense, determine the conditions under which difference between the values of some aerodynamic characteristics of a 3-D projectile and those of a reference projectile is of a higher order than the difference in their shapes. Area rules are used in gasdynamics, particularly for determining drag coefficients of 3-D projectiles under different high velocity flight conditions, provided a body of revolution is chosen as the reference projectile. It is assumed that both projectiles have the same length. The differences between the areas of cross-section and the contours of the projectiles at every cross-section along the longitudinal axis are assumed to be small. The area rule asserts that if the magnitude of these differences is of the order of ε, the magnitude of the difference in the drag forces acting on these projectiles is of the order of ε^2. Area rules have been determined for special ranges of flight conditions for various integral characteristics (not only for the drag force), including the characteristics that can be expressed through functionals of a quite general form. More information on this subject may be found, for example, in Bunimovich and Dubinsky (1995) and Apshtein and Titow (1996).

The significance and usefulness of the area rules is discussed in the following. It may be relatively easy to analyze a reference impactor having a simple shape. The method of area rules allows us to predict the behavior of a projectile with a more complex shape using the results obtained for the reference impactor. This method can essentially simplify the prediction procedures for projectiles having complicated shapes, and it is of particular importance for predicting the consequences of small geometrical changes on the performance of a projectile.

Area rules have also been established for penetrating impactors (Ben-Dor et al., 1997b, 1998c). Using the assumption of the LIT, two versions of such rules are found in the following sections for the BLV, namely, when a body of revolution or a projectile with a polygonal cross-section is considered as the reference impactor.

2. BODY OF REVOLUTION AS A REFERENCE IMPACTOR

2.1 Statement of the problem

Consider the penetration of a rigid sharp 3-D impactor Γ into a shield with a finite thickness and assume that the general LIM is valid. Equations (2-2.11)-(2-2.13) allow us to rewrite the functional D determining the drag force as follows:

$$D[\Phi(x,\vartheta),h,w] = \int_{\theta(h)}^{\Theta(h)} dx \int_0^{2\pi} \Xi(\hat{u},w)\hat{U}d\vartheta, \tag{2.1}$$

where

$$\Xi(\hat{u},w) = \Omega_0(\hat{u},\sqrt{w})/\hat{u}, \tag{2.2}$$

$$\hat{u}(x,\vartheta) = \frac{\Phi\Phi_x}{\sqrt{\Phi^2(\Phi_x^2+1)+\Phi_\vartheta^2}}, \quad \hat{U}(x,\vartheta) = \Phi\Phi_x, \tag{2.3}$$

$$w = v^2. \tag{2.4}$$

If $w = w(h)$ is a solution of the equation of motion of the impactor

$$\frac{m}{2}\frac{dw}{dh} = -D[\Phi(x,\vartheta),h,w] \tag{2.5}$$

with the initial condition

$$w(b+L) = 0, \tag{2.6}$$

then the BLV is determined as $v_{bl} = \sqrt{w_{bl}}$ and $w_{bl} = w(0)$.

Together with a 3-D impactor Γ whose shape is determined by the function $\Phi(x,\vartheta)$ let us consider a reference body of revolution Γ° with the equation of the generatrix $\rho = r(x)$ having the same length and the same cross-sectional area $\sigma(x)$ at every cross-section. The latter implies the relationship:

$$\frac{1}{2}\int_0^{2\pi} \Phi^2(x,\vartheta)d\vartheta = \pi r^2(x) \tag{2.7}$$

Equations (2-2.14), (2-4.1), (2-4.2) and (2.7) show that if the material density of the impactors γ_{imp} is the same, then the impactors have the same mass of the nose. Assuming that their cylindrical parts have the same mass, it may be concluded that the parameter m in Eq. (2.5) is the same for both impactors.

Assume also that the shapes of the impactors Γ and Γ° are close to each other in the sense that

$$\Phi(x,\vartheta) = r(x) + \varepsilon\xi(x,\vartheta) + O(\varepsilon^2), \tag{2.8}$$

where $\xi(x,\vartheta)$ is some function, and ε is a small parameter. Hereafter, a non-numerical subscript will denote differentiation with respect to the corresponding variable. Our goal is to estimate the difference between values of the BLV for impactors Γ and Γ°.

2.2 Derivation of the area rules

Substituting $\Phi(x,\vartheta)$ from Eq. (2.8) to Eq. (2.7), we obtain after some algebra:

$$r(x)\Delta(x)\varepsilon + O(\varepsilon^2) = 0, \tag{2.9}$$

where

$$\Delta(x) = \int_0^{2\pi} \xi(x,\vartheta)d\vartheta, \tag{2.10}$$

and, consequently,

$$\Delta(x) \sim \varepsilon \ . \tag{2.11}$$

Now let us use the general approach of the perturbation theory (see, e.g., Bush, 1992). The solutions of Eq. (2.5) with initial condition given by Eq. (2.6) can be expanded into a series of a small parameter ε:

$$w(h) = w^{[0]}(h) + w^{[1]}(h)\varepsilon + O(\varepsilon^2), \tag{2.12}$$

where functions $w^{[0]}(h)$ and $w^{[1]}(h)$ must be determined. For this purpose substitute $\Phi(x,\vartheta)$ from Eq. (2.8) and $w(h)$ from Eq. (2.12) into Eq. (2.5):

$$\frac{m}{2}\left[w_h^{[0]} + w_h^{[1]}\varepsilon + O(\varepsilon^2)\right] = -D\, r(x) + \varepsilon\xi(x,\vartheta), h, w^{[0]} + w^{[1]}\varepsilon + O(\varepsilon^2) \tag{2.13}$$

Drag coefficient D in Eq. (2.13) can be expanded in Taylor series with respect to a small parameter ε:

$$D = D^{\circ} + D_{\varepsilon}^{\circ}\varepsilon + O(\varepsilon^2), \tag{2.14}$$

where the superscript denotes the value for $\varepsilon = 0$, i.e.,

$$D^{\circ} = D\left[r(x), h, w^{[0]}\right], \quad D_{\varepsilon}^{\circ} = \left.\frac{\partial D}{\partial \varepsilon}\right|_{\varepsilon=0} . \tag{2.15}$$

Then Eq. (2.13) can be rewritten as:

$$(m/2)\left[w_h^{[0]} + w_h^{[1]}\varepsilon + O(\varepsilon^2)\right] = -D^{\circ} - D_{\varepsilon}^{\circ}\varepsilon - O(\varepsilon^2). \tag{2.16}$$

Taking into account Eq. (2.12) let us calculate the derivative D_{ε} of the function determined by Eq. (2.1):

$$D_{\varepsilon} = \int_{\theta(h)}^{\Theta(h)} dx \int_0^{2\pi} \left[\Xi_u \hat{u}_{\varepsilon}\hat{U} + \Xi\,\hat{U}_{\varepsilon} + \Xi_w w_{\varepsilon}\hat{U}\right] d\vartheta, \tag{2.17}$$

where

$$\begin{aligned}\hat{U} &= [r+\varepsilon\xi+O(\varepsilon^2)][r_x+\varepsilon\xi_x+O(\varepsilon^2)] \\ &= rr_x+(\xi r_x+r\xi_x)\varepsilon+O(\varepsilon^2),\end{aligned} \tag{2.18}$$

$$\begin{aligned}\hat{u} &= \frac{\hat{U}}{\sqrt{U^2+[r+\varepsilon\xi+O(\varepsilon^2)]^2+[\varepsilon\xi_\theta+O(\varepsilon^2)]^2}} \\ &= \frac{rr_x+(\xi r_x+r\xi_x)\varepsilon+O(\varepsilon^2)}{\sqrt{r^2(r_x^2+1)+2r[r_x(\xi r_x+r\xi_x)+\xi]\varepsilon+O(\varepsilon^2)}}.\end{aligned} \tag{2.19}$$

Then let us calculate the derivatives $\hat{u}_\varepsilon, \hat{U}_\varepsilon$. After some algebra and using Eq. (2.17), we arrive at the following expression for D_ε°:

$$D_\varepsilon^\circ = 2\pi w^{[1]} I + J, \tag{2.20}$$

where

$$I = \int_{\theta(h)}^{\Theta(h)} rr_x \Xi_w(\hat{u}^\circ, w^{[0]})dx, \quad \hat{u}^\circ = \frac{r_x}{\sqrt{r_x^2+1}}, \tag{2.21}$$

$$J = \int_{\theta(h)}^{\Theta(h)} [\varphi(x,h)\Delta(x)+\psi(x,h)\Delta_x(x)]dx, \tag{2.22}$$

$$\begin{aligned}\varphi(x,h) &= r_x \Xi(\hat{u}^\circ, w^{[0]}(h)), \\ \psi(x,h) &= \frac{r}{r_x}\left[\frac{1}{r_x}(\hat{u}^\circ)^3 \Xi_u(\hat{u}^\circ, w^{[0]}(h)) + \varphi(x,h)\right].\end{aligned} \tag{2.23}$$

Integrating by parts the second term in formula for J in Eqs. (2.22), we find that:

$$\begin{aligned}J &= \int_{\theta(h)}^{\Theta(h)} \left\{\varphi(x,h)\Delta(x)+\frac{\partial}{\partial x}[\psi(x,h)\Delta(x)]-\psi_x(x,h)\Delta(x)\right\}dx \\ &= \psi(\Theta(h),h)\Delta(\Theta(h))-\psi(\theta(h),h)\Delta(\theta(h)) \\ &+ \int_{\theta(h)}^{\Theta(h)} [\varphi(x,h)-\psi_x(x,h)]\Delta(x)dx.\end{aligned} \tag{2.24}$$

Taking into account Eq. (2.11), we obtain the following estimate, $J \sim \varepsilon$. Therefore, Eq. (2.16) can be rewritten in the following form:

$$(m/2)\left[w_h^{[0]} + w_h^{[1]}\varepsilon + O(\varepsilon^2)\right] = -D^{\circ} - \left[2\pi w^{[1]} I\right]\varepsilon - O(\varepsilon^2). \qquad (2.25)$$

Equating the coefficients near the same powers of ε in both sides of Eq. (2.25) and taking into account Eq. (2.6), we obtain two ordinary differential equations (ODEs) for determining $w^{[0]}$ and $w^{[1]}$:

$$(m/2)w_h^{[0]} = -D\left[r(x), h, w^{[0]}\right], \quad w^{[0]}(b+L) = 0, \qquad (2.26)$$

and

$$(m/2)w_h^{[1]} = -2\pi w^{[1]} I(h), \quad w^{[1]}(b+L) = 0. \qquad (2.27)$$

Equation (2.26) describes the motion of the impactor Γ°. A general solution of problem for ODE given by Eq. (2.27) reads:

$$w^{[1]}(h) = C\exp\left(-\frac{4\pi}{m}\int_0^h I(\tilde{h})d\tilde{h}\right), \qquad (2.28)$$

where C is a constant. Since the initial condition given by the second relation in Eq. (2.27) can be satisfied only when $C = 0$, the solution for $w^{[1]}$ reads:

$$w^{[1]}(h) = 0. \qquad (2.29)$$

Equation (2.12) yields the following estimate for the function describing motion of the impactor with impact velocity equal to the BLV:

$$w(h) = w^{[0]}(h) + O(\varepsilon^2). \qquad (2.30)$$

Taking into account Eq. (2.4), we obtain the following estimate for the BLV of the impactor Γ:

$$v_{bl} = \sqrt{w^{[0]}(0) + O(\varepsilon^2)} = v_{bl}^{\circ} + O(\varepsilon^2), \qquad (2.31)$$

where v_{bl}° is the BLV of the impactor Γ°.

Therefore, the difference between the BLVs of the impactors Γ and Γ° is of the order of ε^2, i.e., the area rule for the BLV has been established.

Similar laws can be determined for other characteristics of the penetration. Assume that impactor Γ begins its motion in the shield with the impact velocity v_{imp}, i.e., the initial condition

$$w(0) = v_{imp}^2 \tag{2.32}$$

replaces the initial condition given by Eq. (2.6). Then the initial conditions for $w^{[0]}(h)$ and $w^{[1]}(h)$ read:

$$w^{[0]}(0) = v_{imp}^2, \quad w^{[1]}(0) = 0 . \tag{2.33}$$

Equation (2.28) determines the solution of Eq. (2.27) for $w^{[1]}$ with the initial condition given by Eq. (2.33). Therefore, the estimate given by Eq. (2.30) is valid, where $w^{[0]}$ is the solution of Eq. (2.26) with the initial condition given by Eq. (2.33). Equation (2.30) implies the following estimate of the difference between the velocities of the impactors Γ and Γ° at the same depth, $v(h)$ and $v^{\circ}(h)$:

$$v(h) = v^{\circ}(h) + O(\varepsilon^2) . \tag{2.34}$$

Therefore, if the impact velocities of the impactors Γ and Γ° are the same, the difference between their velocities at any depth $h \le b + L$ inside a shield is of the order of ε^2 until both impactors continue their motion inside the shield.

The latter conclusion is also valid if the impactors penetrate into a SIS. The only difference is that in this case $\theta(h) = 0$ and $\Theta(h)$ is determined by Eq. (2-2.2). It should be noted that the above-given analysis does not yield the estimate for the difference between the maximum DOPs. However, if the impactor Γ° stops its motion in the shield while the impactor Γ continues to move, the velocity of the impactor Γ is of the order of ε^2.

It must be noted that the above-given analysis can be simplified if one replaces Eq. (2.7) by the condition:

$$\Delta(x) = 0 . \tag{2.35}$$

However, the name *area rule* would be inappropriate in this case, and the meaning of such a rule would be less transparent.

2.3 An illustrative example

Let us consider conical impactors and use elliptical cones as 3-D impactors Γ so as to illustrate the above described area rules. The equation of the surface of the impactors Γ reads (see Eq. 2-4.3):

$$\Phi(x,\vartheta)=kx\eta(\vartheta),\, k=\frac{1}{L},\eta(\vartheta)=\frac{a_x a_y}{\sqrt{a_x^2\, sin^2\vartheta+a_y^2\, cos^2\vartheta}}, \tag{2.36}$$

where function $\rho=\eta(\vartheta)$ is the equation of the shank of the cone, a_x and a_y are the lengths of its semi-axes.

For the model that is determined by Eq. (2-8.26) with $\mu_{fr}=0$ and $\alpha=0$ ($\beta=2$), Eqs. (2-8.30) and (2-8.31) yield:

$$C_0=a_0 k\int_0^{2\pi}\eta^2 d\vartheta=2\pi a_0 k a_x a_y, \tag{2.37}$$

$$C_2=a_2 k^3\int_0^{2\pi}\frac{\eta^6}{\eta^2(k^2\eta^2+1)+\eta'^2}d\vartheta=\frac{2\pi a_2 k^3(a_x a_y)^2}{\sqrt{(k^2 a_x^2+1)(k^2 a_y^2+1)}}. \tag{2.38}$$

Taking into account that $k=1/L$, Eq. (2-8.22) with $\beta=2$ reads:

$$v_{bl}^2=\frac{a_0}{a_2}\frac{q}{\overline{a}_x\overline{a}_y}\left[exp\left(\lambda_{16}\frac{(\overline{a}_x\overline{a}_y)^2}{q}\right)-1\right], \tag{2.39}$$

where

$$\lambda_{16}=\frac{2\pi bL^2 a_2}{m},\,\overline{a}_x=\frac{a_x}{L},\,\overline{a}_y=\frac{a_y}{L},\,q=\sqrt{(\overline{a}_x^2+1)(\overline{a}_y^2+1)}. \tag{2.40}$$

Let the impactor Γ° be a cone with radius of the shank R. The cross-sectional area of an elliptical cone may be written as:

$$\sigma(x)=\frac{1}{2}kx\int_0^{2\pi}\eta^2 d\vartheta=\pi kxa_x a_y \tag{2.41}$$

Hence, the condition of equality of the areas of the impactors Γ and Γ° in the same cross-sections may be written as:

$$\bar{a}_x \bar{a}_y = \tau^2, \quad \tau = R/L \,. \tag{2.42}$$

It is convenient to select the parameter ε determining the difference between the shapes of the impactors Γ and Γ° as follows:

$$\varepsilon = \frac{\bar{a}_x}{\bar{a}_y} - 1 = \frac{a_x}{a_y} - 1 \,. \tag{2.43}$$

Then Eq. (2.42) yields:

$$\bar{a}_x = \tau\sqrt{1+\varepsilon}, \quad \bar{a}_y = \tau/\sqrt{1+\varepsilon} \,, \tag{2.44}$$

and Eq. (2.39) can be rewritten in the following form:

$$v_{bl}(\varepsilon) = \frac{1}{\tau}\sqrt{\frac{a_0}{a_2} q\left[\exp\left(\frac{\lambda_{16}\tau^4}{q}\right) - 1\right]} \,, \tag{2.45}$$

where

$$q(\varepsilon) = \sqrt{[\tau^2(\varepsilon+1)+1][\tau^2/(\varepsilon+1)+1]} \,. \tag{2.46}$$

Clearly, $v_{bl}(0) = v_{bl}^\circ$ is the BLV of the impactor Γ°.

Using a Taylor series expansion with respect to ε and substituting $\bar{a}_x$ and $\bar{a}_y$ from Eq. (2.44), allows to obtain the following expression for function $\eta(\vartheta)$ in Eq. (2.36):

$$\frac{\eta(\vartheta)}{R} = \sqrt{\sin^2\vartheta(\varepsilon+1) + \frac{\cos^2\vartheta}{\varepsilon+1}} = 1 + \frac{1}{2}\cos 2\vartheta + O(\varepsilon^2) \,. \tag{2.47}$$

Therefore, the equation of the surface of the impactor Γ is described in the form of Eq. (2.8) with:

$$r(x) = \tau x, \quad \xi(x,\vartheta) = 0.5\tau x \cos 2\vartheta \,, \tag{2.48}$$

for which Eq. (2.31) is satisfied.

The area rule in this case can be proved by considering Taylor series expansion of the function $v_{bl}(\varepsilon)$ with respect to ε. Since

$$v'_{bl}(\varepsilon)=\frac{a_0 q'(\varepsilon)Q(q)}{2\tau^2 a_2 v_{bl}(\varepsilon)}, \tag{2.49}$$

$$q'(\varepsilon)=\frac{\tau^2\varepsilon(\varepsilon+2)}{2q(\varepsilon)(\varepsilon+1)^2},\quad Q(q)=\left(1-\frac{\lambda_{16}\tau^4}{q}\right)exp\left(\frac{\lambda_{16}\tau^4}{q}\right)-1, \tag{2.50}$$

then $q'(0)=0$ and, consequently, $v'_{bl}(0)=0$. Therefore, the area rules given by Eq. (2.31) are valid.

Figure 9-1a-b illustrates the square-law dependence between $\tilde{\delta}=(v_{bl}-v^{\circ}_{bl})/v^{\circ}_{bl}$ and ε. Clearly, the dimensionless parameter $\tilde{\delta}$ does not depend, for given λ_{16} and τ, upon the parameters of the model used. However, to remain approximately in the range of the parameters corresponding to the ballistic impact conditions, we use the restriction $100\le v^{\circ}_{bl}(m/s)\le 900$ applied to shields manufactured from soft steel (the parameters of the model for soft steel are taken from *Table 10-3*). Inspection of these figures shows that the normalized difference between the BLV of the elliptical impactor Γ and the BLV of the reference impactor Γ° is very small and negative. The latter finding means that the BLV of an elliptical-shaped conical impactor is, for small ε, smaller than the BLV of a conical impactor - a body of revolution with the same area of the shank and, consequently, with the same volume.

Let us prove that this advantage is retained for arbitrary ε, i.e., for all elliptical-shaped conical impactors. Let us calculate the derivative:

$$\frac{dQ}{d\varepsilon}=q'(\varepsilon)\frac{dQ}{dq}=q'(\varepsilon)\frac{(\lambda_{16}\tau^4)^2}{q^3}exp\left(\frac{\lambda_{16}\tau^4}{q}\right). \tag{2.51}$$

Equation (2.50) shows that $q'(\varepsilon)>0$, and Eq. (2.51) implies that $dQ/d\varepsilon>0$. Since Q is an increasing function of ε and

$$\lim_{\varepsilon\to+\infty} Q(q(\varepsilon))=0, \tag{2.52}$$

it can be concluded that $Q<0$ for $\varepsilon\ge 0$. Eq. (2.49) implies that $v'_{bl}(\varepsilon)<0$ for finite values of $\varepsilon\ge 0$.

Therefore, $v_{bl}(\varepsilon)<v^{\circ}_{bl}=v_{bl}(0)$ for all $\varepsilon>0$, i.e., the BLV of the impactor Γ is always smaller than the BLV of the impactor Γ°. Moreover, the difference between the BLVs of the impactors Γ and Γ° increases with the increase of ε and tends to a finite value when $\varepsilon \to +\infty$.

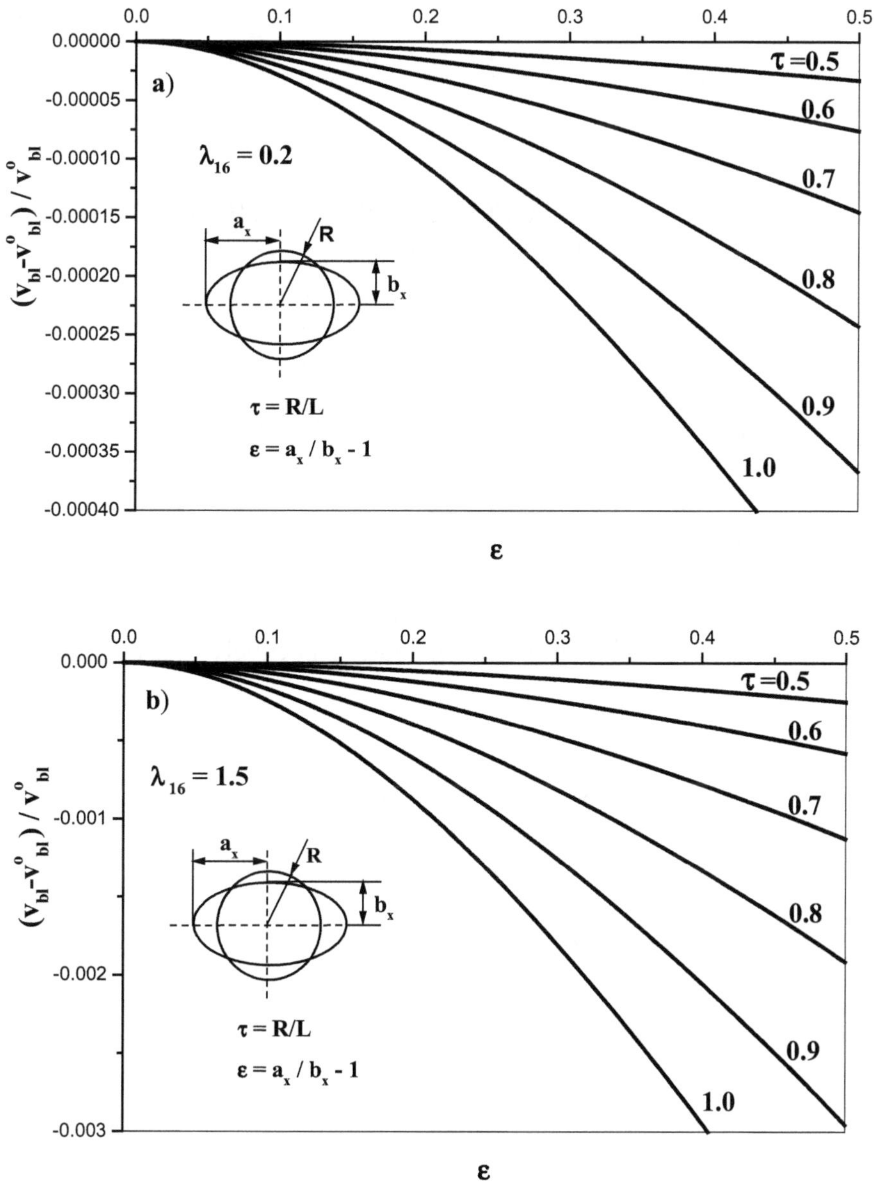

Figure 9-1. Normalized difference between the BLV of the elliptical impactor, v_{bl} and the BLV of the reference impactor, v_{bl}°, vs . the normalized difference in their shapes, ε.

The latter limit can be calculated using L'Hospital's rule:

$$\lim_{\varepsilon\to+\infty} v_{bl}(\varepsilon)=\frac{1}{\tau}\sqrt{\frac{a_0}{a_2}}\sqrt{\lim_{\varepsilon\to+\infty}\frac{exp(\lambda_{16}\tau^4/q)-1}{1/q}}$$
$$=\frac{1}{\tau}\sqrt{\frac{a_0}{a_2}}\sqrt{\lambda_{16}\tau^4\lim_{\varepsilon\to+\infty}\frac{-q^{-2}exp(\lambda_{16}\tau^4/q)}{-q^{-2}}}=\tau\sqrt{\frac{a_0\lambda_{16}}{a_2}}. \tag{2.53}$$

3. BODY WITH A POLYGONAL CROSS SECTION AS A REFERENCE IMPACTOR

In this section, as a reference impactor Γ° we consider a 3-D impactor with polygonal cross-sections. We assume that all cross-sections of this reference impactor are geometrically similar and that an inscribed circle exists for each polygon. The equation of the surface of the reference impactor can be written as follows:

$$\rho=\Phi^\circ(x,\vartheta),\quad \Phi^\circ(x,\vartheta)=r(x)\eta(\vartheta), \tag{3.1}$$

where functions $r(x)$ and $\eta(\vartheta)$ determine the longitudinal contour of the impactor and the shape of its cross-sections, respectively. It is assumed that

$$r(L)=1. \tag{3.2}$$

Therefore, $\eta(\vartheta)$ describes the impactor's surface in the cross-section $x=L$ (see *Figure 9-2*). A polygon with N sides can be determined by the radius of the inscribed circle g and the angles $\alpha_i\ (i=1,...,N)$. Then

$$\eta(\vartheta)=\begin{cases}\eta_1(\vartheta)\ if\ \vartheta_0\le\vartheta\le\vartheta_1\\ \dots\\ \eta_i(\vartheta)\ if\ \vartheta_{i-1}\le\vartheta\le\vartheta_i,\\ \dots\\ \eta_N(\vartheta)\ if\ \vartheta_{N-1}\le\vartheta\le\vartheta_N\end{cases} \tag{3.3}$$

where

$$\eta_i(\vartheta)=g\,sec(\alpha-\alpha_i),\quad i=1,...,N, \tag{3.4}$$

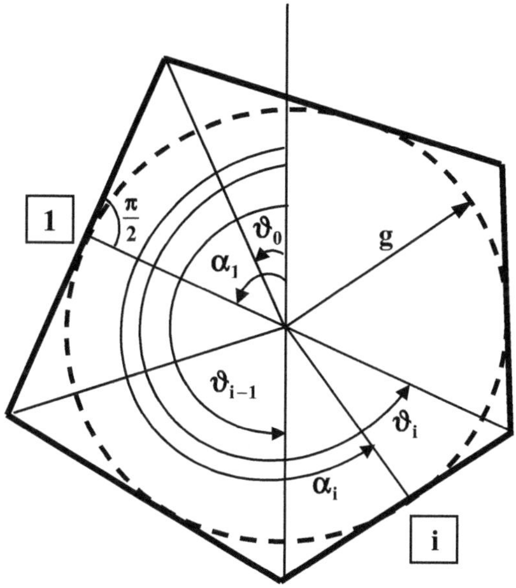

Figure 9-2. Cross-section of the reference impactor.

$$\vartheta_i = \begin{cases} (\alpha_1 + \alpha_N)/2 - \pi & if \quad i = 0 \\ (\alpha_i + \alpha_{i+1})/2 & if \quad 1 \le i \le N-1 \, . \\ (\alpha_1 + \alpha_N)/2 + \pi & if \quad i = N \end{cases} \tag{3.5}$$

Similarly to the derivation in the previous section, let us consider another impactor Γ and assume that both impactors have the same length L, density and areas at every cross-section; therefore, they have the same mass. It is assumed that shapes of both impactors are close to each other in the following sense:

$$\Phi(x,\vartheta) = r(x)\eta(\vartheta) + \varepsilon\xi(x,\vartheta) + O(\varepsilon^2), \tag{3.6}$$

where $\xi(x,\vartheta)$ is some function, ε is a small parameter, and function $\Phi(x,\vartheta)$ determines the shape of the impactor Γ. The requirement of continuity of the surface must be satisfied for both impactors.

From Eqs. (3.3)-(3.6), the equality of the cross-sectional areas of the projectiles Γ and Γ° can be written, for some x, as follows:

$$\frac{1}{2}\sum_{i=1}^{N}\int_{\vartheta_{i-1}}^{\vartheta_i}[gr(x)sec(\alpha-\alpha_i)+\varepsilon\xi+O(\varepsilon^2)]^2 d\vartheta$$
$$=\frac{1}{2}\sum_{i=1}^{N}\int_{\vartheta_{i-1}}^{\vartheta_i}[gr(x)sec(\alpha-\alpha_i)]^2 d\vartheta. \tag{3.7}$$

Equation (3.7) yields the following estimate:

$$\Delta(x)=\sum_{i=1}^{N}\int_{\vartheta_{i-1}}^{\vartheta_i}r(x)sec(\alpha-\alpha_i)d\vartheta\sim\varepsilon. \tag{3.8}$$

Equation (2.1) for the functional D determining the drag force acting at the impactor can be rewritten as follows:

$$D[\Phi(x,\vartheta),h,w]=\sum_{i=1}^{N}D_i[r(x)\eta_i(\vartheta)+\varepsilon\xi(x,\vartheta)+O(\varepsilon^2),h,w], \tag{3.9}$$

where

$$D_i[r(x)\eta_i(\vartheta)+\varepsilon\xi(x,\vartheta)+O(\varepsilon^2),h,w]=\int_{\theta(h)}^{\Theta(h)}dx\int_{\vartheta_{i-1}}^{\vartheta_i}\Xi(\hat{u}_i,w)\hat{U}_i d\vartheta \tag{3.10}$$

In Eq. (3.10), function Ξ determines the model of projectile-shield interaction; variables $\hat{u}_i$ and $\hat{U}_i$ have the same meaning as $\hat{u}$ and $\hat{U}$ in Eq. (2.1), respectively; and the subscript indicates the range of the angle $\vartheta\in[\vartheta_{i-1},\vartheta_i]$. Then the following expressions are valid:

$$\hat{U}_i=[gr\,sec(\alpha-\alpha_i)+\varepsilon\xi+O(\varepsilon^2)][gr_x\,sec(\alpha-\alpha_i)+\varepsilon\xi_x+O(\varepsilon^2)], \tag{3.11}$$

$$[\hat{U}_i/\hat{u}_i]^2=[gr\,sec(\alpha-\alpha_i)+\varepsilon\xi+O(\varepsilon^2)]^2\{[gr_x\,sec(\alpha-\alpha_i)$$
$$+\varepsilon\xi_x+O(\varepsilon^2)]^2+1\}+[gr\,sin(\alpha-\alpha_i)sec^2(\alpha-\alpha_i)+\varepsilon\xi_\vartheta+O(\varepsilon^2)]^2. \tag{3.12}$$

Assume that w in Eq. (3.10) is replaced by Eq. (2.12). Then the expression for the derivative $\partial D/\partial\varepsilon$ reads:

$$\frac{\partial D}{\partial \varepsilon}=\sum_{i=1}^{N}\frac{\partial D_i}{\partial \varepsilon}, \tag{3.13}$$

where formula for $\partial D_i/\partial\varepsilon$ is similar to that given by Eq. (2.17):

$$\frac{\partial D_i}{\partial \varepsilon}=\int_{\theta(h)}^{\Theta(h)}dx\int_{\vartheta_{i-1}}^{\vartheta_i}\left[\Xi_u\hat{u}_\varepsilon\hat{U}+\Xi\,\hat{U}_\varepsilon+\Xi_w w_\varepsilon\hat{U}\right]d\vartheta. \tag{3.14}$$

After some algebra, the expression for $\partial D/\partial\varepsilon$ at $\varepsilon=0$ can be written as:

$$D_\varepsilon^\circ=[\partial D/\partial\varepsilon]_{\varepsilon=0}=\tilde{I}(h)-\tilde{J}(h)+\tilde{K}(h)w^{[1]}(h), \tag{3.15}$$

where

$$\tilde{I}(h)=\int_{\theta(h)}^{\Theta(h)}E(x,h)\Delta_x(x)dx,\quad \tilde{J}(h)=\int_{\theta(h)}^{\Theta(h)}G(x,h)\Psi(x)dx, \tag{3.16}$$

$$\tilde{K}(h)=g^2\sum_{i=1}^{N}\frac{\sin(\vartheta_i-\vartheta_{i-1})}{\cos(\alpha_i-\vartheta_{i-1})\cos(\alpha_i-\vartheta_i)}$$
$$\times\int_{\theta(h)}^{\Theta(h)}rr_x\Xi_w\left(\hat{u}^\circ,w^{[0]}(h)\right)dx,\quad \hat{u}^\circ=\frac{gr_x}{\sqrt{g^2r_x^2+1}}, \tag{3.17}$$

$$E(x,h)=\frac{1}{r_x}G(x,h)+g\Xi\left(\hat{u}^\circ,w^{[0]}(h)\right), \tag{3.18}$$

$$G(x,h)=\frac{(\hat{u}^\circ)^3}{gr_x}\Xi_u\left(\hat{u}^\circ,w^{[0]}(h)\right), \tag{3.19}$$

$$\Psi(x)=\sum_{i=1}^{N}\int_{\vartheta_{i-1}}^{\vartheta_i}\frac{\partial}{\partial\vartheta}[\xi\sin(\vartheta-\alpha_i)]d\vartheta$$
$$=\sum_{i=1}^{N}\xi(x,\vartheta_i)\sin(\vartheta_i-\alpha_i)-\sum_{i=1}^{N}\xi(x,\vartheta_{i-1})\sin(\vartheta_{i-1}-\alpha_i). \tag{3.20}$$

Using Eq. (3.5), we can rewrite Eq. (3.20) in a more convenient form:

$$\Psi(x)=\xi(x,\vartheta_N)\sin(\vartheta_N-\alpha_N)-\xi(x,\vartheta_0)\sin(\vartheta_0-\alpha_1)$$
$$+\sum_{i=1}^{N-1}\xi(x,\vartheta_i)[\sin(\vartheta_i-\alpha_i)-\sin(\vartheta_i-\alpha_{i+1})] \tag{3.21}$$
$$=2\left[\sin(\frac{\alpha_N-\alpha_1}{2})\xi(x,\vartheta_0)+\sum_{i=1}^{N-1}\sin(\frac{\alpha_{i+1}-\alpha_1}{2})\xi(x,\vartheta_i)\right].$$

Note that all u_i for $\varepsilon=0$ do not depend on ϑ. Therefore, the first argument in the expressions for Ξ, Ξ_u and Ξ_w in Eqs. (3.17)-(3.19) does not have a subscript.

Integrating by parts the first integral in Eqs. (3.16), we find that:

$$\tilde{I}=\int_{\theta(h)}^{\Theta(h)}\left\{\frac{\partial}{\partial x}[E(x,h)\Delta(x)]-E_x(x,h)\Delta(x)\right\}dx$$
$$=E(\Theta(h),h)\Delta(\Theta(h))-E(\theta(h),h)\Delta(\theta(h))-\int_{\theta(h)}^{\Theta(h)}E_x(x,h)\Delta(x)dx. \tag{3.22}$$

Taking into account Eq. (3.8), we obtain the following estimate:

$$\tilde{I}\sim\varepsilon. \tag{3.23}$$

Assume now that the following requirement is satisfied:

$$\tilde{J}=0. \tag{3.24}$$

We now list the most important cases when condition (3.24) is valid using Eq. (3.21).

1) For every x, the contour of the cross-section of the impactor Γ passes through the apexes of the polygon at the cross-section of the impactor Γ°. Then all $\xi(x,\vartheta_i)=0$, and Eq. (3.24) is valid.

2) If the cross-section of the impactor Γ° is a regular polygon, then

$$\vartheta_i=\vartheta_0+\frac{2\pi}{N}i,\quad \alpha_i=\gamma_0+\frac{\pi}{N}(2i-1),\quad i=1\div N, \tag{3.25}$$

and Eq. (3.21) implies that

$$\Psi(x) = 2\sin\frac{\pi}{N}\sum_{i=0}^{N-1}\xi(x,\vartheta_i). \tag{3.26}$$

Then from Eqs. (3.6) and (3.26), it can be concluded that Eq. (3.24) is valid when

$$\sum_{i=0}^{N-1}[\Phi(x,\vartheta_i) - r(x)\eta(\vartheta_i)] = 0. \tag{3.27}$$

3) Function $\Psi(x) \to 0$ when $N \to \infty$ (see Eq. 3.26). Then the reference impactor is a body of revolution. This special case was analyzed in *Section 9-2*.

Therefore, Eqs. (3.23) and (3.24) allow us using Eq. (3.15) to obtain the following estimates of D_ε:

$$D_\varepsilon^\circ = \widetilde{K}(h)w^{[1]}(h) + O(\varepsilon). \tag{3.28}$$

As was found in the previous section, we obtain two ODEs for $w^{[0]}$ and $w^{[1]}$:

$$(m/2)w_h^{[0]} = -D\left[r(x)\eta(\vartheta), h, w^{[0]}\right], \quad w^{[0]}(b+L) = 0, \tag{3.29}$$

and

$$(m/2)w_h^{[1]} = -w^{[1]}\widetilde{K}(h), \quad w^{[1]}(b+L) = 0. \tag{3.30}$$

Hence all the conclusions and estimates obtained in *Section 9-2* are also valid in this case.

PART 3: OPTIMIZATION OF NON-HOMOGENEOUS SHIELDS

Chapter 10

OPTIMIZATION OF MULTI-LAYERED AND SPACED DUCTILE SHIELDS

1. INTRODUCTION

Several topics associated with the penetration mechanics of layered (including spaced) shields have been extensively studied in the literature, and some examples of these topics are given below.

1) Comparing the ballistic characteristics of monolithic shields with those of shields composed of several plates with the same total thickness and manufactured from the same material.

The plates may be in contact or there may be air gaps between the plates. Therefore, as alternatives to the monolithic shield, many types of shields are possible with different numbers of plates and different thicknesses of the plates and of the air gaps.

2) Analysis of the effect of the order of plates manufactured from different materials on the ballistic characteristics of the shield.

The simplest case of this problem is interchanging the plates in a two-layered shield. In the general case, the number of plates may vary and they may be manufactured from different materials.

3) Investigating the combined effects, e.g., changing the order of plates and using air gaps, on the ballistic performance of the shield and various problems of optimization of the structure of the shield.

The ballistic performance of a shield depends on its structure, the material properties of its elements, and the shape of the penetrator; e. g., penetration by a striker with bluntness is accompanied by plugging, and penetration into a thin plate causes petalling (see, e.g., Backman and

Goldsmith, 1978; Corbett et al., 1996; Wierzbicki, 1999). Penetration modes may differ from shield to shield. Significant role can be played by the obliquity of the impact, the deformation or fragmentation of the impactor and other factors.

A brief survey of the state-of-the-art presented below (mainly on penetration in metal shields) validates the assessment of Radin and Goldsmith (1988) that: "only limited results for multiple target materials exist in the literature..., and the results obtained cannot easily be correlated since different target and projectile materials, nose shapes, impact geometries and striker speeds were used". Clearly, the latter assessment is not related to the problem of selecting the best shield out of the given set of shields against the impactor with a given shape. This problem can be often solved experimentally, and the obtained results can be explained using relatively simple physical reasoning. The problem is to determine a more-or-less general law that will enable prediction of the change of ballistic characteristics of the shield by varying the structure of the shield. This problem has not been solved as yet, although a number of experimental and theoretical studies have been performed in this direction.

Hurlich (1950) noted that the earliest study on the modern use of spaced armor that he found was performed in 1913 for armor of naval vessels. He presented some qualitative arguments in favor of spaced armor (mostly for tanks), a number of tables with experimental results, some references and curious historical information.

Honda et al. (1930) (see also Goldsmith, 1960) investigated experimentally the impact of steel plates by conical-nosed projectiles. It was found that a shield composed of thin plates had a lower ballistic resistance than a monolithic shield with the same thickness. However, a spaced shield with thicknesses of the plates equal to the half-thickness of a monolithic shield performed better than a monolithic shield. Marom and Bodner (1979) conducted a combined analytical and experimental comparative study of monolithic, layered and spaced thin aluminum shields. They found that the ballistic resistance of a monolithic shield is higher than that of a multi-layered shield with the plates in contact and lower than the ballistic resistance of a spaced shield. The study of Radin and Goldsmith (1988) was also based on semi-empirical models and experimental investigations. They found a monolithic aluminum shield to be superior to a layered shield with the same total thickness for conical-nose and blunt projectiles, while spaced shields were less effective. Corran et al. (1983a,b), using experimental results on penetration of mild steel plates by impactors having "increasingly rounded nose shape", plotted a curve describing dependence of perforation energy vs. plate thickness for all considered variants of the shield and found a "kink" in the curve "at about 3.5 mm total thickness". The occurrence of the kink was explained by the change of character of energy absorption. The authors arrived at the following conclusions. "(1) The order of unequal plate thickness is important... (2) No advance is found in using multilayered targets below the kink... Above this point the best combinations may approach the best-fit line to the single layer tests below the kink... (3) There is an advantage in placing the layers in contact".

Nixdorff (1984a,b; 1987a) compared the ballistic performance of a monolithic metal shield with shields manufactured from the same material, having the same total thickness, and consisting of several plates in contact. Using the theory developed by Awerbuch and Bodner (Awerbuch, 1970; Awerbuch and Bodner, 1974), Nixdorff showed that separation of a homogeneous shield into several layers implies a reduction of the BLV of the shield.

Woodward and Cimpoeru (1998) developed a simple phenomenological model that considers the perforation of laminates as a two-stage process of indentation on the impact side of the shield and either shear or dishing failure on the exit side, depending on the shield configuration. Experimental data for laminated aluminum alloy shields perforated by flat-ended or conical penetrators were used in this investigation.

Zukas (1996) and Zukas and Scheffler (2001) found, on the basis of numerical simulations with metallic shields, that "layering dramatically weakens thin [$b/(2R)<1$] and intermediate [$3<b/(2R)<10$] thickness targets", while "thick targets [$b/(2R)>10$] show small changes in projectile residual properties [residual mass and residual velocity] when compared to their monoblock equivalent".

Madhu et al. (2003) conducted experiments with aluminum plates impacted normally and concluded that "there is no significant change in the ballistic performance due to layering of such intermediate thickness of plates". They compared a monolithic shield with two- and three-layered shields of the same thickness. Gupta and Madhu (1997), using experimental results obtained for aluminum and steel plates, arrived at the same conclusion with respect to "relatively thick plates". For thin shields, "the layered combinations in contact give higher residual velocity [when compared with a monolithic shield] for the plates of both the materials tested". It was also found that for a "spaced target... the residual velocity was higher than the plates in contact", for the same impact velocity.

Weidemaier et al. (1993) conducted experiments and numerical simulations on the perforation of steel barriers by spherical impactors with a diameter of 17 mm. They studied a monolithic shield with a thickness of 43 mm and shields composed of plates in contact having the same total thickness. It was found that the ballistic characteristics of layered shields depended strongly on the order of the plates having different thicknesses and that layering could improve or impair the ballistic performance of the shield.

Almohandes et al. (1996) conducted a comprehensive experimental study on the perforation of mild steel by standard 7.62 mm bullets. They investigated layered in contact, spaced and monolithic shields with total thickness in the range 8-14 mm. The efficiencies of shields were assessed by comparing their residual velocities for the same magnitude of the impact velocity. It was found that single shields were more effective than laminated shields of the same total thickness, regardless of the configuration or striking velocity, and that the difference in performance diminished as the striking velocity increased. Moreover, the effectiveness of laminated shields – whether in contact or spaced – increased as the number of plates comprising each shield decreased. Ballistic performance of laminated shields is further enhanced by using the thickest lamina as the back lamina. The authors also studied shields with different structures in which fiberglass reinforced polyester was used as the filler material, and showed that these shields performed better than weight-equivalent steel shields. The experimental results of Almohandes et al. (1996) were used by Liang et al. (2005) for validating their proposed approximate penetration model. This model was used for comparative analysis of shields with different structures. It was concluded that the ballistic performance was the best for the double shield when the ratio of the first-layer thickness to the total thickness was about 0.75, and the worst performance was obtained when this ratio was equal to 0.5. An air gap slightly influenced the resistance to perforation in multi-layered shields.

Elek et al. (2005) developed a simple model to describe the perforation of monolithic and multi-layered thin metallic plates by a flat-ended cylindrical impactor and used their model for the analysis of the ballistic properties of multi-layered spaced shields. The main results of this study may be summarized as follows. The suggested model predicted that the monolithic shield will have larger resistance than any other multi-layered shield with standoff distance between layers and equivalent total mass. The analysis of penetration in a two-layered shield showed that the maximum resistance could be obtained for very low (<20% of total thickness) or very high (>80% of total thickness) first-layer thickness. The increase of the number of spaced layers of a multi-layered shield, at constant total mass, caused a further decrease of ballistic resistance. Deterioration of the ballistic performance of thin steel shields against flat-ended cylindrical impactors caused by layering had been noticed earlier by Zaid et al. (1973).

Shirai et al. (1997) investigated experimentally and numerically the impact resistance of reinforced concrete plates against projectile impact.

They found that double-layered plates could be expected to have higher impact resistance than standard plates.

Park et al. (2005) suggested a multi-stage procedure for optimization of a two-layered shield. In the first stage, using numerical simulations to describe penetration into shields with different thicknesses of the layers, $b^{(1)}$ and $b^{(2)}$, they determined the average temperature of a shield, T_{ave}, the average equivalent plastic strain σ_{ave} and the maximum equivalent plastic strain in a critical element of the shield σ_{max}. In the second stage, the approximate functions describing the dependencies, T_{ave}, σ_{ave} and σ_{max} vs. $b^{(1)}$ и $b^{(2)}$, were determined. In the third stage, using a reduction to a single-criterion problem by a linear combination of criteria, they solved a two-objective optimization problem. The authors considered two variants of optimization criteria (T_{ave} or σ_{ave} and the weight of a shield); the constraints included the upper bounds or σ_{max}, and constraints on the thicknesses of the plates and the total thickness of a shield.

Aptukov (1985) and Aptukov et al. (1985), using Pontrjagin's maximum principle, determined the optimum distribution of the mechanical characteristics of a non-homogeneous plate. The areal density of the shield along the trajectory of the impactor until it stopped was used as a criterion, and cylindrical and cone-nosed impactors were considered. The two-term LIM was employed, wherein the assumption about a linear dependence between the coefficients of the model was used. Aptukov et al. (1986) solved the discrete problem of optimization of a layered plate when the shield consisted of several layers of material and the material itself could be chosen from a given set of materials. The cylindrical cavity expansion model suggested by Sagomonyan (1988) was used in that study. All these investigations are summarized in the monograph of Aptukov et al. (1992).

Young (1972, 1997), using phenomenological formulas for soil, rock and concrete homogeneous shields, developed a technique for calculating the penetration into layered shields. Bernard (1978) suggested a model describing penetration in soil/rock multi-layered SISs that takes into account simultaneous interaction of a conical impactor with adjacent layers in a shield. The DOP was determined by numerical simulations. Yossifon et al. (2002) also suggested mathematical models and methods for calculating the ballistic characteristics of multi-layered shields. The survey of Corbett et al. (1996) contains a section on multi-layered shields.

Ben-Dor et al. (1997c; 1998a,d,e,f; 1999b,c,d; 2000c) investigated multi-layered ductile shields using LIMs and CCEMs. Systemized and expanded results of these studies are presented in this chapter.

2. LOCALIZED INTERACTION APPROACH FOR NON-HOMOGENEOUS SHIELDS

2.1 Modeling of non-homogeneous shields

The LIT can easily be generalized to the case of non-homogeneous shields by assuming that functions Ω_n and Ω_τ in Eq. (2-1.1) also depend on the location of the impactor-shield contact point in the coordinate system associated with the shield. This dependence can be described directly or through the parameters $a_0, a_1, \ldots$, that depend on the coordinates. We consider a normal impact and normal penetration of a rigid striker into a non-homogeneous shield with mechanical properties that change only along the direction of the impactor's motion.

Therefore, we assume that $a_0, a_1, \ldots$ are functions of the coordinate ξ associated with the shield (*Figure 10-1*), i. e.:

$$\vec{a} = \vec{a}(\xi), \tag{2.1}$$

where $\vec{a}$ is a vector with components $a_0(\xi), a_1(\xi), \ldots$ Then

$$\Omega_n = \Omega_n(\vec{a}(\xi), u, v), \ \ \Omega_\tau = \Omega_\tau(\vec{a}(\xi), u, v). \tag{2.2}$$

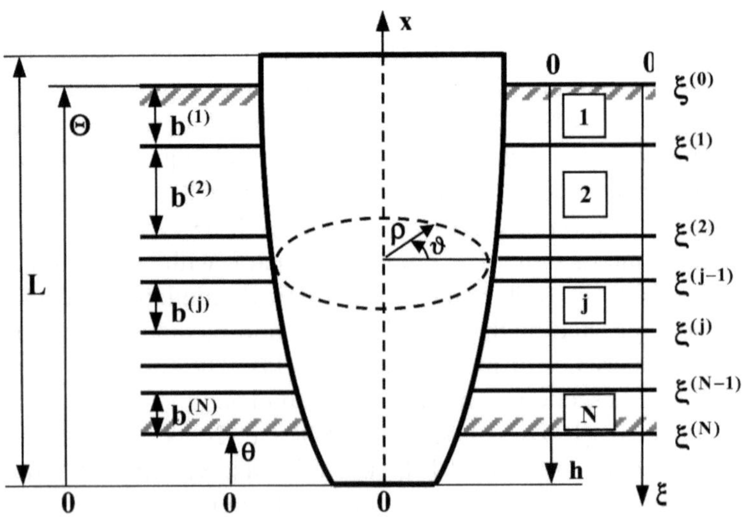

Figure 10-1. Description of LIM for non-homogeneous shields.

Without loss of generality, we can assume that:

$$\Omega_n(0,u,v)=\Omega_\tau(0,u,v)=0\ . \tag{2.3}$$

If this assumption is not valid for the original functions Ω_n and Ω_τ, we can introduce modified functions $\hat{\Omega}_n$ and $\hat{\Omega}_\tau$:

$$\hat{\Omega}_n=a_+\Omega_n,\quad \hat{\Omega}_\tau=a_+\Omega_\tau\ , \tag{2.4}$$

where a_+ is an additional auxiliary parameter.

Definitions of the coordinates x,h and ξ imply the following relationship for every impactor-shield contact point:

$$\xi=h-x\ . \tag{2.5}$$

Hereafter we consider the case of a shield with a finite thickness. Techniques used in this chapter and the derivation procedures employed in *Chapter 2* allow us to derive formulas for a SIS as well.

Equation (2-2.13) can be rewritten as:

$$D(h,v)=\Omega_n(\vec{a}(h),1,v)\sigma(0)\delta(h) + \int_{\theta(h)}^{\Theta(h)} dx \int_0^{2\pi}\Omega_0(\vec{a}(h-x),u(x,\vartheta),v)u_0(x,\vartheta)d\vartheta, \tag{2.6}$$

where a formula similar to Eq. (2-1.6) is valid for Ω_0,

$$\Omega_0(\vec{a}(\xi),u,v)=u\Omega_n(\vec{a}(\xi),u,v)+\sqrt{1-u^2}\,\Omega_\tau(\vec{a}(\xi),u,v), \tag{2.7}$$

and references to other notations are given in *Table 10-1.*

Table 10-1. Reference to the equations defining a LIM for a non-homogeneous shield

Function	Definition
$\sigma(x)$	Eq. (2-2.14)
$\delta(h)$	Eq. (2-2.16)
$u_0(x,\vartheta)$, $u_1(x,\vartheta)$	Eq. (2-2.12)
$\theta(h)$	Eq. (2-2.6)
$\Theta(h)$	Eq. (2-2.2)
$u(x,\vartheta)$	Eq. (2-2.11)

In the case of *3-D conical impactor* when the impactor's shape is described by Eq. (2-4.3), Eq. (2.6) can be rewritten as follows:

$$D(h,v)=\Omega_n(\vec{a}(h),1,v)\sigma(0)\delta(h)+\int_{\theta(h)}^{\Theta(h)}(k_0+kx)\widetilde{G}(\vec{a}(h-x),v)dx, \quad (2.8)$$

where

$$\widetilde{G}(\vec{a}(\xi),v)=\int_0^{2\pi}\Omega_0(\vec{a}(\xi),u(\vartheta),v)\hat{U}(\vartheta)d\vartheta, \quad (2.9)$$

$$\widetilde{G}(0,v)=0, \quad (2.10)$$

$$\hat{U}(\vartheta)=\sqrt{\eta^2(k^2\eta^2+1)+\eta'^2}, \quad u(\vartheta)=k\eta^2/\hat{U}(\vartheta). \quad (2.11)$$

Equation (2-2.20) describing the motion of the impactor and Eq. (2-2.25) are used for determining the BLV.

2.2 A model for multi-layered shields

For a layered shield consisting of N non-interacting plates with thicknesses $b^{(1)},b^{(2)},...,b^{(N)}$ (see *Figure 10-1*), $\vec{a}(\xi)$ is a step-function:

$$\vec{a}(\xi)=\begin{cases}\vec{a}^{(1)}=(a_1^{(1)},a_2^{(1)},...) \text{ if } \xi^{(0)}\le\xi<\xi^{(1)}\\ \dots\\ \vec{a}^{(j)}=(a_1^{(j)},a_2^{(j)},...) \text{ if } \xi^{(j-1)}\le\xi<\xi^{(j)}\\ \dots\\ \vec{a}^{(N)}=(a_1^{(N)},a_2^{(N)},...) \text{ if } \xi^{(N-1)}\le\xi\le\xi^{(N)}\end{cases}, \quad (2.12)$$

where $\xi^{(j)}$ ($j=1,2,...,N$) is the total thickness of the first j layers (in the direction of the axis ξ),

$$\xi^{(j)}=\sum_{\kappa=1}^{j}b^{(\kappa)}, \quad \xi^{(0)}=0, \quad (2.13)$$

and a vector $\vec{a}^{(j)}$ with components $a_1^{(j)},a_2^{(j)},...$ represents the parameters that determine the properties of the plate with number j, i.e., a superscript in parentheses denotes the number of the plate in a multi-layered shield, while a subscript denotes the number of the parameter in the set of parameters that determine the properties of this plate. *Figure 10-2* shows, in the coordinates h and x, the borders between neighboring plates.

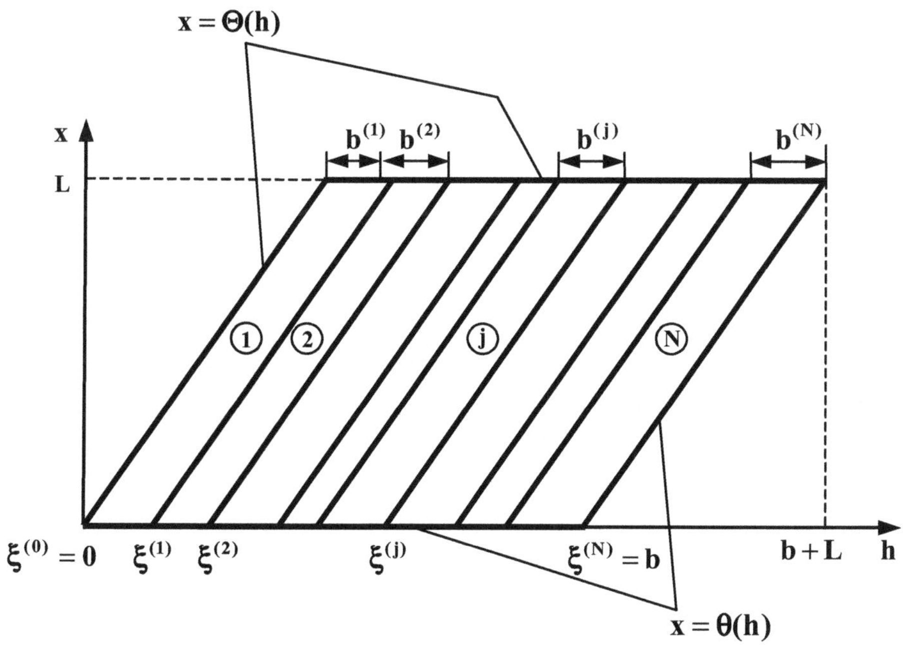

Figure 10-2. Description of the LIM for multi-layered shields.

If the plate with the number j is an air gap, then

$$\vec{a}^{(j)} = 0 \tag{2.14}$$

and Eq. (2.3) is valid.

2.3 Some properties of the model

Let us prove an identity that is often used in the remainder of the chapter. Let $\Xi(x)$ be an arbitrary function, and $a_i(\xi)$ a step function describing variation of a parameter of a multi-layered shield. Then, the following identity is valid:

$$\int_{\theta(h)}^{\Theta(h)} \Xi(x)a_i(h-x)dx = \sum_{j=1}^{N} a_i^{(j)} \int_{x_*\left(h-\xi^{(j)}\right)}^{x_*\left(h-\xi^{(j-1)}\right)} \Xi(x)dx\,, \tag{2.15}$$

where $x_*(z)$ is determined by Eq. (2-2.10).

First let us prove the following identity for any function $\widetilde{\Xi}(x)$ and any $0 \le h_0 \le b+L$:

$$\int_{\theta(h_0)}^{\Theta(h_0)} \widetilde{\Xi}(x)dx = \sum_{j=1}^{N} \int_{x_*\left(h_0-\xi^{(j)}\right)}^{x_*\left(h_0-\xi^{(j-1)}\right)} \widetilde{\Xi}(x)dx\,. \tag{2.16}$$

The simplest method to prove this identity is to consider four possible locations of the straight line $h = h_0$ with respect to the borders of the domain $0 \le h \le b+L$, $\theta(h) \le x \le \Theta(h)$ (see *Figure 10-3a-d*)

Let us consider the case $0 \le h_0 \le min(L,b)$ when $\theta(h)=0, \Theta(h)=h$. This case is shown in *Figure 10-3a* and, in more detail, in *Figure 10-4*. Denote by j_{max} such a number j so that $\xi^{(j_{max}-1)} \le h_0 \le \xi^{(j_{max})}$. Then (see *Figure 10-4*):

$$\int_{\theta(h_0)}^{\Theta(h_0)} \widetilde{\Xi}(x)dx = \int_{0}^{h_0-\xi^{(j_{max}-1)}} \widetilde{\Xi}(x)dx + \sum_{j=j_{max}-1}^{1} \int_{h_0-\xi^{(j)}}^{h_0-\xi^{(j-1)}} \widetilde{\Xi}(x)dx\,. \tag{2.17}$$

Since $x_*\left(h_0-\xi^{(j_{max})}\right)=0$ for $j_{max} \le j \le N$ and $x_*\left(h_0-\xi^{(j)}\right)=h_0-\xi^{(j)}$ for $1 \le j \le j_{max}-1$, we can rewrite Eq. (2.17) as:

$$\begin{aligned} \int_{\theta(h_0)}^{\Theta(h_0)} \widetilde{\Xi}(x)dx &= \int_{x_*\left(h_0-\xi^{(j_{max})}\right)}^{x_*\left(h_0-\xi^{(j_{max}-1)}\right)} \widetilde{\Xi}(x)dx + \sum_{j=j_{max}-1}^{1} \int_{x_*\left(h_0-\xi^{(j)}\right)}^{x_*\left(h_0-\xi^{(j-1)}\right)} \widetilde{\Xi}(x)dx \\ &+ \sum_{j=j_{max}}^{N} \int_{x_*\left(h_0-\xi^{(j)}\right)}^{x_*\left(h_0-\xi^{(j-1)}\right)} \widetilde{\Xi}(x)dx = \sum_{j=1}^{N} \int_{x_*\left(h_0-\xi^{(j)}\right)}^{x_*\left(h_0-\xi^{(j-1)}\right)} \widetilde{\Xi}(x)dx. \end{aligned} \tag{2.18}$$

Therefore, Eq. (2.16) is valid if $0 \le h_0 \le min(L,b)$. The cases shown in *Figure 10-3b-d* can be considered in a similar manner.

Equation (2.16) can be written for $\widetilde{\Xi}(x) = \Xi(x)a_i(h-x)$:

$$\int_{\theta(h)}^{\Theta(h)} \Xi(x) a_i(h-x) dx = \sum_{j=1}^{N} \int_{x_*\left(h-\xi^{(j)}\right)}^{x_*\left(h-\xi^{(j-1)}\right)} \Xi(x) a_i(h-x) dx , \tag{2.19}$$

a) $\mathbf{0 \le h_0 \le \min(L, b)}$

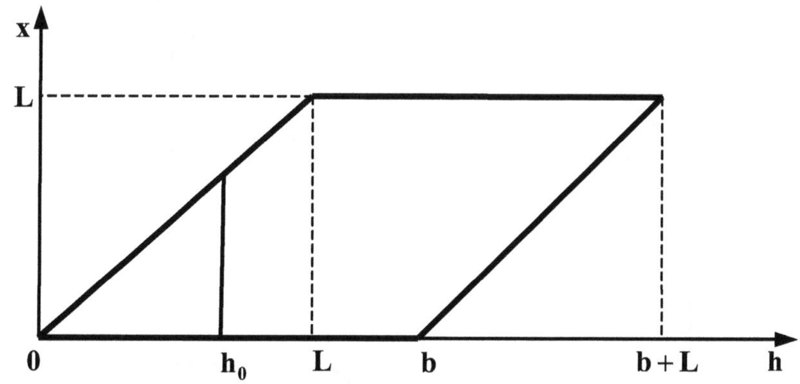

b) $\mathbf{L < b, L \le h_0 \le b}$

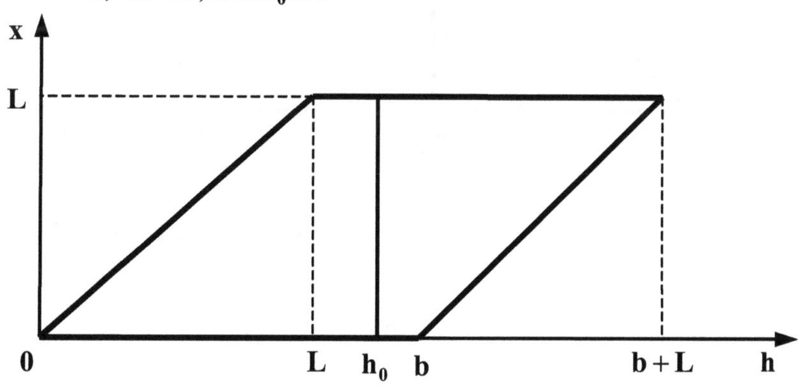

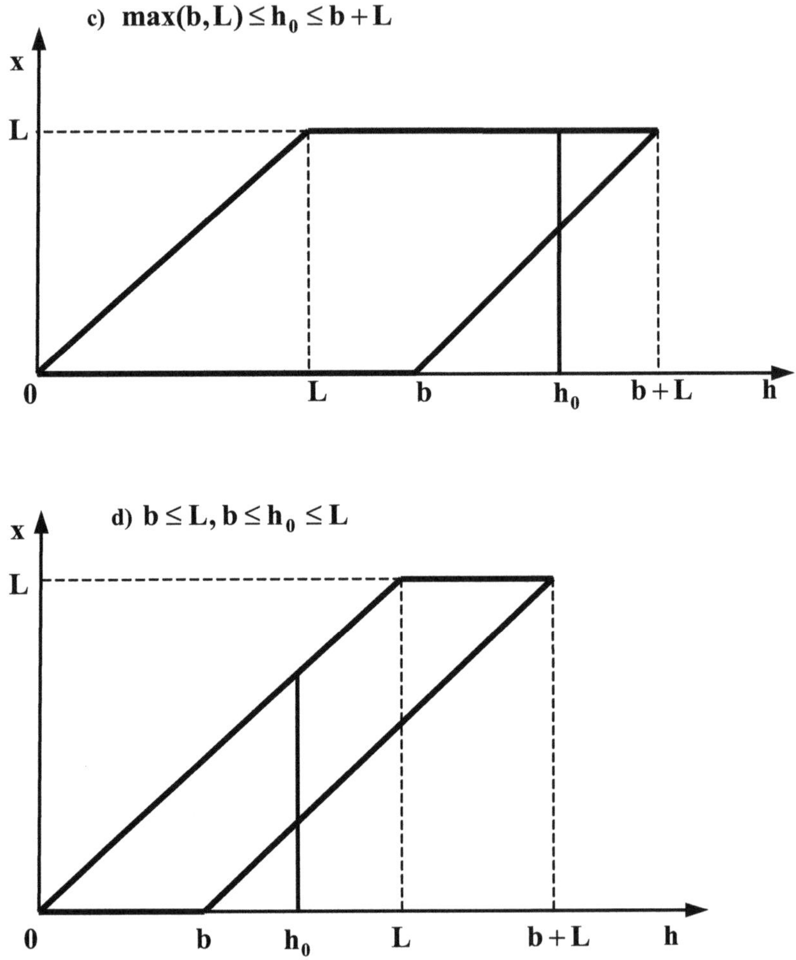

Figure 10-3. Possible locations of the straight line $h = h_0$ with respect to the borders of the domain $0 \le h \le b+L$, $\theta(h) \le x \le \Theta(h)$.

where h_0 is replaced by h. Since $a_i(h-x) = a_i^{(j)}$ if $x_*\left(h-\xi^{(j)}\right) \le x \le x_*\left(h-\xi^{(j-1)}\right)$ and $x_*\left(h-\xi^{(j)}\right) = x_*\left(h-\xi^{(j-1)}\right)$ for $x \le x_*\left(h-\xi^{(j)}\right)$ and for $x \ge x_*\left(h-\xi^{(j-1)}\right)$, we obtain:

$$\int_{x_*\left(h-\xi^{(j)}\right)}^{x_*\left(h-\xi^{(j-1)}\right)} \Xi(x)\, a_i(h-x)\, dx = a_i^{(j)} \int_{x_*\left(h-\xi^{(j)}\right)}^{x_*\left(h-\xi^{(j-1)}\right)} \Xi(x)\, dx . \qquad (2.20)$$

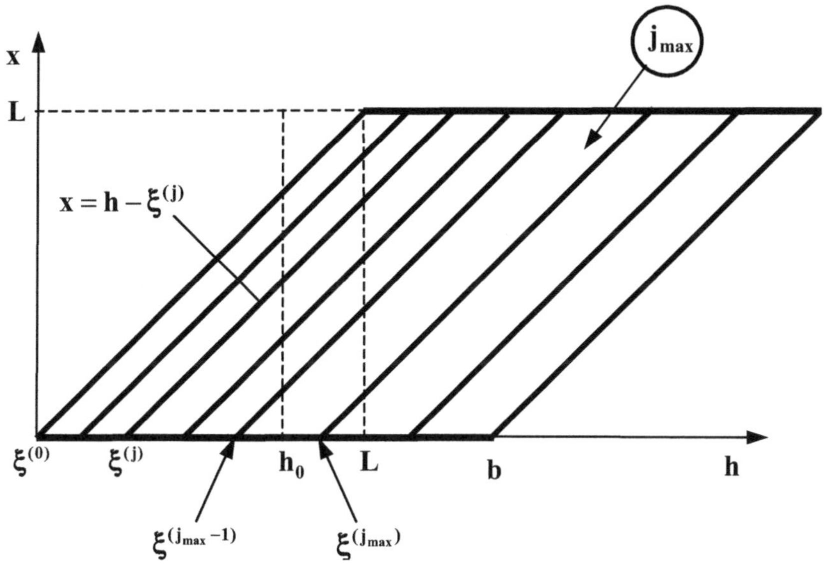

Figure 10-4. Case $0 \le h_0 \le min(L,b)$.

Substitution of the integral in the left-hand side of Eq. (2.20) into Eq. (2.19) completes the proof of validity of Eq. (2.15).

Another useful relationship can be easily proved for some functions $\Xi(x)$ and $\hat{\Xi}(h)$ by interchanging the order of integration and changing the variables $x \to x$, $h - \xi^{(j)} - x \to h$ in the integrals:

$$\int_0^{b+L} dh \hat{\Xi}(h) \int_{x_*\left(h-\xi^{(j)}\right)}^{x_*\left(h-\xi^{(j-1)}\right)} dx \Xi(x) = \int_0^L dx \Xi(x) \int_0^{b^{(j)}} dh \hat{\Xi}(h + \xi^{(j)} + x). \tag{2.21}$$

3. TWO-TERM LOCALIZED INTERACTION MODELS

3.1 Three-dimensional impactors

First, let us consider the following model:

$$\Omega_n(\vec{a}(\xi),u,v) = A_2(\vec{a}(\xi),u)v^2 + A_0(\vec{a}(\xi),u)v^\alpha, \tag{3.1}$$

$$\Omega_\tau\left(\vec{a}(\xi),u,v\right)=\mu_{fr}\Omega_n\left(\vec{a}(\xi),u,v\right). \tag{3.2}$$

This model is a generalization of the model considered in *Section 2-8* for non-homogeneous shields. Equation (2.6) implies Eq. (2-8.2), with $i=0,2$:

$$f_i(h)=A_i\left(\vec{a}(h),1\right)\sigma(0)\delta(h)+\int_{\theta(h)}^{\Theta(h)}dx\int_0^{2\pi}A_i\left(\vec{a}(h-x),u(x,\vartheta)\right)U(x,\vartheta)\,d\vartheta, \tag{3.3}$$

where the notations are given in *Table 10-4*, and U is defined by Eq. (2-3.4). Equations (2-8.4)-(2-8.8) and Eqs. (2-8.11)-(2-8.14) remain valid, taking into account that $f_i(h)$ are defined by Eq. (3.3).

When

$$A_i\left(\vec{a}(\xi),u\right)=a_i(\xi)\,\omega_i(u),\; i=0,2\;, \tag{3.4}$$

where ω_i are some positive functions, then Eq. (3.1) becomes:

$$\Omega_n\left(\vec{a}(\xi),u,v\right)=a_2(\xi)\,\omega_2(u)\,v^2+a_0(\xi)\,\omega_0(u)\,v^\alpha. \tag{3.5}$$

Substituting A_i $(i=0,2)$ given by Eq. (3.4) into Eq. (3.3) we obtain:

$$f_i(h)=a_i(h)\,\omega_i(1)\,\sigma(0)\,\delta(h) + \int_{\theta(h)}^{\Theta(h)}a_i(h-x)dx\int_0^{2\pi}\omega_i\left(u(x,\vartheta)\right)U(x,\vartheta)\,d\vartheta, \tag{3.6}$$

Clearly, in this case Eqs. (2-8.4)-(2-8.8) and Eqs. (2-8.11)-(2-8.14) and references in *Table 10-4* remain valid.

Let us consider this model for a *sharp 3-D impactor* $[\sigma(0)=0]$ penetrating into a multi-layered shield in more detail. Using Eq. (2.15), we can transform Eq. (3.6) to the following form:

$$f_i(h)=\int_{\theta(h)}^{\Theta(h)}a_i(h-x)e_i(x)dx=\sum_{j=1}^{N}a_i^{(j)}\int_{x_*\left(h-\xi^{(j)}\right)}^{x_*\left(h-\xi^{(j-1)}\right)}e_i(x)\,dx\,, \tag{3.7}$$

where $i=0,2,$

$$e_i(x) = \int_0^{2\pi} \omega_i\big(u(x,\vartheta)\big)U(x,\vartheta)\,d\vartheta\,, \tag{3.8}$$

$$\frac{m}{\beta} v_{bl}^{\beta} = \int_0^{b+L} f_0(h)Q(h)\,dh, \quad Q(h) = exp\left(\frac{\beta}{m}\int_0^h f_2(\tilde{h})\,d\tilde{h}\right). \tag{3.9}$$

3.2 Conical three-dimensional impactors

3.2.1 General formulas

In the case of a *3-D conical impactor*, using formulas from *Table 2-2*, we can write Eq. (3.6) as follows:

$$f_i(h) = \frac{k_0^2}{2} a_i(h)\,\omega_i(1)\,\delta(h) \int_0^{2\pi} \eta^2(\vartheta)\,d\vartheta + c_i \int_{\theta(h)}^{\Theta(h)} (k_0 + k\,x)\,a_i(h-x)\,dx\,, \tag{3.10}$$

where $i = 0, 2$,

$$c_i = \int_0^{2\pi} \omega_i\big(u(\vartheta)\big)\tilde{u}(\vartheta)\,d\vartheta\,, \tag{3.11}$$

$$\tilde{u}(\vartheta) = k\eta + \mu_{fr}\sqrt{\eta^2 + \eta'^2}\ , \tag{3.12}$$

and function $u(\vartheta)$ is determined by Eq. (2.11).

When the impactor's nose is a *sharp 3-D conical body* ($k_0 = 0$), Eq. (3.10) can be rewritten as:

$$f_i(h) = c_i k \int_{\theta(h)}^{\Theta(h)} x\,a_i(h-x)\,dx\,, \quad i = 0, 2. \tag{3.13}$$

Note that for a 3-D conical impactor penetrating into non-homogeneous shields a simple analytical solution similar to that given by Eq. (2-8.22) cannot be obtained.

3.2.2 3-D sharp cone

Since the model given by Eqs. (3.2) and (3.5) for a 3-D sharp conical impactor penetrating into a multi-layered shield will be often used in this book, we present the relevant relationships in more detail.

Equation (2.15) allows us to present functions $f_i(h)$ in Eq. (3.13) in the following form:

$$f_2(h)=c_2 k\sum_{j=1}^{N} a_2^{(j)} \int\limits_{x_*\left(h-\xi^{(j)}\right)}^{x_*\left(h-\xi^{(j-1)}\right)} xdx=\frac{m}{\beta}\sum_{j=1}^{N}\psi^{(j)}(h), \tag{3.14}$$

$$f_0(h)=c_0 k\sum_{j=1}^{N} a_0^{(j)} \int\limits_{x_*\left(h-\xi^{(j)}\right)}^{x_*\left(h-\xi^{(j-1)}\right)} xdx=\frac{c_0}{c_2}\frac{m}{\beta}\sum_{j=1}^{N}\chi^{(j)}\psi^{(j)}(h), \tag{3.15}$$

where $j=1,2,...,N$,

$$\varphi^{(j)}(h)=c^{(j)}\int\limits_{0}^{h} d\tilde{h} \int\limits_{x_*\left(\tilde{h}-\xi^{(j)}\right)}^{x_*\left(\tilde{h}-\xi^{(j-1)}\right)} xdx\ , \tag{3.16}$$

$$\psi^{(j)}(h)=\frac{d\varphi^{(j)}}{dh}=c^{(j)} \int\limits_{x_*\left(h-\xi^{(j)}\right)}^{x_*\left(h-\xi^{(j-1)}\right)} xdx\ , \tag{3.17}$$

$$\chi^{(j)}=\frac{a_0^{(j)}}{a_2^{(j)}}\ , \tag{3.18}$$

$$c^{(j)}=\frac{\beta}{m}c_2 k a_2^{(j)}\ , \tag{3.19}$$

and parameters c_i and k depend on the shape of the impactor (see Eq. 3.11), $0<\beta\leq 2$.

Substituting $f_2(h)$ and $f_0(h)$ from Eqs. (3.14) and (3.15) into Eq. (3.9), we obtain:

$$\frac{c_2}{c_0}v_{bl}^{\beta}=\sum_{j=1}^{N}\chi^{(j)}I^{(j)}\ , \tag{3.20}$$

where

$$I^{(j)} = \int_{0}^{b+L} \psi^{(j)} \exp(\varphi(h))\, dh\,, \quad j = 1,2,\ldots,N\,, \tag{3.21}$$

$$\varphi(h) = \sum_{j=1}^{N} \varphi^{(j)}(h)\,. \tag{3.22}$$

For the convenience of the reader, the references to the equations that determine the model for a 3-D sharp conical impactor are summarized in *Table* 10-2.

Table 10-2. References to the equations determining two-term LIMs for a sharp 3-D conical impactor penetrating into a layered shield

Function/parameter	Definition
Ω_n, Ω_τ	Eqs. (3.5), (3.2)
$u, \tilde{u}$	Eq. (3.12)
Φ	Eq. (2-4.3)
v_{bl}	Eq. (3.20)
$\chi^{(j)}$	Eq. (3.18)
$I^{(j)}$	Eq. (3.21)
$\psi^{(j)}$	Eq. (3.17)
φ	Eq. (3.22)
$a_0^{(j)}, a_2^{(j)}$	Eqs. (3.5), (2.12)
c_0, c_2	Eq. (3.11)
$c^{(j)}$	Eq. (3.19)

Let us now prove a useful formula. Equations (3.16), (3.17), (3.21) and (3.22) yield:

$$\sum_{j=1}^{N} I^{(j)} = \int_{0}^{b+L} \frac{d\varphi(h)}{dh} \exp(\varphi)\, dh = \int_{0}^{b+L} \frac{d\exp(\varphi)}{dh} dh = \left[\exp(\varphi)\right]_{h=0}^{h=b+L} \tag{3.23}$$

Using Eq. (2.20) for $\hat{\Xi}(h) = 1$ and $\Xi(x) = x$, we obtain:

$$\varphi^{(j)}(b+L) = c^{(j)} \int_{0}^{b+L} dh \int_{x_*(h-\xi^{(j)})}^{x_*(h-\xi^{(j-1)})} x dx = c^{(j)} \int_{0}^{L} dx\, x \int_{0}^{b^{(j)}} dh = \frac{c^{(j)} b^{(j)} L^2}{2}\,. \tag{3.24}$$

Taking into account that $\varphi^{(j)}(0)=0$ for all j, Eqs. (3.18), (3.22), (3.23) and (3.24) yield:

$$\sum_{j=1}^{N} I^{(j)} = \hat{I}, \quad \hat{I} = exp\left(\frac{\beta L^2 c_2 k}{2m} \sum_{j=1}^{N} a_2^{(j)} b^{(j)} \right) - 1. \tag{3.25}$$

3.2.3 3-D sharp cone of revolution: Newton's model

Consider an impactor having the shape of a cone of revolution. The model given by Eqs. (3.2) and (3.5) with

$$\omega_0(u)=1, \omega_2(u)=u^2, \mu_{fr}=0 \tag{3.26}$$

yields the following equations:

$$v_{bl}^{\beta} = \frac{k^2+1}{k^2} \lambda_{17} \int_0^{\bar{b}+1} d\bar{h}\, exp\left(\lambda_{17} \tilde{P}(\bar{h})\right) \sum_{j=1}^{N} a_0^{(j)} \bar{K}^{(j)}(\bar{h}), \tag{3.27}$$

where

$$\begin{aligned} \tilde{P}(\bar{h}) &= \sum_{i=1}^{N} a_2^{(i)} \int_0^{\bar{h}} \bar{K}^{(i)}(\tilde{h}) d\tilde{h} = \frac{1}{2} \sum_{i=1}^{N} a_2^{(i)} \left[\tilde{\Psi}(\bar{h}, \bar{\xi}^{(i-1)}) - \tilde{\Psi}(\bar{h}, \bar{\xi}^{(i)}) \right] \\ &= \frac{1}{2} \left[a_2^{(1)} \tilde{\Psi}(\bar{h}, 0) - a_2^{(N)} \tilde{\Psi}(\bar{h}, \bar{b}) + \sum_{1 \le i \le N-1} \left(a_2^{(i+1)} - a_2^{(i)} \right) \tilde{\Psi}(\bar{h}, \bar{\xi}^{(i)}) \right], \end{aligned} \tag{3.28}$$

$$\tilde{\Psi}(\bar{h}, \bar{c}) = \int_0^{\bar{h}} X_\diamond^2(\tilde{h} - \bar{c}) d\tilde{h} = \begin{cases} 0 & if \quad \bar{h} \le \bar{c} \\ \frac{1}{3}(\bar{h} - \bar{c})^3 & if \quad \bar{c} \le \bar{h} \le \bar{c}+1 \\ \bar{h} - \bar{c} - 2/3 & if \quad \bar{h} \geqq c+1 \end{cases}, \tag{3.29}$$

$$\bar{K}^{(j)}(\bar{h}) = \int_{X_\diamond(\bar{h}-\bar{\xi}^{(j)})}^{X_\diamond(\bar{h}-\bar{\xi}^{(j-1)})} \bar{x} d\bar{x} = \frac{1}{2} \left[X_\diamond^2 \left(\bar{h} - \bar{\xi}^{(j-1)} \right) - X_\diamond^2 \left(\bar{h} - \bar{\xi}^{(j)} \right) \right], \tag{3.30}$$

$$\lambda_{17} = \frac{2\pi \beta k^4 L^3}{(k^2+1)m}, \tag{3.31}$$

$$\bar{h} = \frac{h}{L}, \quad \bar{\xi}^{(j)} = \frac{\xi^{(j)}}{L}, \quad \bar{x} = \frac{x}{L}, \quad \bar{b} = \frac{b}{L}. \tag{3.32}$$

4. EFFECT OF AIR GAPS ON THE BALLISTIC RESISTANCE OF A SHIELD

4.1 Multi-layered shield with the layers of the same material: general localized interaction model

Let us consider a shield consisting of μ_{max} plates manufactured from the same material with air gaps between the plates (*Figure 10-5*). The shield is perforated by a 3-D sharp conical impactor. There are no constraints on the widths of the air gaps, so that the impactor can interact with several plates simultaneously.
For a sharp impactor, $k_0 = 0$, and Eq. (2.8) can be rewritten as:

$$D(h,v) = k \int_{\theta(h)}^{\Theta(h)} x\widetilde{G}(\vec{a}(h-x),v)dx\,, \tag{4.1}$$

where the notations are explained in *Section 10-2.1*.

Let us define a function $\varepsilon(\xi)$, which is equal to 1 if the point with the coordinate ξ (see *Figure 10-5*) is located in any plate and is equal to 0 if this point is located in an air gap:

$$\begin{aligned} &\varepsilon(\xi) = \varepsilon^{(j)} \quad \text{if} \quad \xi^{(j-1)} \le \xi \le \xi^{(j)}, \quad j = 1,2,\ldots,\mu_{max} - 1, \\ &\varepsilon^{(1)} = \ldots = \varepsilon^{(2\mu-1)} = \ldots = \varepsilon^{(2\mu_{max}-1)} = 1, \\ &\varepsilon^{(2)} = \ldots = \varepsilon^{(2\mu)} = \ldots = \varepsilon^{(2\mu_{max}-2)} = 0. \end{aligned} \tag{4.2}$$

Then

$$\widetilde{G}(\vec{a}(\xi),v) = \varepsilon(\xi)\widetilde{G}(\vec{a}^{(1)},v) = \varepsilon(\xi)\int_0^{2\pi} \Omega_0(\vec{a}^{(1)},u(\vartheta),v)\widehat{U}(\vartheta)d\vartheta\,, \tag{4.3}$$

where vector $\vec{a}^{(1)}$ with constant components $a_1^{(1)}, a_2^{(1)},\ldots$ represents the parameters determining the material properties of the plates. Using Eq. (4.3), we can write the expression for drag force (Eq. 4.1) as:

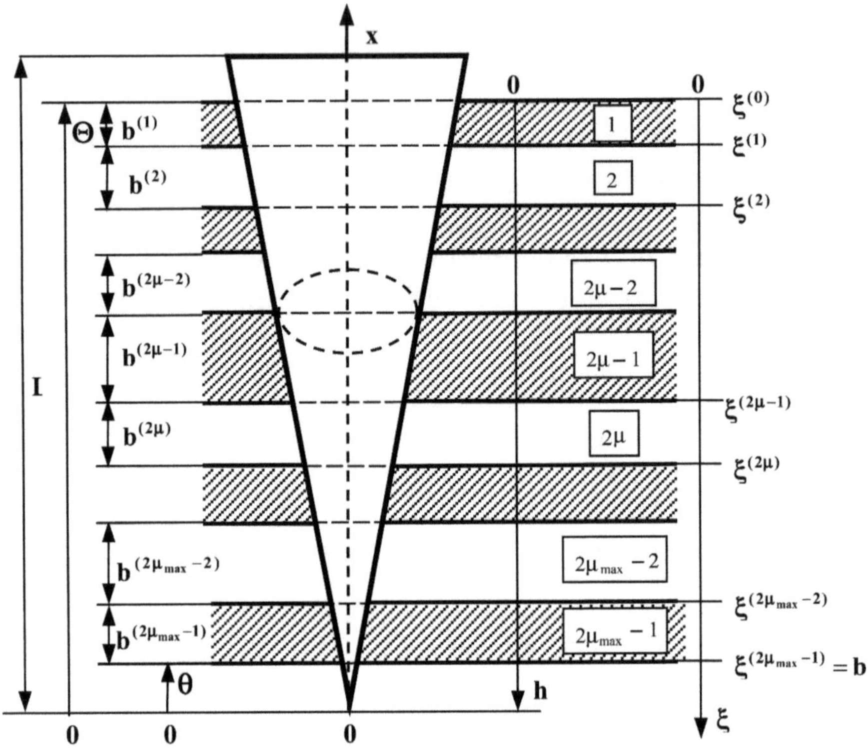

Figure 10-5. Structure of a shield with air gaps.

$$D(h,v)=k\,\widetilde{G}\left(\vec{a}^{(1)},v\right)\int_{\theta(h)}^{\Theta(h)} x\varepsilon(h-x)dx\,. \tag{4.4}$$

The equation of motion of the impactor reads:

$$mv\frac{dv}{dh}=-k\,\widetilde{G}\left(\vec{a}^{(1)},v\right)\int_{\theta(h)}^{\Theta(h)} x\varepsilon(h-x)dx\,, \tag{4.5}$$

or

$$\frac{mv}{k\,\widetilde{G}\left(\vec{a}^{(1)},v\right)}dv=-\left(\int_{\theta(h)}^{\Theta(h)} x\varepsilon(h-x)dx\right)dh\,. \tag{4.6}$$

Integrating the left-hand side of this equation over v from $v = v_{imp}$ to $v = v_{res}$ and right-hand side over h from $h = 0$ to $h = b + L$, we obtain the relationship between the impact velocity and the residual velocity of the impactor:

$$\psi(v_{imp}) - \psi(v_{res}) = \widetilde{J}\ , \tag{4.7}$$

where

$$\psi(z) = \frac{m}{k}\int_0^z \frac{vdv}{\widetilde{G}\left(\vec{a}^{(1)}, v\right)}\ , \tag{4.8}$$

$$\widetilde{J} = \int_0^{b+L} dh \int_{\theta(h)}^{\Theta(h)} x\varepsilon(h - x)dx\ . \tag{4.9}$$

Equation (4.7) yields an equation for the BLV by substituting $v_{imp} = v_{bl}$ and $v_{res} = 0$:

$$\psi(v_{bl}) = \widetilde{J}. \tag{4.10}$$

Integral in Eq. (4.9) can be transformed by changing the variables, $x = x, h = x + \xi$ (see *Figure 10-6a-b*):

$$\widetilde{J} = \int_0^L xdx \int_0^b \varepsilon(\xi)d\xi = \int_0^L xdx \left(\sum_{\mu=1}^{\mu_{max}} \int_{\xi^{(2\mu-2)}}^{\xi^{(2\mu-1)}} \varepsilon(\xi)d\xi + \sum_{\mu=1}^{\mu_{max}-1} \int_{\xi^{(2\mu-1)}}^{\xi^{(2\mu)}} \varepsilon(\xi)d\xi \right). \tag{4.11}$$

The second sum in Eq. (4.11) equals 0, while the integrals in the first sum yield the thicknesses of the corresponding plates. After some algebra, Eq. (4.11) yields:

$$\widetilde{J} = \int_0^L xdx \sum_{\mu=1}^{\mu_{max}} \left(\xi^{(2\mu-1)} - \xi^{(2\mu-2)} \right) = \int_0^L b_{sum} xdx = 0.5L^2 b_{sum}\ , \tag{4.12}$$

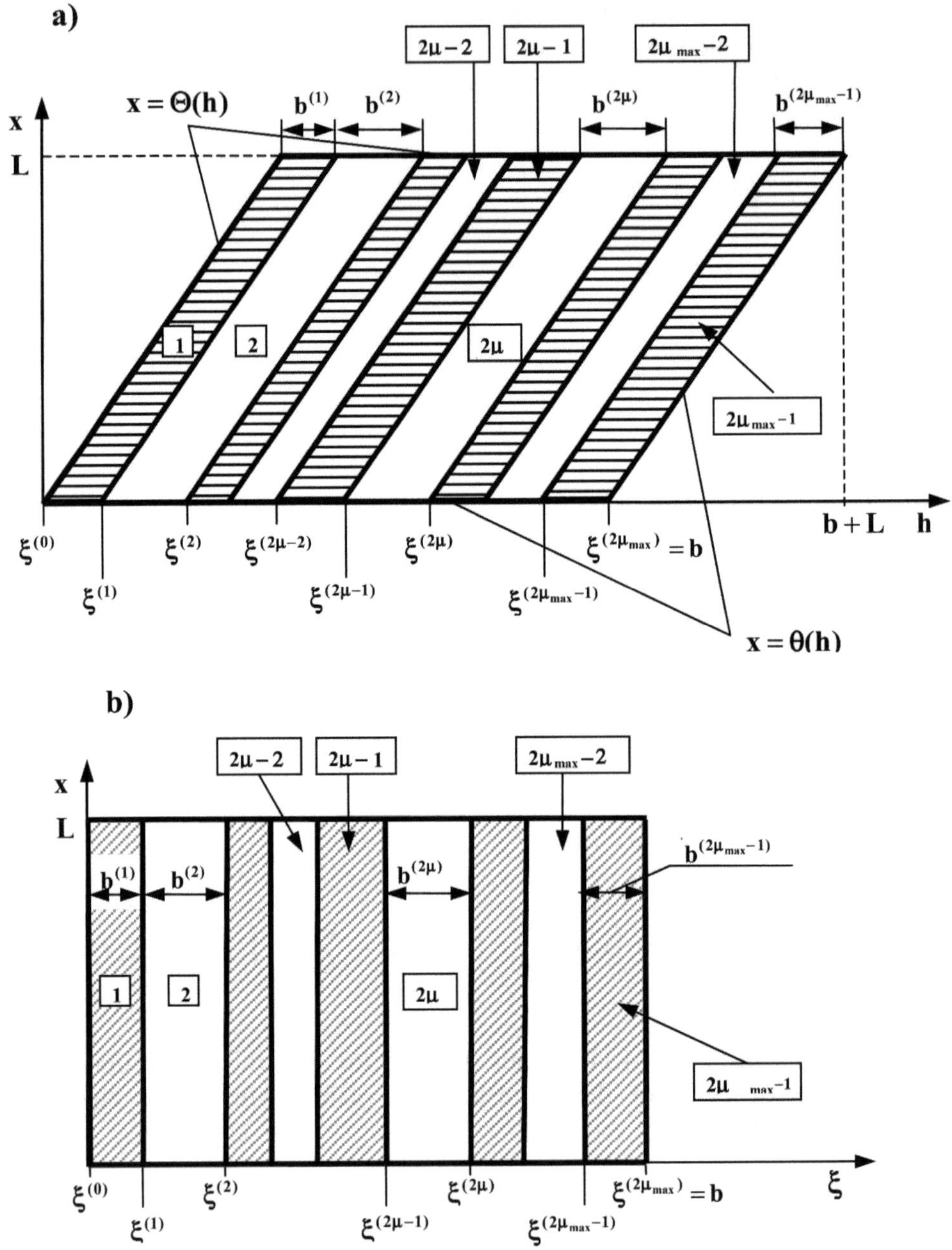

Figure 10-6. Coordinates used in the description of the impactor-shield interaction.

where b_{sum} is the total thickness of all the plates:

$$b_{sum} = \sum_{\mu=1}^{\mu_{max}} b^{(2\mu-1)} . \tag{4.13}$$

Therefore, Eqs. (4.7), (4.10) and (4.12) imply that:

$$v_{bl} = \psi^{-1}(0.5L^2 b_{sum}) , \tag{4.14}$$

$$v_{res} = \psi^{-1}\left(\psi(v_{imp}) - 0.5L^2 b_{sum}\right), v_{imp} \geq v_{bl} . \tag{4.15}$$

Equations (4.14) and (4.15) show that the residual velocity for a given impact velocity and the BLV are the same for all shields (spaced and non-spaced) with the same total thickness b_{sum}. It must be emphasized that the latter result was obtained by using a general LIM. Note that the model used does not account for the difference between the resistance properties of a monolithic shield and a shield consisting of several plates that are in-contact and have the same total thickness; it also neglects interaction between plates in contact.

Experimental data on ballistic penetration of conical impactors into ductile shields with air gaps are quite scarce. Radin and Goldsmith (1988) investigated the impact response of shields composed of soft aluminum multi-layered plates penetrated by a hard-steel 60-grad conical-nosed projectile. Two of their experiments are relevant to our study. In the first experiment, the BLVs for a shield consisting of two plates, 1.6 mm thick, were found to be 93.2 m/s and 90.6 m/s for adjacent and spaced plates, respectively. In the second experiment, the plates of thickness 3.2 mm were used; the measured BLVs were 160.4 m/s and 153.4 m/s, respectively. The values of BLV for adjacent and spaced plates were thus close. The increase in the BLV for plates in contact may be associated with friction between the layers (Corran et al., 1983b) that has not been taken into account in the used models.

4.2 Two-layered shield with different materials of the plates: two-term localized interaction model

4.2.1 Statement of the problem

Consider penetration of a 3-D rigid conical impactor into two-layered armor with an air gap between the plates. The notations used in the analysis of this problem are shown in *Figure 10-7*. The shield consists of two plates with thicknesses $b^{(1)}$ and $b^{(3)}$ manufactured, in general, from different materials. The first (front) plate is located between the cross-sections $\xi = \xi^{(0)} = 0$ and $\xi = \xi^{(1)} = b^{(1)}$, and the second plate is located between the cross-sections $\xi = \xi^{(2)} = b^{(1)} + b^{(2)}$ and $\xi = \xi^{(3)} = b^{(1)} + b^{(2)} + b^{(2)}$, where $b^{(2)}$ is the thickness of the air gap.

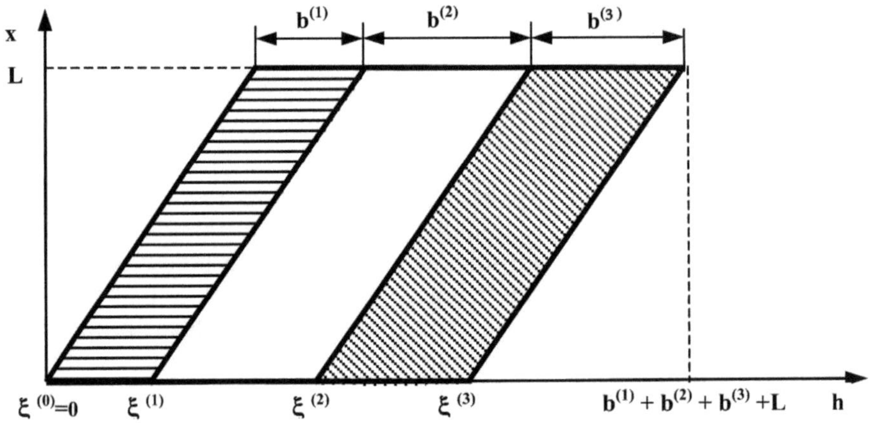

Figure 10-7. Two-layered shield with air gap.

We use the impactor-shield interaction model defined in *Table 10-5*. In the considered case ($j = 1,3$):

$$\varphi^{(j)}(h) = c^{(j)} \int_0^h d\tilde{h} \int_{x_*\left(\tilde{h}-\xi^{(j)}\right)}^{x_*\left(\tilde{h}-\xi^{(j-1)}\right)} x dx \, , \tag{4.16}$$

$$\frac{c_2}{c_0} v_{bl}^{\beta} = \chi^{(1)} I^{(1)} + \chi^{(3)} I^{(3)} \, , \tag{4.17}$$

$$I^{(j)} = \int_0^{b+L} \frac{d\varphi^{(j)}}{dh} exp\left(\varphi^{(1)} + \varphi^{(3)}\right) dh \, , \tag{4.18}$$

where $\chi^{(j)}$ and $c^{(j)}$ are determined by Eq. (3.18) and Eq. (3.19), respectively (with $j = 1,3$).

First let us consider the dependence of the BLV v_{bl} on the thickness of the air gap $b^{(2)}$.

4.2.2 Investigation of the problem

Using Eq. (3.25), we can write:

$$I^{(3)} = \hat{I} - I^{(1)} , \tag{4.19}$$

where

$$\hat{I} = exp\left[\frac{\beta}{2m} c_2 k\left(a_2^{(1)} b^{(1)} + a_2^{(3)} b^{(3)}\right)\right]. \tag{4.20}$$

Substituting $I^{(3)}$ from Eq. (4.19) into Eq. (4.17), we obtain:

$$\frac{c_2}{c_0} v_{bl}^{\beta} = \chi^{(3)} \hat{I} + \left(\chi^{(1)} - \chi^{(3)}\right) I^{(1)} , \tag{4.21}$$

where only the integral $I^{(1)}$ depends on the thickness of the air gap $b^{(2)}$.

Consider the integral $I^{(1)}$ in more detail. From Eq. (4.16), this integral can be transformed, by changing variables, as follows:

$$\begin{aligned} I^{(1)} &= c^{(1)} \int_0^{b+L} dh \, exp\left(\varphi^{(1)} + \varphi^{(3)}\right) \int_{x_*\left(h-\xi^{(1)}\right)}^{x_*(h)} x dx \\ &= c^{(1)} \int_0^L x dx \int_x^{x+b^{(1)}} dh \, exp\left(\varphi^{(1)}\right) exp\left(\varphi^{(3)}\right), \end{aligned} \tag{4.22}$$

where

$$\varphi^{(1)}(h) = c^{(1)} \int_0^h d\tilde{h} \int_{x_*\left(\tilde{h}-b^{(1)}\right)}^{x_*\left(\tilde{h}\right)} x dx , \tag{4.23}$$

$$\varphi^{(3)}(h) = c^{(3)} \hat{\varphi}^{(3)}(h, \xi^{(2)}), \tag{4.24}$$

$$\hat{\varphi}^{(3)}(h,\xi^{(2)})=\int_0^h d\tilde{h}\int_{x_*(\tilde{h}-\xi^{(3)})}^{x_*(\tilde{h}-\xi^{(2)})} x\,dx=\int_0^h d\tilde{h}\int_{x_*(\tilde{h}-\xi^{(3)})}^{x_*(\tilde{h}-\xi^{(2)})} x\,dx=\frac{1}{6}(I_+-I_-), \tag{4.25}$$

$$I_+=3\int_0^h x_*^2\left(\tilde{h}-\xi^{(2)}\right)d\tilde{h}=\begin{cases}0 & if \quad h\le\xi^{(2)}\\ \left(h-\xi^{(2)}\right)^3 & if \quad \xi^{(2)}\le h\le\xi^{(2)}+L\\ 3L^2\left(h-\xi^{(2)}\right)-2L^3 & if \quad h\ge\xi^{(2)}+L\end{cases}, \tag{4.26}$$

$$I_-=3\int_0^h x_*^2\left(\tilde{h}-\xi^{(3)}\right)d\tilde{h}=3\int_0^h x_*^2\left(\tilde{h}-\xi^{(2)}-b^{(3)}\right)d\tilde{h}$$

$$=\begin{cases}0 & if \quad h\le\xi^{(2)}+b^{(3)}\\ \left(h-\xi^{(2)}-b^{(3)}\right)^3 & if \quad \xi^{(2)}+b^{(3)}\le h\le\xi^{(2)}+b^{(3)}+L\\ 3L^2\left(h-\xi^{(2)}-b^{(3)}\right)-2L^3 & if \quad h\ge\xi^{(2)}+b^{(3)}+L\end{cases} \tag{4.27}$$

and $\hat{\varphi}^{(3)}$ is now considered as a function of two arguments, h and $\xi^{(2)}=b^{(1)}+b^{(2)}$. Equations (4.22)-(4.27) show that the BLV depends on the thickness of the air gap $b^{(2)}$ only through the function $\hat{\varphi}^{(3)}$.

Let us derive a formula for the function $\hat{\varphi}^{(3)}(h,\xi^{(2)})$. Equations (4.26) and (4.27) yield:

$$6\hat{\varphi}^{(3)}(h,\xi^{(2)})$$

$$=\begin{cases}0 & if \quad h\le\xi^{(2)}\\ \left(h-\xi^{(2)}\right)^3 & if \quad \xi^{(2)}\le h\le\xi^{(3)}\\ \left(h-\xi^{(2)}\right)^3-\left(h-\xi^{(3)}\right)^3 & if \quad \xi^{(3)}\le h\le\xi^{(2)}+L\\ 3L^2\left(h-\xi^{(2)}\right)-2L^3-\left(h-\xi^{(3)}\right)^3 & if \quad \xi^{(2)}+L\le h\le\xi^{(3)}+L\\ 3L^2b^{(3)} & if \quad h\ge\xi^{(3)}+L\end{cases} \tag{4.28}$$

for $b^{(3)}\le L$, and

$$6\hat{\varphi}^{(3)}(h,\xi^{(2)})$$
$$=\begin{cases} 0 & \text{if} \quad h \le \xi^{(2)} \\ \left(h-\xi^{(2)}\right)^3 & \text{if} \quad \xi^{(2)} \le h \le \xi^{(2)}+L \\ 3L^2\left(h-\xi^{(3)}\right)-2L^3 & \text{if} \quad \xi^{(2)}+L \le h \le \xi^{(3)} \\ 3L^2\left(h-\xi^{(2)}\right)-2L^3-\left(h-\xi^{(3)}\right)^3 & \text{if} \quad \xi^{(3)} \le h \le \xi^{(3)}+L \\ 3L^2 b^{(3)} & \text{if} \quad h \ge \xi^{(2)}+L+b^{(3)} \end{cases} \tag{4.29}$$

for $b^{(3)} \ge L$, where $\xi^{(3)} = \xi^{(2)} + b^{(3)}$.

The expression for the derivative of $\hat{\varphi}^{(3)}(h,\xi^{(2)})$ with respect to $\xi^{(2)}$ reads:

$$2\frac{\partial\hat{\varphi}^{(3)}(h,\xi^{(2)})}{\partial\xi^{(2)}}$$
$$=\begin{cases} 0 & \text{if} \quad h \le \xi^{(2)} \\ -\left(h-\xi^{(2)}\right)^2 & \text{if} \quad \xi^{(2)} \le h \le \xi^{(3)} \\ -b^{(3)}\left[2\left(h-\xi^{(2)}\right)-b^{(3)}\right] & \text{if} \quad \xi^{(3)} \le h \le \xi^{(2)}+L \\ \left(h-\xi^{(3)}-L\right)\left(h-\xi^{(3)}+L\right) & \text{if} \quad \xi^{(2)}+L \le h \le \xi^{(3)}+L \\ 0 & \text{if} \quad h \ge \xi^{(2)}+L+b^{(3)} \end{cases} \tag{4.30}$$

for $b^{(3)} \le L$, and

$$2\frac{\partial\hat{\varphi}^{(3)}(h,\xi^{(2)})}{\partial\xi^{(2)}}$$
$$=\begin{cases} 0 & \text{if} \quad h \le \xi^{(2)} \\ -\left(h-\xi^{(2)}\right)^2 & \text{if} \quad \xi^{(2)} \le h \le \xi^{(2)}+L \\ -L^2 & \text{if} \quad \xi^{(2)}+L \le h \le \xi^{(3)} \\ \left(h-\xi^{(3)}-L\right)\left(h-\xi^{(3)}+L\right) & \text{if} \quad \xi^{(3)} \le h \le \xi^{(3)}+L \\ 0 & \text{if} \quad h \ge \xi^{(3)}+L \end{cases} \tag{4.31}$$

for $b^{(3)} \ge L$.

We included the boundary points between the neighboring intervals in the formula for the derivative of $\hat{\varphi}^{(3)}(h,\xi^{(2)})$ with respect to $\xi^{(2)}$ because the left and right derivatives in these points coincide, i.e., function $\hat{\varphi}^{(3)}(h,\xi^{(2)})$ has a continuous first partial derivative with respect to $\xi^{(2)}$.

Inspection of Eqs. (4.28)-(4.31) allows us to draw the following conclusions.

Assume that $\xi^{(2)} \geq b^{(1)} + L\,(b^{(2)} \geq L)$. Since $\hat{\varphi}^{(3)} = 0$ for $h \leq \xi^{(2)}$, then $\hat{\varphi}^{(3)} = 0$ for $h \leq b^{(1)} + L$ as well. Consequently, $\hat{\varphi}^{(3)} = 0$ in the domain of integration in Eq. (4.22), $0 \leq x \leq L, x \leq h \leq x + b^{(1)}$. Hence, $I^{(1)}$ and the BLV do not depend on the thickness of the air gap. Clearly, this conclusion is valid because the plates are perforated consecutively.

If $b^{(1)} \leq \xi^{(2)} < b^{(1)} + L$, then $\hat{\varphi}^{(3)}$ is a non-increasing function of $\xi^{(2)}$ at every sub-interval, and there is always the sub-interval of h where $\hat{\varphi}^{(3)}$ is a decreasing function of $\xi^{(2)}$. Therefore, the integral $I^{(1)}$ decreases when $\xi^{(2)}$ increases.

Thus, Eq. (3.5) implies the following claim. When $\chi^{(1)} > \chi^{(3)}$ $\left(\chi^{(1)} < \chi^{(3)}\right)$, then the BLV of the shield decreases (increases) with increasing air gap thickness from zero to the length of the impactor, and remains constant with a further increase of the air gap thickness. If $\chi^{(1)} = \chi^{(3)}$, the BLV does not depend on the thickness of the air gap.

5. EFFECT OF THE ORDER OF THE PLATES ON THE BALLISTIC RESISTANCE OF A SHIELD

5.1 Two-layered shield consisting of plates in contact

5.1.1 Formulation of the problem

Consider the penetration of a 3-D rigid conical impactor into two-layered armor. In the following exposition, we use the notations of the *Section 10-1.2.* Some details pertinent to the considered problem are shown in *Figure 10-8a-b*. The shield consists of two plates in contact with thicknesses $b^{(1)}$ and $b^{(2)}$ in the direction of the impactor's motion (reference position of the plates). The plates, in general, are manufactured from different materials. All superscripts in the notations of the parameters are associated with this order of the plates in the shield. We consider the reference order of the plates in the armor (*Figure 10-8a*) and the inverse order of the plates (*Figure 10-8b*) with the goal of determining the effect of the order of the plates on the BLV of the shield.

In the analysis, we use the model described in *Table 10-5*. For purposes of convenience, we rewrite some relationships for the two-term case $(j = 1,2)$ and replace $\xi^{(0)}, \xi^{(1)}, \xi^{(2)}$ by $0, b^{(1)}, b^{(1)} + b^{(2)}$. After these manipulations, we obtain:

$$\varphi^{(1)}(h)=c^{(1)}\int_0^h d\tilde{h}\int_{x_*(\tilde{h}-b^{(1)})}^{x_*(\tilde{h})} xdx,\ \varphi^{(2)}(h)=c^{(2)}\int_0^h d\tilde{h}\int_{x_*(\tilde{h}-b^{(1)}-b^{(2)})}^{x_*(\tilde{h}-b^{(1)})} xdx\ , \tag{5.1}$$

$$\frac{c_2}{c_0}v_{bl}^{\beta}=\chi^{(1)}I^{(1)}+\chi^{(2)}I^{(2)}, \tag{5.2}$$

$$I^{(j)}=\int_0^{b+L}\frac{d\varphi^{(j)}}{dh}exp(\varphi(h))\,dh\,,\ \varphi=\varphi^{(1)}+\varphi^{(2)},\ j=1,2\,, \tag{5.3}$$

where $\chi^{(j)}$ and $c^{(j)}$ are determined by Eq. (3.18) and Eq. (3.19), respectively.

The model for the shield with the inverse positions of the plates (subscript *inv* indicates the parameters of this model) may be written as:

$$\varphi_{inv}^{(1)}(h)=c^{(2)}\int_0^h d\tilde{h}\int_{x_*(\tilde{h}-b^{(2)})}^{x_*(\tilde{h})} xdx,\ \ \varphi_{inv}^{(2)}(h)=c^{(1)}\int_0^h d\tilde{h}\int_{x_*(\tilde{h}-b^{(1)}-b^{(2)})}^{x_*(\tilde{h}-b^{(2)})} xdx\ , \tag{5.4}$$

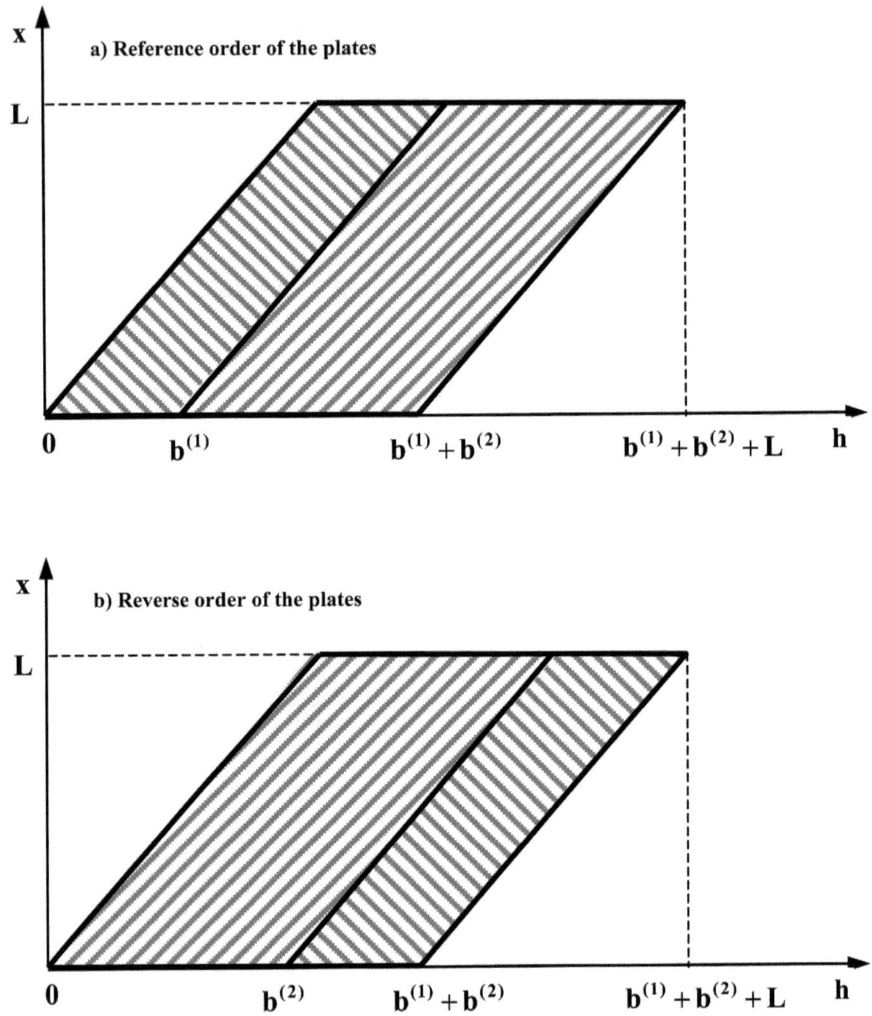

Figure 10-8. Two variants of a two-layered shield.

$$\frac{c_2}{c_0} v^{\beta}_{bl\ inv} = \chi^{(2)} I^{(1)}_{inv} + \chi^{(1)} I^{(2)}_{inv}, \tag{5.5}$$

$$I^{(j)}_{inv} = \int_0^{b+L} \frac{d\varphi^{(j)}_{inv}}{dh} exp(\varphi_{inv}(h))\, dh\,, \quad \varphi_{inv} = \varphi^{(1)}_{inv} + \varphi^{(2)}_{inv}, \quad j = 1,2. \tag{5.6}$$

5.1.2 Investigation of the problem

Using Eq. (3.25) we can write:

$$I^{(2)} = \hat{I} - I^{(1)}, \quad I^{(2)}_{inv} = \hat{I} - I^{(1)}_{inv}, \tag{5.7}$$

where

$$\hat{I} = exp\left[\frac{\beta}{2m} c_2 k \left(a_2^{(1)} b^{(1)} + a_2^{(2)} b^{(2)}\right)\right] \tag{5.8}$$

does not depend on the order of the plates in the shield. Substituting $I^{(2)}$ and $I^{(2)}_{inv}$ from Eq. (5.7) into Eqs. (5.2) and (5.5), respectively, we obtain:

$$\frac{c_2}{c_0} v^{\beta}_{bl} = \chi^{(1)} I^{(1)} + \chi^{(2)} \left(\hat{I} - I^{(1)}\right), \tag{5.9}$$

$$\frac{c_2}{c_0} v^{\beta}_{bl\ inv} = \chi^{(2)} I^{(1)}_{inv} + \chi^{(1)} \left(\hat{I} - I^{(1)}_{inv}\right), \tag{5.10}$$

and

$$\begin{aligned} \frac{c_2}{c_0} (v^{\beta}_{bl\ inv} - v^{\beta}_{bl}) &= \left(\chi^{(1)} - \chi^{(2)}\right)\left(\hat{I} - I^{(1)} - I^{(1)}_{inv}\right) \\ &= \left(\chi^{(1)} - \chi^{(2)}\right)\left(I^{(2)} - I^{(1)}_{inv}\right). \end{aligned} \tag{5.11}$$

Let us now prove that $I^{(2)} - I^{(1)}_{inv} > 0$.

We interchange the variables h and x in the integrals $I^{(2)}$ and $I^{(1)}_{inv}$ and then change the variable $h - b^{(1)} \to h$. The result reads:

$$I^{(2)} = c^{(2)} \int_0^b dh\, exp(\varphi(h)) \int_{x_*(h-b^{(1)}-b^{(2)})}^{x_*(h-b^{(1)})} x dx = c^{(2)} \int_0^L x dx \int_{x+b^{(1)}}^{x+b^{(1)} \; b^{(2)}} exp(\varphi(h)) dh$$

$$= c^{(2)} \int_0^L x dx \int_x^{x+b^{(2)}} exp(\varphi(h+b^{(2)})) dh, \tag{5.12}$$

$$I_{inv}^{(1)} = c^{(2)} \int_0^L x dx \int_x^{x+b^{(2)}} exp(\varphi_{inv}(h)) dh. \tag{5.13}$$

Then

$$\begin{aligned} \frac{1}{c^{(2)}} \left(I^{(2)} - I_{inv}^{(1)} \right) &= \int_0^L x dx \int_x^{x+b^{(2)}} \left[exp\left(\varphi(h+b^{(2)})\right) - exp(\varphi_{inv}(h)) \right] dh \\ &= \int_0^L x dx \int_x^{x+b^{(2)}} \left[exp\left(\hat{G}(h)\right) - 1 \right] exp(\varphi_{inv}(h)) dh, \end{aligned} \tag{5.14}$$

where

$$\hat{G}(h) = \varphi(h+b^{(2)}) - \varphi_{inv}(h) = c^{(1)} g^{(1)}(h) + c^{(2)} g^{(2)}(h), \tag{5.15}$$

$$g^{(j)}(h) = \frac{\varphi^{(j)}(h+b^{(1)}) - \varphi_{inv}^{(3-j)}(h)}{c^{(j)}}, \; j = 1,2, \tag{5.16}$$

$$g^{(1)}(h) = \int_0^{h+b^{(1)}} d\tilde{h} \int_{x_*(\tilde{h}-b^{(1)})}^{x_*(\tilde{h})} x dx - \int_0^h d\tilde{h} \int_{x_*(\tilde{h}-b^{(1)}-b^{(2)})}^{x_*(\tilde{h}-b^{(2)})} x dx \;, \tag{5.17}$$

$$g^{(2)}(h) = \int_0^{h+b^{(1)}} d\tilde{h} \int_{x_*(\tilde{h}-b^{(1)}-b^{(2)})}^{x_*(\tilde{h}-b^{(1)})} x dx - \int_0^h d\tilde{h} \int_{x_*(\tilde{h}-b^{(2)})}^{x_*(\tilde{h})} x dx. \tag{5.18}$$

Let us now investigate the sign of $g^{(1)}(h)$. After the change of variables $\tilde{h} - b^{(2)} \to \tilde{h}$ in the second integral in Eq. (5.17), the equation reads:

$$g^{(1)}(h) = \int_{0}^{h+b^{(1)}} d\widetilde{h} \int_{x_*(\widetilde{h}-b^{(1)})}^{x_*(\widetilde{h})} x dx - \int_{-b^{(2)}}^{h-b^{(2)}} d\widetilde{h} \int_{x_*(\widetilde{h}-b^{(1)})}^{x_*(\widetilde{h})} x dx . \quad (5.19)$$

Since $x_*(\widetilde{h}) = x_*(\widetilde{h} - b^{(1)}) = 0$ for $\widetilde{h} \le 0$, the upper bound $\widetilde{h} = -b^{(2)}$ in the integral in the second term in Eq. (5.19) can be replaced by $\widetilde{h} = 0$. Then

$$g^{(1)}(h) = \int_{0}^{h+b^{(1)}} d\widetilde{h} \int_{x_*(\widetilde{h}-b^{(1)})}^{x_*(\widetilde{h})} x dx - \int_{0}^{h-b^{(2)}} d\widetilde{h} \int_{x_*(\widetilde{h}-b^{(1)})}^{x_*(\widetilde{h})} x dx = \int_{h-b^{(2)}}^{h+b^{(1)}} d\widetilde{h} \int_{x_*(\widetilde{h}-b^{(1)})}^{x_*(\widetilde{h})} x dx \ge 0 . \quad (5.20)$$

Similarly, after the change of variables $\widetilde{h} - b^{(1)} \to \widetilde{h}$ in the first integral in Eq. (5.18), the equation reads:

$$g^{(2)}(h) = \int_{-b^{(1)}}^{h} d\widetilde{h} \int_{x_*(\widetilde{h}-b^{(2)})}^{x_*(\widetilde{h})} x dx - \int_{0}^{h} d\widetilde{h} \int_{x_*(\widetilde{h}-b^{(2)})}^{x_*(\widetilde{h})} x dx = \int_{-b^{(1)}}^{0} d\widetilde{h} \int_{x_*(\widetilde{h}-b^{(2)})}^{x_*(\widetilde{h})} x dx = 0 . \quad (5.21)$$

Therefore, $\hat{G}(h) \ge 0$, and Eq. (5.14) implies that:

$$I^{(2)} - I^{(1)}_{inv} > 0 \quad (5.22)$$

and Eq. (5.11) allows us to conclude that:

$$sign(v_{bl\,inv} - v_{bl}) = sign\left(\chi^{(1)} - \chi^{(2)}\right). \quad (5.23)$$

We have shown that the maximum BLV of the shield is attained when the plates are arranged in order of the increasing values of parameters χ.

5.2 Two-layered spaced shield

Let us consider four shields (see *Figure 10-9a-d*). The first shield (*Figure 10-9a*) is denoted by 1-2 ($v_{bl}^{[1-2]}$ denotes the corresponding BLV). This is a two-layer shield without an air gap, which consists of two plates with the thicknesses $b^{(1)}$ (the front plate) and $b^{(2)}$ (the rear plate). The second shield (*Figure 10-9b*) is denoted by 1-0-2 ($v_{bl}^{[1-0-2]}$ is its BLV), and it consists of the same plates in the same order with an air gap. The third shield (*Figure 10-9c*) and the fourth shield (*Figure 10-9d*) are similar to the shields 1-2 and 1-0-2, respectively, but the plates are arranged in the opposite order. These shields are denoted by 2-1 and 2-0-1 (with BLVs $v_{bl}^{[2-1]}$ and $v_{bl}^{[2-0-1]}$, respectively).

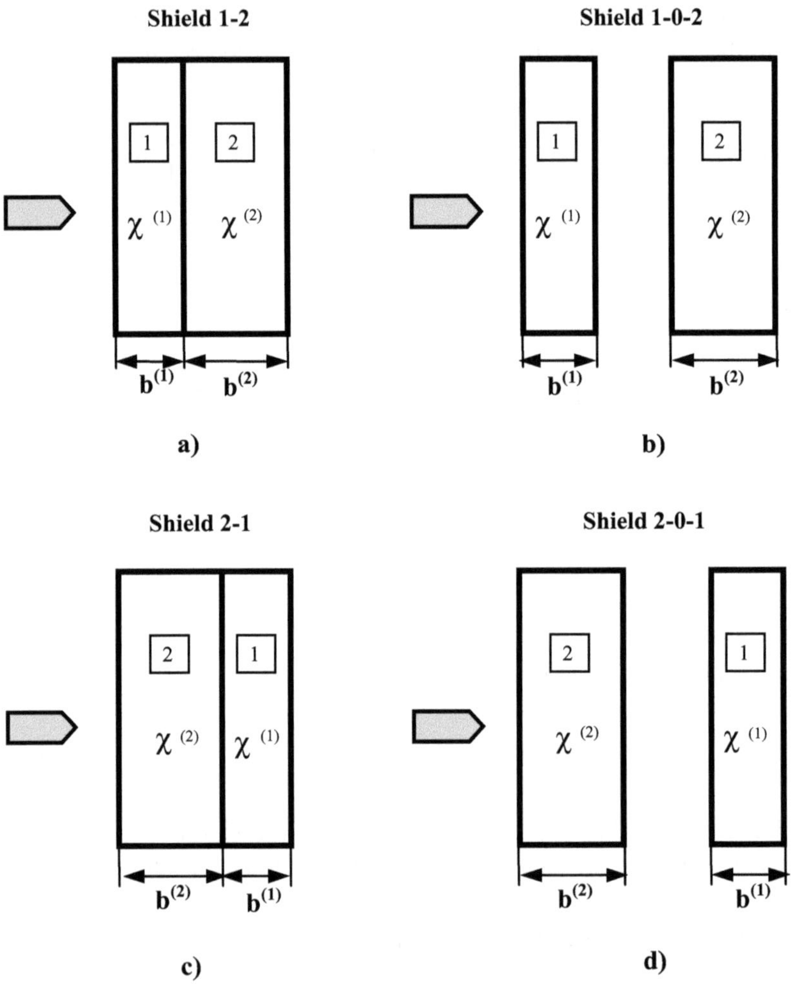

Figure 10-9. Compared shields.

Assume that $\chi^{(1)} > \chi^{(2)}$, where parameter $\chi^{(j)}$ characterizes the material properties of the j-th plate. Then, the results of *Section 10-5.1* imply that:

$$v_{bl}^{[2-1]} > v_{bl}^{[1-2]}, \tag{5.24}$$

while the results of *Section 10-4.2* imply that:

$$v_{bl}^{[1-\infty-2]} \le v_{bl}^{[1-0-2]} < v_{bl}^{[1-2]}, \quad v_{bl}^{[2-\infty-1]} \ge v_{bl}^{[2-0-1]} > v_{bl}^{[2-1]}. \tag{5.25}$$

Here, $v_{bl}^{[1-\infty-2]}$ and $v_{bl}^{[2-\infty-1]}$, respectively, are the magnitudes of the BLVs of shields 1-0-2 and 2-0-1 with air gaps larger than L. Eqs. (5.24)-(5.25) yield:

$$v_{bl}^{[2-\infty-1]} \ge v_{bl}^{[2-0-1]} > v_{bl}^{[2-1]} > v_{bl}^{[1-2]} > v_{bl}^{[1-0-2]} \ge v_{bl}^{[1-\infty-2]}. \tag{5.26}$$

Therefore the best ballistic performance of a two-layered shield is attained when the plate with the smaller χ is a front plate, and the air gap between the plates is larger than L. In contrast, the minimum BLV is attained when the plates are arranged in the reverse order, and the air gap is larger than L.

5.3 Multi-layered shield with large air gaps

5.3.1 Basic relationship

In this section, we study the case, in which an N-layered shield is perforated sequentially, i.e., the impactor does not interact simultaneously with two or more plates. The latter assumption is approximately valid if the length of the impactor is much smaller than the thicknesses of the plates. In the framework of the LIA this assumption corresponds to a spaced shield when the widths of the air gaps are larger than the length of the impactor L. We employ the LIM described in *Table 10-5* for a 3-D conical impactor and will use the corresponding general formulas. However, in the following exposition, we employ simple method to derive the relationship for the BLV. Since the plates in the shield are perforated independently, a one-layer shield model can be used for each plate.

Let us rewrite the formulas of the model of the *Section 2-8.2.2* for the plate with number j (see *Figure 10-10*). To this end, formulas for $A_2(u)$ and $A_0(u)$ in Eq. (2-8.1) can be written as follows:

$$A_2(u) = a_2^{(j)}\omega_2(u), \quad A_0(u) = a_0^{(j)}\omega_0(u), \quad j = 1,...,N, \tag{5.27}$$

where parameters $a_0^{(j)}$ and $a_2^{(j)}$ ($j = 1,...,N$) characterizing material properties of the plates are determined by step functions $a_0(\xi)$ and $a_2(\xi)$, respectively. Note that functions $a_0(\xi), a_2(\xi)$, $\omega_0(u)$ and $\omega_2(u)$ that are used in this section and in Eq. (3.4) have the same meaning. The analog of

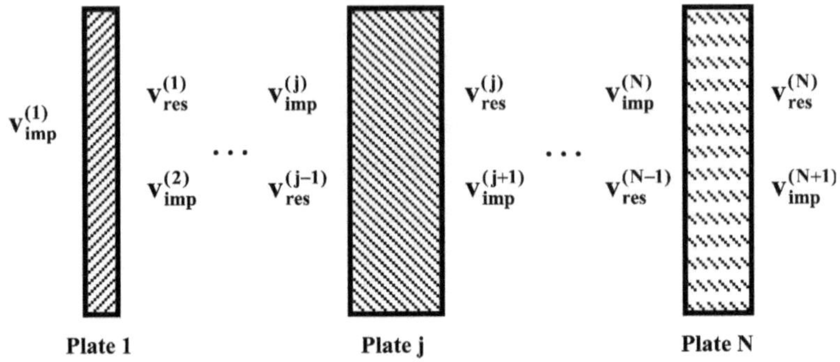

Figure 10-10. Multi-layered shield with large air gaps.

Eq. (2-8.24), the correlation between the impact velocity $v_{imp}^{(j)}$ and the residual velocity of the j-th ($j = 1,...,N$) plate $v_{res}^{(j)}$, reads:

$$v_{res}^{(j)\beta} = E^{(j)} v_{imp}^{(j)\beta} - F^{(j)}, \tag{5.28}$$

where

$$E^{(j)} = \frac{1}{T^{(j)}}, \tag{5.29}$$

$$F^{(j)} = \frac{C_0^{(j)}}{C_2^{(j)}} \left[1 - \frac{1}{T^{(j)}} \right], \tag{5.30}$$

$$T^{(j)} = exp\left(\frac{\beta k L^2 C_2^{(j)} b^{(j)}}{2m} \right). \tag{5.31}$$

Parameters $C_0^{(j)}$ and $C_2^{(j)}$ are determined similarly to Eq. (2-8.16). Using Eq. (5.27), we can rewrite equation Eq. (2-8.16) as follows:

$$C_i^{(j)} = a_i^{(j)} c_i, \quad i = 0,2, \quad j = 1 \div N, \tag{5.32}$$

where coefficients c_i are determined by Eq. (3.11). Substituting $C_i^{(j)}$ from Eq. (5.32) into Eq. (5.30), we obtain:

$$F^{(j)} = \chi^{(j)} \frac{c_0}{c_2} \left[1 - \frac{1}{T^{(j)}} \right], \quad \chi^{(j)} = \frac{a_0^{(j)}}{a_2^{(j)}}. \tag{5.33}$$

5.3.2 Formula for the BLV of the shield

Since the plates are perforated independently (see *Figure 10-10*), the residual velocity of the j-th plate, $v_{res}^{(j)}$, is equal to the impact velocity of the $(j+1)$-th plate, $v_{imp}^{(j+1)}$, for $j = 1,\ldots,N-1$. In addition, the impact velocity of the shield v_{imp} is equal to $v_{imp}^{(1)}$, and the residual velocity of the shield v_{res} is equal to $v_{res}^{(N)} = v_{imp}^{(N+1)}$ (notation $v_{imp}^{(N+1)}$ is introduced here for convenience). Then, Eq. (5.28) can be rewritten in the following form:

$$v_{imp}^{(j+1)\beta} = E^{(j)} v_{imp}^{(j)\beta} - F^{(j)}, \quad j = 1 \div N. \tag{5.34}$$

Eliminating $v_{imp}^{(j)}$ $(j = 2,\ldots,N)$ from these equations, we can obtain the correlation between the impact velocity of the shield, $v_{imp} = v_{imp}^{(1)}$, and the residual velocity of the shield, $v_{res} = v_{imp}^{(N+1)}$. The result is given by the following formula:

$$v_{imp}^{(N+1)\beta} = \left[\prod_{j=1}^{N} E^{(j)} \right] v_{imp}^{(1)\beta} - \sum_{j=1}^{N-1} F^{(j)} \prod_{v=j+1}^{N} E^{(v)} - F^{(N)}, \quad N \geq 2. \tag{5.35}$$

Equation (5.35) can be proved by induction. Indeed, let us rewrite Eq. (5.34) for $j = 1$ and $j = 2$:

$$v_{imp}^{(2)\beta} = E^{(1)} v_{imp}^{(1)\beta} - F^{(1)}, \; v_{imp}^{(3)\beta} = E^{(2)} v_{imp}^{(2)\beta} - F^{(2)}. \tag{5.36}$$

Eliminating $v_{imp}^{(2)}$, we obtain:

$$v_{imp}^{(3)\beta} = E^{(1)} E^{(2)} v_{imp}^{(1)\beta} - E^{(2)} F^{(1)} - F^{(2)} \tag{5.37}$$

Since Eq. (5.35) for $N = 2$ yields the same formula, Eq. (5.35) is valid for $N = 2$.

Assume that Eq. (5.35) is valid for an *N*-layered shield ($N > 2$ is an integer number). Now, let us prove the validity of this equation for a shield consisting of $N+1$ plates. In other words, we assume that Eq. (5.35) is valid for a sub-shield consisting of the first N plates and prove, using Eq. (5.34), that Eq. (5.35) is valid for the whole shield. Eq. (5.34) for $j = N+1$ reads:

$$v_{imp}^{(N+2)\beta} = E^{(N+1)} v_{imp}^{(N+1)\beta} - F^{(N+1)}, \quad j = 1 \div N. \tag{5.38}$$

Substituting $v_{imp}^{(N+1)}$ from Eq. (5.35) to Eq. (5.38), we find that:

$$v_{imp}^{(N+2)\beta} = E^{(N+1)} \left[\prod_{j=1}^{N} E^{(j)} \right] v_{imp}^{(1)\beta} - E^{(N+1)} \sum_{j=1}^{N-1} F^{(j)} \prod_{v=j+1}^{N} E^{(v)}$$
$$- E^{(N+1)} F^{(N)} - F^{(N+1)} = \left[\prod_{j=1}^{N+1} E^{(j)} \right] v_{imp}^{(1)\beta} - \sum_{j=1}^{N} F^{(j)} \prod_{v=j+1}^{N+1} E^{(v)} - F^{(N+1)}. \tag{5.39}$$

Since the formula for $v_{imp}^{(N+2)}$ (Eq. 5.39) coincides with Eq. (5.35) after replacing N by $N+1$, it can be concluded that Eq. (5.35) is valid for arbitrary values of N.

Substituting $v_{imp}^{(N+1)} = 0$ and $v_{imp}^{(1)} = v_{bl}$ into Eq. (5.35), we obtain a formula for the BLV of the shield, v_{bl}:

$$v_{bl}^{\beta} = \left[\prod_{j=1}^{N} \frac{1}{E^{(j)}} \right] \cdot \left[\sum_{j=1}^{N-1} F^{(j)} \prod_{v=j+1}^{N} E^{(v)} + F^{(N)} \right] = \sum_{j=1}^{N} F^{(j)} \prod_{v=1}^{j} \frac{1}{E^{(v)}}. \tag{5.40}$$

Using Eqs. (5.29) and (5.33) for $E^{(j)}$ and $F^{(j)}$, we can rewrite Eq. (5.40) as follows:

$$\frac{c_2}{c_0} v_{bl}^{\beta} = \sum_{j=1}^{N} \chi^{(j)} [T^{(j)} - 1] \prod_{v=0}^{j-1} T^{(v)}, \quad T^{(0)} = 1. \tag{5.41}$$

In the case of an impactor having the shape of a cone of revolution and using the model given by Eqs. (3.2), (3.5) and (3.26), we obtain:

$$\frac{c_2}{c_0} = \frac{k^2}{k^2 + 1}, \quad T^{(j)} = exp\left(\frac{\lambda_{17} b^{(j)} a_2^{(j)}}{2L} \right), \tag{5.42}$$

where λ_{17} is defined by Eq. (3.31).

5.3.3 Effect of the order of the plates on the ballistic limit velocity

Let us investigate how the BLV changes when two plates with numbers s and $s+1$ are interchanged. First, let us rewrite Eq. (5.41) so as to separate the terms with superscripts s and $s+1$ from the sum:

$$\frac{c_2}{c_0} v_{bl}^{\beta} = U_{1\div s-1} + U_{s\div s+1} + U_{s+2} + U_{s+3\div N}, \tag{5.43}$$

where

$$U_{1\div s-1} = \begin{cases} \sum_{j=1}^{s-1} \chi^{(j)} [T^{(j)} - 1] \prod_{v=0}^{j-1} T^{(v)} & \text{if } \quad s > 1 \\ 0 & \text{if } \quad s = 1 \end{cases}, \tag{5.44}$$

$$U_{s\div s+1} = \{ \chi^{(s)} [T^{(s)} - 1] + \chi^{(s+1)} [T^{(s+1)} - 1] T^{(s)} \} \prod_{v=0}^{s-1} T^{(v)}, \tag{5.45}$$

$$U_{s+2} = \begin{cases} \chi^{(s+2)} [T^{(s+2)} - 1] T^{(s)} T^{(s+1)} \prod_{v=0}^{s-1} T^{(v)} & \text{if } \quad s \le N-2 \\ 0 & \text{if } \quad s = N-1 \end{cases}, \tag{5.46}$$

$$U_{s+3\div N} = \begin{cases} T^{(s)} T^{(s+1)} \sum_{j=s+3}^{N} \chi^{(j)} [T^{(j)} - 1] \prod_{\substack{0 \le v \le s-1 \\ s+2 \le v \le j-1}} T^{(v)} & \text{if } \quad s \le N-3 \\ 0 & \text{if } \quad s > N-3 \end{cases}. \tag{5.47}$$

Consider now a shield with the reverse order of the plates with numbers s and $s+1$. The formula for the BLV, $v_{bl}^{[s+1,s]}$, can be obtained from Eqs. (5.43)-(5.47) by interchanging $T^{(s)} \leftrightarrow T^{(s+1)}$ and $\chi^{(s)} \leftrightarrow \chi^{(s+1)}$. Therefore,

$$\frac{c_2}{c_0} v_{bl}^{[s+1,s]\,\beta} = \widetilde{U}_{1\div s-1} + \widetilde{U}_{s\div s+1} + \widetilde{U}_{s+2} + \widetilde{U}_{s+3\div N}, \tag{5.48}$$

where

$$\widetilde{U}_{1\div s-1}=U_{1\div s-1}\ ,\ \widetilde{U}_{s+2}=U_{s+2},\ \widetilde{U}_{s+3\div N}=U_{s+3\div N}, \tag{5.49}$$

$$\widetilde{U}_{s\div s+1}=\{\chi^{(s+1)}[T^{(s+1)}-1]+\chi^{(s)}[T^{(s)}-1]T^{(s+1)}\}\prod_{\nu=0}^{s-1}T^{(\nu)}. \tag{5.50}$$

Equations (5.43), (5.48) and (5.49) yield:

$$\frac{c_2}{c_0}(v_{bl}^{[s+1,s]\,\beta}-v_{bl}^{\beta})=\widetilde{U}_{s\div s+1}-U_{s\div s+1}\,. \tag{5.51}$$

Substituting Eqs. (5.45) and (5.50) to Eq. (5.51), we obtain after some algebra:

$$\frac{c_2}{c_0}(v_{bl}^{[s+1,s]\,\beta}-v_{bl}^{\beta})=\left\{[T^{(s)}-1][T^{(s+1)}-1]\prod_{\nu=0}^{s-1}T^{(\nu)}\right\}[\chi^{(s)}-\chi^{(s+1)}]. \tag{5.52}$$

Since $T^{(\nu)}>1$ for all ν and $c_0>0, c_2>0, \beta>0$, Eq. (5.52) implies that:

$$sign(v_{bl}^{[s+1,s]}-v_{bl})=sign(\chi^{(s)}-\chi^{(s+1)}). \tag{5.53}$$

Thus, we have proved that if two adjacent plates in a multi-layered shield with large air gaps are arranged such that the value of the parameter χ for the first plate is larger than that for the second plate, the BLV of the shield can be increased by interchanging these plates.

It can also be easily shown that the maximum BLV of the shield is obtained when the plates are arranged in the order of increasing values of χ. To prove the latter claim, let us assume that this claim is not valid, i.e., the ballistic limit for some sequence of the plates is maximum but for some $0<s<N$, $\chi^{(s)}>\chi^{(s+1)}$. Interchanging the order of these plates results in an increase of the BLV that contradicts the initial assumption. Therefore, the claim is proved. It must be noted that the optimal order of the plates is independent of their thicknesses. Clearly, the minimum BLV is achieved when the plates are arranged in the shield in the inverse order.

The model given by Eq. (5.38), with $\beta=2$ and coefficients that have different meanings depending on the adopted physical model of perforation have been employed in a number of studies of multi-layered shields. Nixdorff (1984b) used this model and derived an equation similar to Eq. (5.41). Baranov et al. (1991) determined the criterion given by Eq. (5.53) for the case of a shield consisting of two layers. Seliverstov (2001) attempted to determine a general rule for the optimal order of plates in a multi-layered shield.

5.4 Non-conical impactor: a special case

In this section, we consider penetration in a multi-layered shield consisting of N plates in contact under the following assumptions: (i) the impactor is, generally, a 3-D sharp *non-conical projectile*, (ii) the plates have the same thickness, and (iii) the impactor-shield interaction model determined by Eqs. (3.2) and (3.5) is valid, where the parameter a_2 is the same for all the plates, i.e.,

$$a_2(\xi)=a_2=const \tag{5.54}$$

in Eq. (3.5). This model together with assumption (5.54) is applicable when the plates are manufactured from different types of functionally graded material such as the plates have approximately the same density but their strength characteristics are different.

Taking into account Eq. (5.54) and using the model for sharp 3-D impactor determined by Eqs. (3.7)-(3.9), we arrive at the following equations:

$$\frac{m}{\beta}v_{bl}^{\beta} = \int_0^{b+L} dh\, Q(h) \sum_{j=1}^{N} a_0^{(j)} \int_{x_*\left(h-\xi^{(j)}\right)}^{x_*\left(h-\xi^{(j-1)}\right)} e_0(x)\, dx\,, \tag{5.55}$$

$$Q(h)=exp\left(\frac{\beta a_2}{m}\int_0^h d\tilde{h} \int_{\theta(\tilde{h})}^{\Theta(\tilde{h})} e_2(x)dx\right), \tag{5.56}$$

where $e_i(x)$ is determined by Eq. (3.6). Using Eq. (2.21), we can rewrite Eq. (5.55) as follows:

$$\frac{m}{\beta}v_{bl}^{\beta} = \int_0^{L} dx e_0(x) \sum_{j=1}^{N} a_0^{(j)} \int_0^{b^{(j)}} Q(h+\xi^{(j-1)}+x)dh\,. \tag{5.57}$$

Now let us investigate how the BLV of the shield changes when two plates with numbers s and t $(t>s)$ having the same thickness are interchanged.

Equation (5.57) can be rewritten as:

$$\frac{m}{\beta}v_{bl}^{\beta} = \int_0^L dx e_0(x) \left[\sum_{\substack{0 \le j \le N \\ j \ne s, j \ne t}} a_0^{(j)} \int_0^{b^{(j)}} Q(h + \xi^{(j-1)} + x) dh \right. \\ \left. + a_0^{(s)} \int_0^{b^{(s)}} Q(h + \xi^{(s-1)} + x) dh + a_0^{(t)} \int_0^{b^{(t)}} Q(h + \xi^{(t-1)} + x) dh \right]. \tag{5.58}$$

Consider now a shield with the reverse order of the plates with numbers s and t. The formula for the BLV, $v_{bl}^{[t,s]}$, can be obtained from Eqs. (5.58) by interchanging $a_0^{(s)} \leftrightarrow a_0^{(t)}$ and $b^{(s)} \leftrightarrow b^{(t)}$:

$$\frac{m}{\beta}v_{bl}^{[t,s]\beta} = \int_0^L dx e_0(x) \left[\sum_{\substack{0 \le j \le N \\ j \ne s, j \ne t}} a_0^{(j)} \int_0^{b^{(j)}} Q(h + \xi^{(j-1)} + x) dh \right. \\ \left. + a_0^{(t)} \int_0^{b^{(t)}} Q(h + \xi^{(s-1)} + x) dh + a_0^{(s)} \int_0^{b^{(s)}} Q(h + \xi^{(t-1)} + x) dh \right]. \tag{5.59}$$

Taking into account that $b^{(s)} = b^{(t)}$, Eqs. (5.58) and (5.59) yield:

$$\frac{m}{\beta}\left(v_{bl}^{[t,s]\beta} - v_{bl}^{\beta}\right) \\ = \left(a_0^{(s)} - a_0^{(t)}\right) \int_0^L dx e_0(x) \int_0^{b^{(s)}} Q(h + \xi^{(t-1)} + x) - Q(h + \xi^{(s-1)} + x)\, dh. \tag{5.60}$$

Since $Q(h)$ is an increasing function and $\xi^{(t-1)} > \xi^{(s-1)}$, then $Q(h + \xi^{(t-1)} + x) > Q(h + \xi^{(s-1)} + x)$, and Eq. (5.60) implies that:

$$sign\left(v_{bl}^{[t,s]} - v_{bl}\right) = sign\left(a_0^{(s)} - a_0^{(t)}\right). \tag{5.61}$$

Since in Eq. (5.61) $a_0^{(j)} \propto \chi^{(j)}$, we arrive at the following conclusion. If the value of the parameter χ for the first of two plates having the same thickness is larger than that for the second plate, then the BLV of the shield can be increased by interchanging the plates. The method used in *Section 10-5.3* allows to prove that the maximum BLV is attained when the plates are arranged in the order of increasing values of a_0, and the minimum BLV is achieved when the plates are arranged in the shield in the inverse order.

6. SOME PROPERTIES OF LAYERED SHIELDS WITH A GIVEN AREAL DENSITY

6.1 A class of optimal multi-layered shields

The problem under consideration can be formulated as follows. There are several materials with different properties that can be used for manufacturing the plates in a multi-layered shield. It is assumed that the areal density of the shield, namely, its mass per unit surface area, is given. The goal is to determine the structure of the shield, i.e., the order and the thicknesses of the plates manufactured from different materials that provide the maximum BLV of the shield against a normal impact by a conical-nosed 3-D impactor.

We use the model described in *Table 10-5*, assuming that the parameter $a_2^{(j)}$ is density. Then, the areal density of the shield A is given by the following formula:

$$A = \sum_{j=1}^{N} a_2^{(j)} b^{(j)} . \tag{6.1}$$

Without a loss of generality, it can be assumed that parameters $\chi^{(j)}$ for all plates are different. Denote by $\chi^{(j_*)}$ the maximum value of $\chi^{(j)}$ that is attained for the plate with number j_*. Then Eqs. (3.20) and Eq. (3.25) can be rewritten as follows:

$$\frac{c_2}{c_0} v_{bl}^{\beta} = \sum_{\substack{1 \le j \le N \\ j \ne j_*}} \chi^{(j)} I^{(j)} + \chi^{(j_*)} I^{(j_*)} , \tag{6.2}$$

$$\sum_{\substack{1 \le j \le N \\ j \ne j_*}} I^{(j)} + I^{(j_*)} = \hat{I}, \quad \hat{I} = exp\left(\frac{\beta L^2 c_2 k}{2m} A \right) - 1 . \tag{6.3}$$

Substituting $I^{(j_*)}$ from Eq. (6.3) into Eq. (6.2), we obtain:

$$\frac{c_2}{c_0} v_{bl}^{\beta} = \chi^{(j_*)} \hat{I} - \sum_{\substack{1 \le j \le N \\ j \ne j_*}} \left(\chi^{(j_*)} - \chi^{(j)} \right) I^{(j)}, \tag{6.4}$$

where

$$I^{(j)} = c^{(j)} \int_0^L dx x \int_0^{b^{(j)}} dh \exp\left(\int_0^h d\tilde{h} \sum_{i=1}^N c^{(i)} \int_{x_*(\tilde{h} - \xi^{(i)})}^{x_*(\tilde{h} - \xi^{(i-1)})} x dx \right). \tag{6.5}$$

Equation (6.5) shows that $I^{(j)} > 0$ if $b^{(j)} > 0$ and $I^{(j)} = 0$ only when $b^{(j)} = 0$. Since $\hat{I} = const$ and $\chi^{(j_*)} - \chi^{(j)} > 0$ for $j \ne j_*$, Eq. (6.4) implies that the maximum BLV is attained when $b^{(j)} = 0$ for all $j \ne j_*$.

Therefore, the shield with the maximum BLV consists of one plate or several adjacent plates (these cases are equivalent in the model) manufactured from the material with the maximum value of parameter χ, regardless of the shape of 3-D conical-nosed impactor. The thickness of the optimum shield equals $A / a_2^{(j_*)}$. Similarly, the shield with the minimum BLV consists of one plate or several adjacent plates manufactured from the material with the minimum value of parameter χ. The BLVs of different shields with a given areal density vary in the range between these two limiting values.

6.2 Efficiency of changing the order of the plates in a two-layered shield

Consider the ballistic performance of a two-layer shield with large air gaps against a conical impactor. Formulas for the BLV of the shield with the direct (original) order of the plates (superscript $[1-2]$) and with the inverse order of the plates (superscript $[2-1]$) are given by Eq. (5.41):

$$\frac{c_2}{c_0} v_{bl}^{[1-2]\beta} = \chi^{(1)} [T^{(1)} - 1] + \chi^{(2)} [T^{(2)} - 1] T^{(1)}, \tag{6.6}$$

$$\frac{c_2}{c_0} v_{bl}^{[2-1]\beta} = \chi^{(2)} [T^{(2)} - 1] + \chi^{(1)} [T^{(1)} - 1] T^{(2)}. \tag{6.7}$$

Here, according to Eqs. (5.31) and (5.32),

$$T^{(j)} = exp\left(\overline{A}^{(j)}\right), \quad \overline{A}^{(j)} = \frac{\beta k L^2 c_2}{2m} a_2^{(j)} b^{(j)}, \tag{6.8}$$

the superscript in parentheses corresponds to the number of the plate in the original shield and $\overline{A}^{(j)}$ is the dimensionless areal density of the j-th plate ($a_2^{(j)}$ is the density of a plate).

Assume that:

$$\chi^{(1)} > \chi^{(2)}. \tag{6.9}$$

Then $v_{bl}^{[2-1]} > v_{bl}^{[1-2]}$ and we can consider the ratio

$$\widetilde{\varepsilon} = v_{bl}^{[2-1]} / v_{bl}^{[1-2]} \tag{6.10}$$

as a characteristic of the efficiency of interchanging the plates in the shield.

Let us determine the maximum magnitude of this parameter, i.e., the maximum $\widetilde{\varepsilon}$ that can be attained for a given areal density of the shield:

$$\overline{A} = \overline{A}^{(1)} + \overline{A}^{(2)}. \tag{6.11}$$

This problem can be reduced to the maximization of

$$\widetilde{\varepsilon}^{\beta} = \frac{\chi^{(2)}[T^{(2)} - 1] + \chi^{(1)}[T^{(1)} - 1]T^{(2)}}{\chi^{(1)}[T^{(1)} - 1] + \chi^{(2)}[T^{(2)} - 1]T^{(1)}} \tag{6.12}$$

under the following restrictions:

$$T^{(1)}T^{(2)} = g, \quad 1 \leq T^{(j)} \leq g, \; j = 1,2, \tag{6.13}$$

where

$$g = exp\left(\overline{A}\right). \tag{6.14}$$

Substituting $T^{(2)} = g / T^{(1)}$ into Eq. (6.12) we obtain, after simple algebra, the following problem:

$$\widetilde{\varepsilon}^{\beta} = f_0\left(T^{(1)}\right) = \frac{g(1-\xi) + (g\xi - 1)T^{(1)}}{(\xi - 1)T^{(1)2} + (g - \xi)T^{(1)}} \to max, \tag{6.15}$$

where

$$\xi = \chi^{(1)} / \chi^{(2)} > 1\,, \tag{6.16}$$

g is known, and

$$1 \le T^{(1)} \le g\,. \tag{6.17}$$

To investigate this problem, we calculate the first derivative:

$$f_0'\left(T^{(1)}\right) = \frac{(1-\xi) f_1\left(T^{(1)}\right)}{\left[(\xi-1)T^{(1)2} + (g-\xi)T^{(1)}\right]^2}\,, \tag{6.18}$$

where

$$f_1\left(T^{(1)}\right) = (g\xi - 1)T^{(1)2} - 2g(\xi-1)T^{(1)} - g(g-\xi)\,. \tag{6.19}$$

The equation $f_1\left(T^{(1)}\right) = 0$ has two roots:

$$T_*^{(1)} = \frac{\sqrt{g\xi} + g}{\sqrt{g\xi} + 1}, \quad T_{**}^{(1)} = 1 - \frac{g-1}{\sqrt{g\xi} - 1}. \tag{6.20}$$

Since $g > 1$ and $\xi > 1$, it can be concluded that the root $T_{**}^{(1)} < 1$, i.e., it is located outside the domain determined by Eq. (6.17). The formula for the first root can be written in two forms:

$$T_*^{(1)} = 1 + \frac{g-1}{\sqrt{g\xi} + 1}, \quad T_*^{(1)} = g - \frac{(g-1)\sqrt{g\xi}}{\sqrt{g\xi} + 1}\,. \tag{6.21}$$

Clearly, this root satisfies the constraints given by Eq. (6.17). Since at the boundaries of the interval $[1, g]$

$$f_1(1) = -(g-1)^2 < 0, \quad f_1(g) = g\xi(g-1)^2 > 0\,, \tag{6.22}$$

then $f_1\left(T^{(1)}\right) < 0$ when $1 \le T^{(1)} < T_*^{(1)}$ and $f_1\left(T^{(1)}\right) > 0$ when $T_*^{(1)} < T^{(1)} \le g$. Hence, as shown by Eq. (6.18), function $f_0\left(T^{(1)}\right)$ increases in the interval $1 \le T^{(1)} < T_*^{(1)}$ and decreases for $T_*^{(1)} < T^{(1)} \le g$. The maximum of this function is reached at the point:

$$T_{max}^{(1)} = T_*^{(1)} = \frac{\sqrt{g\xi} + g}{\sqrt{g\xi} + 1}. \tag{6.23}$$

Equation (6.23) yields a formula for the maximum value of the ratio $\tilde{\varepsilon}$, $\tilde{\varepsilon}_{max}$, which can be obtained by substituting the maximum $T^{(1)}$ from Eq. (6.23) into Eq. (6.15):

$$\tilde{\varepsilon}_{max} = \left[\frac{\sqrt{g\xi} + 1}{\sqrt{g} + \sqrt{\xi}}\right]^{2/\beta}. \tag{6.24}$$

Equation (6.24) for $\tilde{\varepsilon}_{max}$ can be rewritten as:

$$\tilde{\varepsilon}_{max} = \left[\sqrt{\xi} - \frac{\xi - 1}{\sqrt{g} + \sqrt{\xi}}\right]^{2/\beta}. \tag{6.25}$$

Equation (6.25) shows that $\tilde{\varepsilon}_{max}$ increases with an increase of the areal density of the shield, $\overline{A} = ln(g)$. It is interesting to note that there is a limiting value of $\tilde{\varepsilon}_{max}$, which can be obtained when $g \to \infty$, i.e., always

$$\tilde{\varepsilon}_{max} \leq \xi^{1/\beta}. \tag{6.26}$$

Expanding $\tilde{\varepsilon}_{max}(\beta = 2)$ in a Taylor series of small parameter $\overline{A}$, we obtain the following approximate formula:

$$\tilde{\varepsilon}_{max} \approx 1 + \frac{\sqrt{\xi} - 1}{2(\sqrt{\xi} + 1)}\overline{A}. \tag{6.27}$$

7. OPTIMIZATION OF IMPACTORS AGAINST MULTI-LAYERED SHIELDS

Let us consider penetration of a 3-D sharp conical impactor into a non-homogeneous shield by using the impactor-shield interaction model summarized in *Table 10-5* with:

$$\omega_0(u) = 1, \mu_{fr} = 0. \tag{7.1}$$

The equation of the impactor's surface, $\rho = \Phi(x,\vartheta)$, reads:

$$\rho = kx\eta(\vartheta), \quad k=1, \tag{7.2}$$

where $\rho = L\eta(\vartheta)$ is the equation of the shank of the cone. Substituting Eqs. (7.1) and (7.2) into the equations listed in *Table 10-5*, we obtain the following formula for the BLV:

$$v_{bl}^{\beta} = \frac{\beta}{m}\int_0^{b+L} dh\, f_0(h)\exp\left(\frac{\beta}{m}\int_0^h f_2(\tilde{h})\,d\tilde{h}\right), \tag{7.3}$$

where

$$f_i(h) = c_i \int_{\theta(h)}^{\Theta(h)} x\, a_i(h-x)dx, \tag{7.4}$$

$$c_0 = J_0[\eta(\vartheta)], \quad J_0[\eta(\vartheta)] = \int_0^{2\pi} \eta^2(\vartheta)d\vartheta, \tag{7.5}$$

$$c_2 = J_2[\eta(\vartheta)], \quad J_2[\eta(\vartheta)] = \int_0^{2\pi} \omega_2\left(\frac{\eta^2}{\sqrt{\eta^2(\eta^2+1)+\eta'^2}}\right)\eta^2(\vartheta)d\vartheta. \tag{7.6}$$

Similarly, we can obtain a formula for the DOP:

$$\frac{\beta}{m}\int_0^H dh\, f_0(h)\exp\left(\frac{\beta}{m}\int_0^h f_2(\tilde{h})\,d\tilde{h}\right) = v_{imp}^{\beta}. \tag{7.7}$$

Clearly,

$$c_0 = \overline{\sigma}(L), \quad \overline{\sigma}(L) = \frac{2}{L^2}\sigma(L), \tag{7.8}$$

where $\sigma(L)$ is the area of the impactor's shank.

Using these formulas, Eqs. (7.3) and (7.7) can be rewritten as follows:

$$v_{bl}^{\beta} = \frac{\beta}{mL^2} J_0[\eta(\vartheta)] \int_0^{b+L} \Psi_0(h)[\Psi_2(h)]^{\frac{\beta}{m}J_2[\eta(\vartheta)]} dh , \tag{7.9}$$

$$\frac{\beta}{mL^2} J_0[\eta(\vartheta)] \int_0^{H} \Psi_0(h)[\Psi_2(h)]^{\frac{\beta}{m}J_2[\eta(\vartheta)]} dh = v_{imp}^{\beta} , \tag{7.10}$$

where

$$\Psi_0(h) = \int_{\theta(h)}^{\Theta(h)} x\, a_0(h-x)dx, \Psi_2(h) = exp\left(\int_0^h d\tilde{h} \int_{\theta(\tilde{h})}^{\Theta(\tilde{h})} x\, a_2(\tilde{h}-x)dx \right). \tag{7.11}$$

Assume that parameters $L,b,m,\beta>0$, $\overline{\sigma}(L)=J_0[\eta(\vartheta)]$ and functions $a_0(\xi)$ and $a_2(\xi)$ are given and consider the problem of finding the shape of cross-section of the 3-D conical impactor with minimum BLV. Since $\Psi_0(h)>0, \Psi_2(h)>0, J_2[\eta(\vartheta)]>0$, Eq. (7.9) shows that the solution of the variational problem:

$$J[\eta(\vartheta)] \to min, \quad J_0[\eta(\vartheta)] = \overline{\sigma}(L) \tag{7.12}$$

provides the solution to problem of finding a 3-D conical impactor with minimum BLV.

The latter conclusion is also valid when some additional constraints must be taken into account, e.g., when the contour of the projectile is bounded by a circular ring:

$$\eta(0)=\eta(2\pi), \quad \overline{R}_{min} \le \eta(\vartheta) \le \overline{R}_{max} , \tag{7.13}$$

where $\overline{R}_{min}$ and $\overline{R}_{max}$ are the dimensionless radii. It is important to note that a different choice of $L,b,m,\beta>0, a_0(\xi), a_2(\xi)$ yields the same variational problem. Therefore, the shape of the optimal impactor (in dimensionless variables using L as a reference length) does not depend on these parameters and functions, i.e., it does not depend on the distribution of the material properties of the shield along the trajectory of the projectile.

Similar conclusions are valid when we consider the problem of obtaining the maximum DOP for a given impact velocity, v_{imp}. Indeed, let us assume that $L,b,m,\beta>0, v_{imp}, \overline{\sigma}(L)=J_0[\eta(\vartheta)]$, $a_0(\xi)$ and $a_2(\xi)$ are given. Then Eq. (7.10) can be considered as a definition of the dependence H vs. $J_2[\eta(\vartheta)]$, where H increases with a decrease of $J_2[\eta(\vartheta)]$. Therefore, the solution of the variational problem given by Eq. (7.12) provides the solution of the DOP maximization problem.

To elucidate the meaning of the functional $J_2[\eta(\vartheta)]$, we consider a hypothetical situation in which an impactor moves with a constant velocity $\hat{v}$ in a homogeneous shield. Assume also that impactor is completely immersed in the shield and that the impactor-shield interaction model is determined by Eqs. (3.5) and (7.1) with constant parameters $\hat{a}_0$ and $\hat{a}_2$:

$$\Omega_n = \hat{a}_2\, \omega_2(u)\hat{v}^2 + \hat{a}_0\, \hat{v}^\alpha, \quad \Omega_\tau = 0\,. \tag{7.14}$$

Substituting Eq. (7.14) and $\sigma(0)=0$, $\theta(h)=0$, $\Theta(h)=L$, $uu_0 = x\eta^2$ into Eq. (2.6) and taking into account Eq. (2.7), we obtain the following expression for the drag force, D:

$$\begin{aligned} D &= \int_0^L x dx \int_0^{2\pi} [\,\hat{a}_2\, \omega_2(u)\hat{v}^2 + \hat{a}_0\, \hat{v}^\alpha\,]\eta^2 d\vartheta \\ &= 0.5L^2\{\,\hat{a}_2\hat{v}^2 J_2[\eta(\vartheta)] + \hat{a}_0\, \hat{v}^\alpha \overline{\sigma}(L)\}. \end{aligned} \tag{7.15}$$

Therefore, the solution of the variational problem given by Eq. (7.12) also provides the solution to the problem of determining the shape of the cross-section of the impactor with the minimum drag force among the impactors with a given area of the shank.

It is interesting to note that for a widely used model

$$\omega_2(u) = u^2 \tag{7.16}$$

Eq. (7.15) with $\hat{a}_0 = 0$ and $\hat{a}_2 = \gamma_{sh}$ coincides with Newton's model (see, e.g., Hayes and Probstein, 1959) for calculating the drag force acting on a hypersonic projectile. Therefore, the optimal shapes of penetrating impactors and those of flying projectile are the same, and solutions obtained in gasdynamics can be used in this case. In particular, this enables us to extend the results by Gonor and Chernyi (Miele, 1965) concerning the optimal properties of star-shaped flying projectiles to the case of penetrating impactors.

We have thus shown that among the 3-D conical strikers there are universal optimal impactors penetrating normally into a non-homogenous (layered) SIS or into a SFT. The impactor subjected to the minimum drag force during its motion inside a homogenous shield with a constant velocity penetrates to the maximum depth into a SIS. This impactor has the minimum BLV when it penetrates into a shield with a finite thickness, regardless of the distribution of the material properties of the shield along its penetration path, number of layers, etc. Therefore, there is a useful analogy between optimization problems encountered in gasdynamics of flying projectiles and optimization problems for impactors penetrating into non-homogeneous shields.

Similar results can also be obtained for thin non-conical impactors (see Ben-Dor et al., 1997c).

8. NUMERICAL ANALYSIS AND ILLUSTRATION OF THE RESULTS BASED ON LOCALIZED INTERACTION MODELS

In numerical calculations, we used the model given by Eqs. (3.27)-(3.31) with $\beta = 2$, where parameters a_0 and a_2 are "dynamical hardness" and material density of the shield, correspondingly. This model was proposed and validated in comprehensive experimental studies conducted by Vitman and Stepanov (1959) with conical impactors penetrating into shields manufactured from different metals. The values of the "dynamical hardness" in *Table 10-3* are adopted from the study of Vitman and Ioffe (1948). The materials are ordered according to increasing values of parameter χ. It must be noted that the remark by Recht (1990) concerning similar semi-empirical models is confirmed in this case, i.e., parameter a_0 is significantly larger than the compressive yield strength.

All the calculations presented below were performed for impactors having the shape of a sharp body of revolution with a conical-nose. Although the results of the analysis are usually presented in the dimensionless form, all calculations are performed taking into account the following constraint:

$$v_{bl} \frac{k}{\sqrt{k^2 + 1}} = v_{bl} \sin \upsilon \le 700 \, m/s \ , \tag{8.1}$$

where 2υ is the angle of the conical nose of the impactor. This constraint approximately determines the range in which this model is valid.

The goal of the first series of calculations was to estimate the influence of air gaps on the BLV for real magnitudes of the parameters describing the ballistic impact. It was found that air gaps changed the BLV by only 0.1%. Therefore, in the framework of the considered model the above-described formulas and conclusions for the case of large air gaps are indeed valid for shields with arbitrary air gaps and, in particular, for shields with plates in contact.

Table 10-3. Parameters of the model (adopted from Vitman and Ioffe, 1948)

Number	Material	Dynamical hardness, a_0, N/m^2	Density, a_2, kg/m^3	$\chi = a_0/a_2$, m^2/s^2
1	Aluminum	$350 \cdot 10^6$	2765	$0.127 \cdot 10^6$
2	Soft steel	$1850 \cdot 10^6$	7830	$0.236 \cdot 10^6$
3	Copper	$910 \cdot 10^6$	2700	$0.337 \cdot 10^6$
4	Duraluminum	$1330 \cdot 10^6$	2765	$0.481 \cdot 10^6$

Figure 10-11a-d shows the efficiency of interchanging the order of the plates in the shield depending on the thicknesses of the plates in the case of two plates for different values of λ_{17} (see Eq. 3.31). The following numbers are assigned to the materials of the plates. The superscript $[1-2]$ means that the order of the plates in the shield is such that the plate manufactured from material #1 is perforated before the plate manufactured from material #2. Since parameter χ increases with an increase in the number of the material, $v_{bl}^{[2-1]} \geq v_{bl}^{[1-2]}$. Some curves in *Figure 10-11a-d* were not continued because of the constraint given by Eq. (8.1). Inspection of these figures shows that the arrangement of the plates in increasing order of parameter χ can result in significant increase of the BLV in comparison with the reverse order of the plates, especially for relatively large values of λ_{17}.

Figure 10-12 shows a quantitative estimate of the results obtained in Section 10-6.2, i.e., it shows the maximum effect on the BLV that can be achieved by changing the order of the plates in the shield. This effect is characterized by the parameter $\tilde{\varepsilon}_{max}$ in Eq. (6.24), where $\overline{A}$ is the dimensionless areal density of the shield (Eq. 6.8) and ξ depends on the plates materials (Eq. 6.16). The effect is especially large when the difference between the values of the parameter χ for the plates is large.

Theoretically predicted structures of the three-layered shield arranged in increasing order of the BLV are shown in *Figure 10-13*, in which the transition to the structure with the larger BLV is performed by interchanging the neighboring plates when the magnitude of the parameter χ for the first plate (in the direction of penetration) is larger than the value of parameter χ for the second plate. In this manner, we obtain two series of structures of the shield:

$$v_{bl}^{[3-2-1]} < v_{bl}^{[2-3-1]} < v_{bl}^{[2-1-3]} < v_{bl}^{[1-2-3]} \tag{8.2}$$

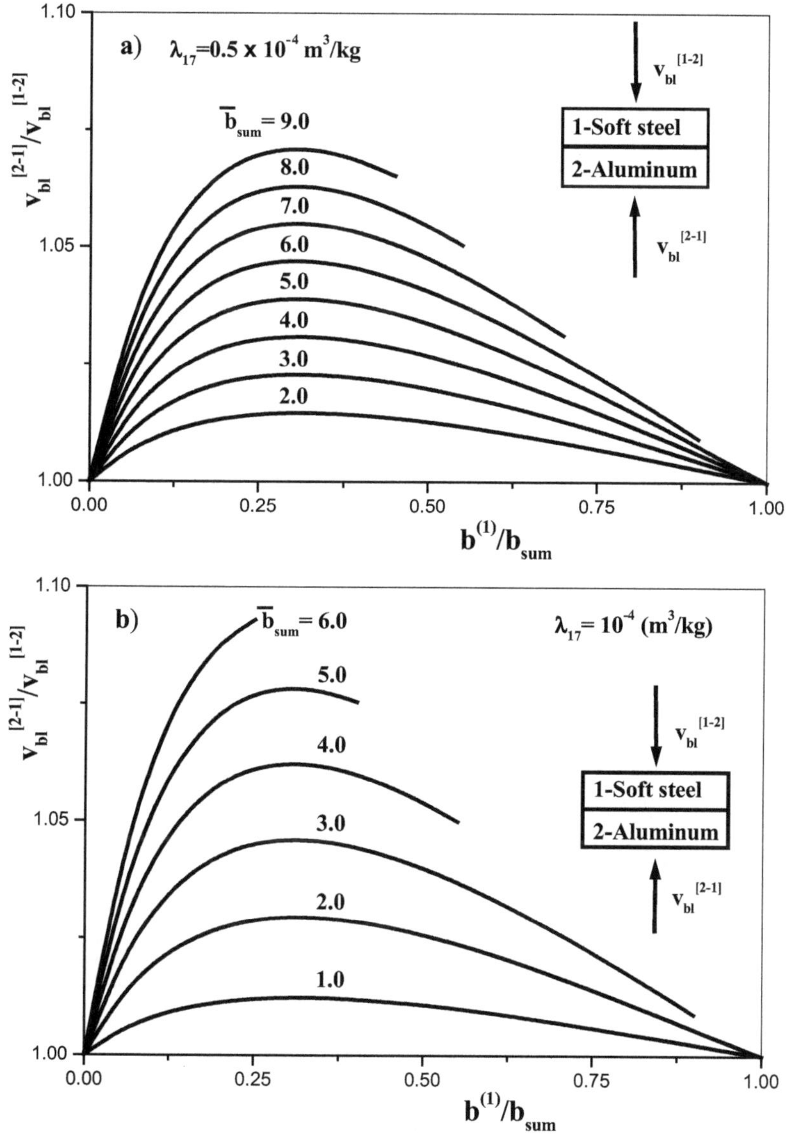
a) λ_{17}=0.5 x 10^{-4} m^3/kg
$\bar{b}_{sum}$= 9.0
8.0
7.0
6.0
5.0
4.0
3.0
2.0
1-Soft steel
2-Aluminum
$v_{bl}^{[1-2]}$
$v_{bl}^{[2-1]}$
$v_{bl}^{[2-1]}/v_{bl}^{[1-2]}$
$b^{(1)}/b_{sum}$
1.10
1.05
1.00
0.00
0.25
0.50
0.75
1.00
b) $\bar{b}_{sum}$= 6.0
5.0
4.0
3.0
2.0
1.0
λ_{17}= 10^{-4} (m^3/kg)
1-Soft steel
2-Aluminum
$v_{bl}^{[1-2]}$
$v_{bl}^{[2-1]}$
$v_{bl}^{[2-1]}/v_{bl}^{[1-2]}$
$b^{(1)}/b_{sum}$
1.10
1.05
1.00
0.00
0.25
0.50
0.75
1.00

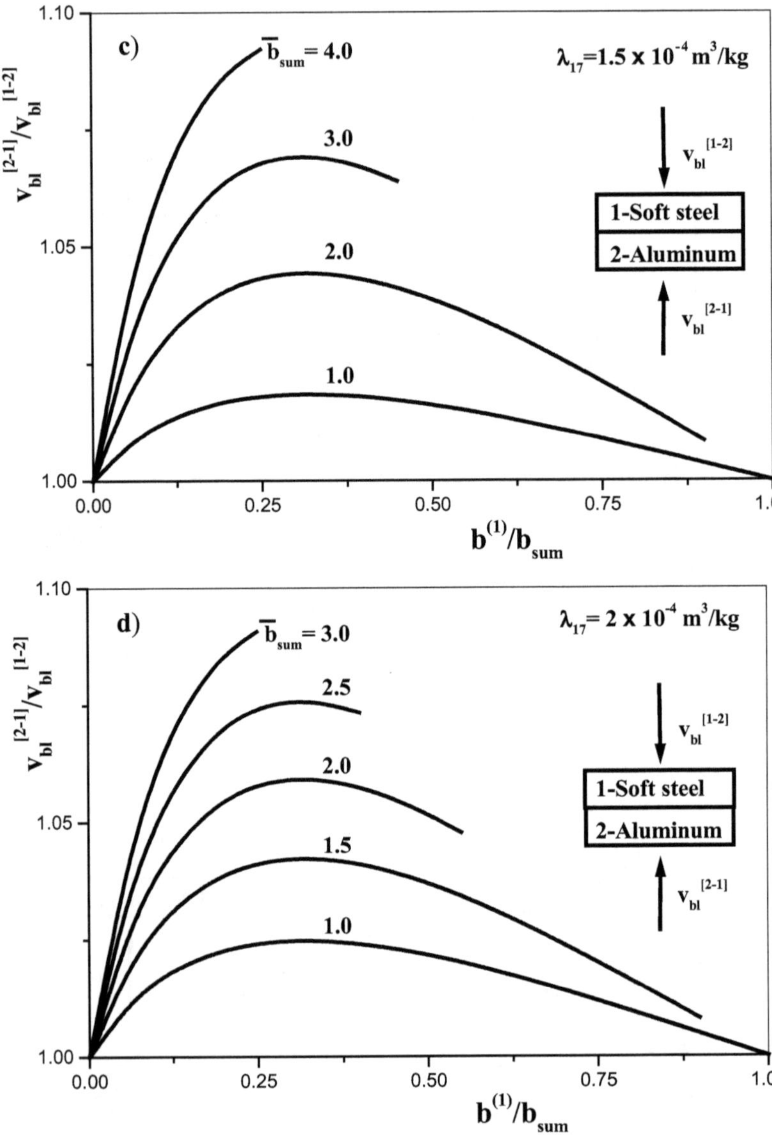

Figure 10-11. Efficiency of the optimum arrangement of plates in the shield. [1-2] and [2-1] denote the arrangements of the plates with the minimum and maximum BLV, respectively; $b^{(1)}$ is the thickness of the steel plate, $\bar{b}_{sum}$ is the dimensionless (normalized by L) thickness of the shield.

and

$$v_{bl}^{[3-2-1]} < v_{bl}^{[3-1-2]} < v_{bl}^{[1-3-2]} < v_{bl}^{[1-2-3]} . \quad (8.3)$$

Figure 10-14a-d shows the dependencies of parameter $\overline{v}_{bl}^{[...]} = v_{bl}^{[...]} / v_{bl}^{[3-2-1]}$ vs. the dimensionless (normalized by L) thickness of the shield, $\overline{b}_{sum}$, for different given ratios between the thicknesses of the plates. Here $v_{bl}^{[...]}$ denotes the BLV of the shield with the corresponding structure, while $v_{bl}^{[3-2-1]}$ is the BLV of the shield with the smallest BLV. In all calculations, $\lambda_{17} = 0.0001 m^3/kg$ was used. The relative locations of the curves are in accordance with the inequalities given by Eqs. (8.2) and (8.3), in which the order of the plates can have a strong affect on the BLV of the shield.

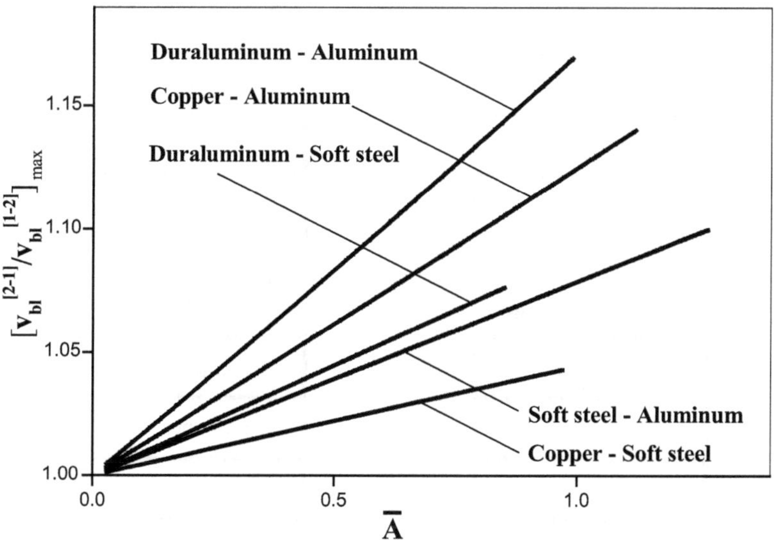

Figure 10-12. Maximum effect on the BLV achieved by changing the order of the plates vs. dimensionless areal density of the shield $\overline{A}$.

9. APPLICATION OF A CYLINDRICAL CAITY EXPANSION APPROXIMATION FOR STUDY OF SPACED SHIELDS

9.1 Model for shield with large air gaps

Consider the case in which an N-layered shield is perforated sequentially by an impactor having the shape of a cone of revolution, i.e., the impactor does not interact with two or more plates simultaneously. The latter requirement was used in *Section 10-5.3* in the analysis based on LIM. In this section we use the CCEA model given by Eq. (3-3.30). For a multi-layered shields, this model can be written as follows:

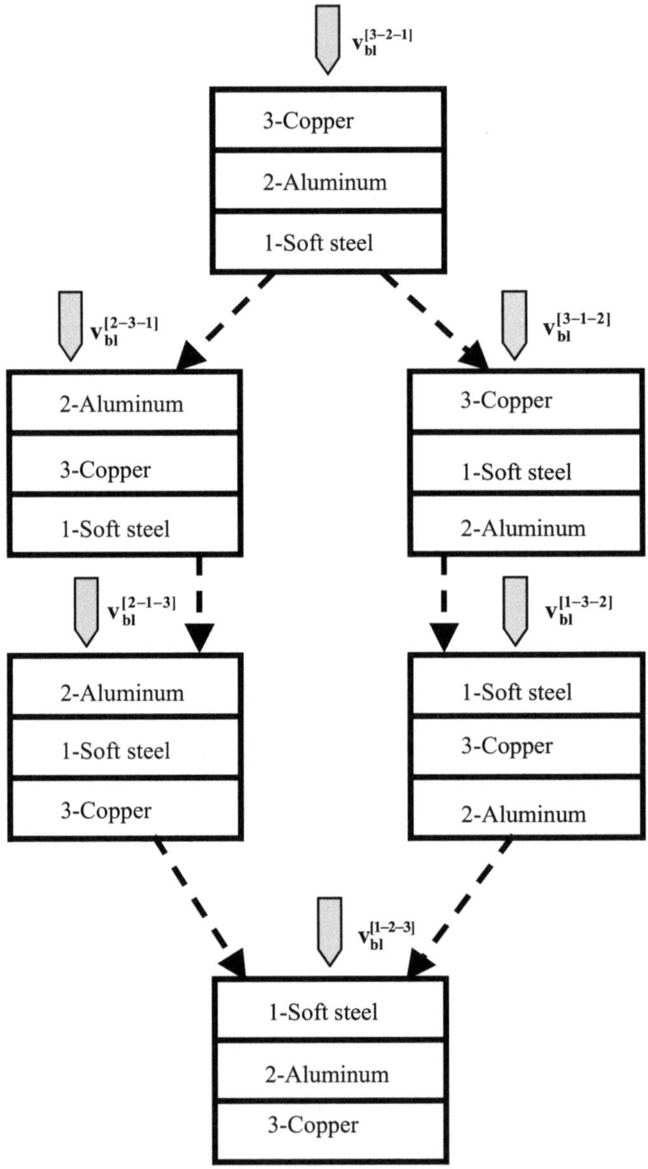

Figure 10-13. Structures of the shield in increasing order of the BLV.

$$p = a_2^{(j)} \dot{y}^2 + a_1^{(j)} y\ddot{y} + a_0^{(j)} \,. \tag{9.1}$$

For the j-th plate, Eqs. (3-4.19) can be rewritten as:

$$w_{res}^{(j)} = E^{(j)} w_{imp}^{(j)} - F^{(j)}, \tag{9.2}$$

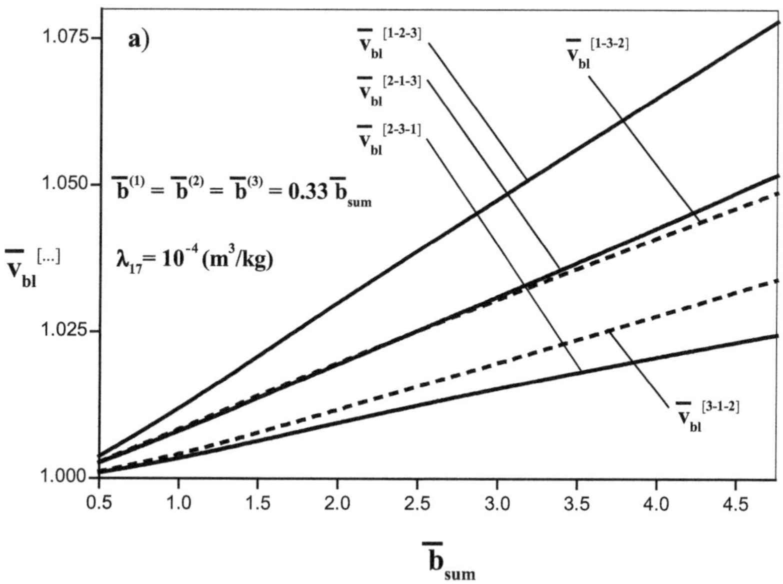

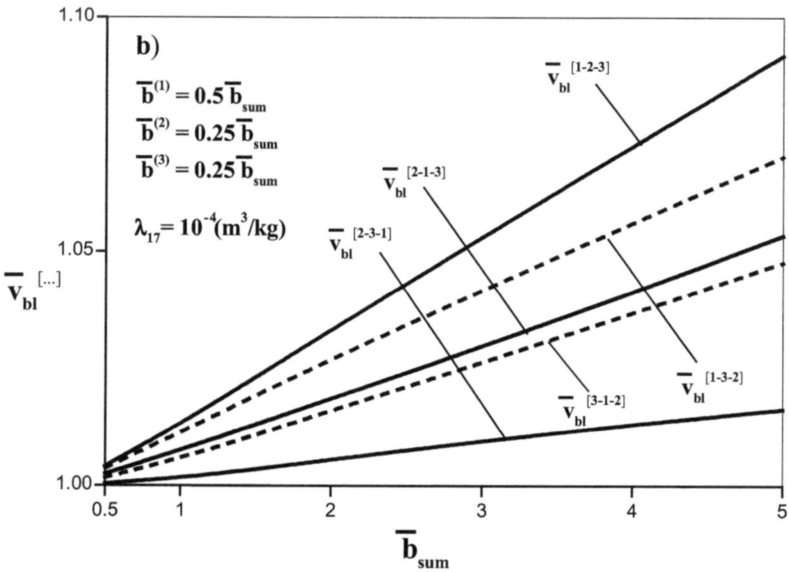

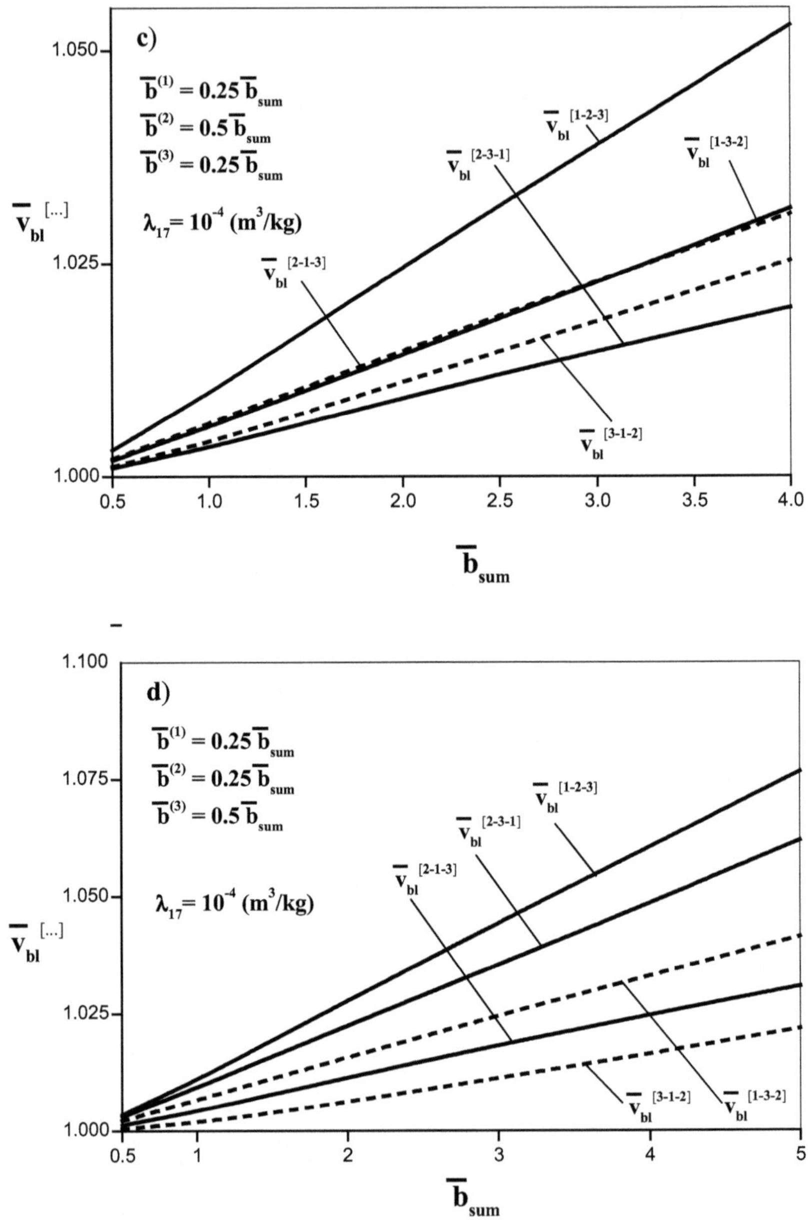

Figure 10-14. Effect of the order of the plates on the BLV of the shield.

where

$$E^{(j)} = \frac{1}{\hat{T}^{(j)}}, \quad F^{(j)} = \frac{\chi^{(j)}}{k^2}\left[1 - \frac{1}{T^{(j)}}\right], \quad \chi^{(j)} = \frac{a_0^{(j)}}{a_2^{(j)}} \tag{9.3}$$

and $\hat{T}^{(j)} > 1$ are determined by Eqs. (3-3.36), (3-4.13)-(3-4.14) and (3-4.20) written for the j-th plate.

Clearly, the model described by Eqs. (9.2)-(9.3) differs from the model determined by Eqs. (5.28), (5.29) and (5.33) only in terms of the coefficient in the formulas for $F^{(j)}$ and in that notation $\hat{T}^{(j)}$ is used instead of $T^{(j)}$. Therefore, we can write the relation for the BLV of the shield in a manner similar to Eq. (5.41):

$$k^2 w_{bl} = \sum_{j=1}^{N} \chi^{(j)} [\hat{T}^{(j)} - 1] \prod_{v=0}^{j-1} \hat{T}^{(v)}, \quad \hat{T}^{(0)} = 1. \tag{9.4}$$

Consequently, the following conclusions of *Section 10-5.3.3* remain valid in the considered case. For two adjacent plates arranged such that the value of the parameter χ for the first plate is larger than that for the second plate, the BLV of the shield can be increased by interchanging the plates. The maximum BLV of the shield is achieved when the plates are arranged in increasing order of parameter χ. The minimum BLV is obtained when the plates are arranged in the shield in the inverse order.

Since $\chi^{(j)}$ in the CCEA model given by Eq. (9.1) does not depend on the coefficient $a_1^{(j)}$ we conclude that the above-described ballistic properties of the shield are valid for an arbitrary choice of this coefficient.

If all $\chi^{(j)}$ are the same, i.e., all the plates are manufactured from the same material, then Eq. (9.4) implies:

$$\frac{1}{\lambda_0} w_{bl} = S_1 - S_2, \tag{9.5}$$

where

$$S_1 = \sum_{j=1}^{N} \prod_{v=0}^{j} \hat{T}^{(v)} = \sum_{j=1}^{N-1} \prod_{v=0}^{j} \hat{T}^{(v)} + \prod_{v=0}^{N} \hat{T}^{(v)}, \tag{9.6}$$

$$S_2 = \sum_{j=1}^{N} \prod_{v=0}^{j-1} \hat{T}^{(v)} = \sum_{j=2}^{N} \prod_{v=0}^{j-1} \hat{T}^{(v)} + 1 = \sum_{j=1}^{N-1} \prod_{v=0}^{j} \hat{T}^{(v)} + 1. \tag{9.7}$$

Consequently,

$$\frac{1}{\lambda_0} w_{bl} = \prod_{v=1}^{N} \hat{T}^{(v)} - 1 = exp\left[\lambda_{18} \sum_{j=1}^{N} G(\bar{b}^{(j)})\right] - 1, \tag{9.8}$$

where λ_0, λ_1 and function G are determined by Eqs. (3-4.16), (3-4.29) and (3-4.28), respectively,

$$\lambda_{18} = 3a_2\lambda_1^2 / a_1, \tag{9.9}$$

and $\bar{b}^{(j)} = b^{(j)}/L$ is the dimensionless thickness of the j-th plate. Hereafter, the argument λ_1 of the function G is not shown when a shield consists of plates manufactured from the same material.

Equation (9.8) shows that the BLV does not depend on the order of the plates in the multi-layered shield.

In *Section 10-4.1*, it is shown that LIM enables us to arrive at the following conclusion. When the plates in a spaced shield are manufactured from the same material, the BLV of the shield against a conical impactor depends only on the total thickness of the plates. In the case of the CCEA model, the BLV depends upon the widths of the air gaps and thicknesses of the plates in the shield.

9.2 Comparison of monolithic and spaced shields

9.2.1 Formulation of the problem

Let us compare the BLV of monolithic shields, $v_{bl\,mon} = \sqrt{w_{bl\,mon}}$, with the BLV of spaced shields, $v_{bl} = \sqrt{w_{bl}}$, using a model that is based on the CCEA and determined by Eq. (9.1). Assume that monolithic shield and N plates of the spaced shield are manufactured from the same material and the total thickness of the plates equals to the thickness of the monolithic shield, i.e.,

$$\sum_{j=1}^{N} \bar{b}^{(j)} = \bar{b}_{sum}. \tag{9.10}$$

We also assume that air gaps between the plates are wider than the length of the conical nose of the impactor.

For the monolithic shield, Eq. (9.8) can be written as:

$$\frac{1}{\lambda_0} w_{bl\,mon} = exp\left[\lambda_{18} G(\overline{b}_{sum})\right] - 1 . \tag{9.11}$$

Equations (9.8) and (9.11) yield:

$$w_{bl} - w_{bl\,mon} = \lambda_0 \, exp\left[\lambda_{18} G(\overline{b}_{sum})\right] \cdot \left[exp(\lambda_{18} \Gamma) - 1\right] , \tag{9.12}$$

where

$$\Gamma = \Gamma(\overline{b}^{(1)}, \overline{b}^{(2)}, ..., \overline{b}^{(N)}) = \sum_{j=1}^{N} G(\overline{b}^{(j)}) - G\left(\sum_{j=1}^{N} \overline{b}^{(j)}\right) . \tag{9.13}$$

Clearly,

$$sign(w_{bl} - w_{bl\,mon}) = sign(\Gamma) . \tag{9.14}$$

Therefore, provided that Eq (9.10) is satisfied, function $\Gamma(\overline{b}^{(1)}, \overline{b}^{(2)}, ..., \overline{b}^{(N)})$ characterizes the differences between the BLV of the monolithic and spaced shields. In the following, we investigate the behavior and perform minimization of this function. In *Section 10-9.2*, we show that the spaced shield has superior ballistic characteristics than a monolithic shield with the same total thickness of the plates. In *Section 10-9.3* we determine the best partition of a monolithic shield into plates such that the difference between the ballistic performances of monolithic and spaced shields is maximum.

9.2.2 Some particular cases

Before proceeding to the general case, we consider two particular cases.

Let all $\overline{b}^{(j)} \geq 1$ ($b^{(j)} \geq L$). Then $\overline{b}_{sum} \geq 1$ ($b_{sum} \geq L$), $G(z) = \Psi(\lambda_1) + \lambda_1(z-1)$ in the considered range of variation of the argument, and:

$$\Gamma=\sum_{j=1}^{N}\left[\Psi(\lambda_1)+\lambda_1(\bar{b}^{(j)}-1)\right]-\Psi(\lambda_1)-\lambda_1\left[\sum_{j=1}^{N}\bar{b}^{(j)}-1\right]$$

$$=(N-1)[\Psi(\lambda_1)-\lambda_1]=(N-1)\left[\int_0^{\lambda_1}\frac{dz}{1-z^3}-\int_0^{\lambda_1}dz\right] \tag{9.15}$$

$$=(N-1)\int_0^{\lambda_1}\frac{z^3dz}{1-z^3}>0.$$

Consider the second limiting case when all $\bar{b}^{(j)}\le 1$ and $\bar{b}_{sum}\le 1$. Then $G(z)=\Psi(\lambda_1)-\Psi(\lambda_1(1-z))$. Using the identity

$$\Psi(\lambda_1(1-z))=\Psi(\lambda_1)+\int_{\lambda_1}^{\lambda_1(1-z)}\frac{dt}{1-t^3} \tag{9.16}$$

for $z=\bar{b}^{(j)}$ and $z=\sum_{j=1}^{N}\bar{b}^{(j)}$ we obtain:

$$\Gamma=\sum_{j=1}^{N}\int_{\lambda_1(1-\bar{b}^{(j)})}^{\lambda_1}\frac{dt}{1-t^3}-\int_{\lambda_1(1-\bar{\xi}^{(N)})}^{\lambda_1}\frac{dt}{1-t^3}, \tag{9.17}$$

where

$$\bar{\xi}^{(j)}=\sum_{\nu=0}^{j}\bar{b}^{(\nu)},\quad \bar{\xi}^{(0)}=\bar{b}^{(0)}=0,\quad j=0\div N. \tag{9.18}$$

The second integral in Eq. (9.17) can be written as:

$$\int_{\lambda_1(1-\bar{\xi}^{(N)})}^{\lambda_1}\frac{dt}{1-t^3}=\sum_{j=0}^{N}\int_{\lambda_1(1-\bar{\xi}^{(j)})}^{\lambda_1(1-\bar{\xi}^{(j-1)})}\frac{dt}{1-t^3}=\sum_{j=0}^{N}\int_{\lambda_1(1-\bar{b}^{(j)})}^{\lambda_1}\frac{d\tilde{t}}{1-\left(\tilde{t}-\lambda_1\bar{\xi}^{(j-1)}\right)^3}. \tag{9.19}$$

In Eq. (9.19) the j-th term in the last sum is obtained from the corresponding term in the previous sum by changing the variables, $t=\tilde{t}-\lambda_1\bar{\xi}^{(j-1)}$. Substituting Eq. (9.19) into Eq. (9.17), we obtain after simple algebra:

$$\Gamma = \lambda_1 \sum_{j=1}^{N} \overline{\xi}^{(j-1)} \int_{\lambda_1(1-\overline{b}^{(j)})}^{\lambda_1} \frac{t^2 + t\left(t - \lambda_1 \overline{\xi}^{(j-1)}\right) + \left(t - \lambda_1 \overline{\xi}^{(j-1)}\right)^2}{(1-t^3)\left[1 - \left(t - \lambda_1 \overline{\xi}^{(j-1)}\right)^3\right]} dt \,. \quad (9.20)$$

Since the variable t varies in the range from $\lambda_1\left(1-\overline{b}^{(j)}\right)$ to λ_1,

$$\lambda_1\left(1-\overline{\xi}^{(j)}\right) \le t - \lambda_1 \overline{\xi}^{(j-1)} \le \lambda_1\left(1-\overline{\xi}^{(j-1)}\right). \quad (9.21)$$

Taking into account that $\lambda_1 < 1$ and $0 < \overline{\xi}^{(\nu)} \le \overline{b}_{sum} \le 1$ for $\nu = 1 \div N$, Eq. (9.21) implies that $0 \le t - \lambda_1 \overline{\xi}^{(j-1)} < 1$. Therefore, each integral in Eq. (9.20) is positive, and $\Gamma > 0$.

We have proved that in the two examples under consideration a spaced shield is superior to a monolithic shield.

The remaining cases may by analyzed in a similar manner. However, we will use another, more general approach.

9.2.3 General case

Using Eqs. (3-4.26) and (3-4.28), we can calculate the derivatives of the function G over z:

$$G'(z) = \begin{cases} \dfrac{\lambda_1}{1-(\lambda_1(1-z))^3} & if \quad z \le 1 \\ \lambda_1 & if \quad z \ge 1 \end{cases}, \quad (9.22)$$

$$G''(z) = \begin{cases} -\dfrac{3\lambda_1^4(1-z)^2}{\left[1-(\lambda_1(1-z))^3\right]^2} & if \quad z \le 1 \\ 0 & if \quad z \ge 1 \end{cases}. \quad (9.23)$$

Inspection of formulas for $G(z)$, $G'(z)$ and $G''(z)$ shows that these functions are continuous at point $z = 1$. Therefore, it can be concluded that continuous function $G(z)$ has continuous first and second derivatives. Consequently, $\Gamma(\overline{b}^{(1)}, \overline{b}^{(2)}, ..., \overline{b}^{(N)})$ is a sufficiently smooth function that can be expanded in a Taylor series:

$$\Gamma(\overline{b}^{(1)}, \overline{b}^{(2)}, ..., \overline{b}^{(N)}) = \Gamma(0,0,...,0) + \sum_{\nu=1}^{N} \Gamma_\nu(\kappa\overline{b}^{(1)}, \kappa\overline{b}^{(2)}, ..., \kappa\overline{b}^{(N)})\overline{b}^{(\nu)}, \quad (9.24)$$

where $0<\kappa<1$ and:

$$\Gamma_\nu(\bar{b}^{(1)},\bar{b}^{(2)},...,\bar{b}^{(N)})=\frac{\partial\Gamma}{\partial\bar{b}^{(\nu)}}=G'(\bar{b}^{(\nu)})-G'\left(\sum_{j=1}^{N}\bar{b}^{(j)}\right). \quad (9.25)$$

Note that in a Taylor series (Eq. 9.24) the remainder term is written in the Lagrange form (Korn and Korn, 1968). Equation (9.23) implies that $G''(z),\leq 0$ i.e., $G'(z)$ is non-increasing function. Since $\bar{b}^{(\nu)}<\bar{b}_{sum}=\bar{b}^{(1)}+\bar{b}^{(2)}+...+\bar{b}^{(N)}$, then $G'(\kappa\bar{b}^{(\nu)})-G'(\kappa\bar{b}_{sum})\geq 0$ and

$$\Gamma_\nu(\kappa\bar{b}^{(1)},\kappa\bar{b}^{(2)},...,\kappa\bar{b}^{(N)})\geq 0\,. \quad (9.26)$$

Since $\Gamma(0,0,...,0)=0$ and $\bar{b}^{(j)}>0$, Eq. (9.26) allows us to conclude that:

$$\Gamma(\bar{b}^{(1)},\bar{b}^{(2)},...,\bar{b}^{(N)})\geq 0\,. \quad (9.27)$$

Let us now prove that equality in Eq. (9.27) is not possible. Assume that $\Gamma(\bar{b}_*^{(1)},\bar{b}_*^{(2)},...,\bar{b}_*^{(N)})=0$ for any positive $\bar{b}_*^{(1)},\bar{b}_*^{(2)},...,\bar{b}^{(N)}$. Then Eq. (9.24) with $\Gamma(0,0,...,0)=0$ implies that all Γ_ν $(\kappa b^{(\bar{1})}$, $\kappa\bar{b}^{(2)}$,..., $\kappa\bar{b}^{(N)})=0$, i.e., $G'(\kappa\bar{b}_*^{(\nu)})=G'(\kappa\bar{b}_*)$, where $\bar{b}_*=\bar{b}_*^{(1)}+\bar{b}_*^{(2)}+...+\bar{b}_*^{(N)}$. Since $\kappa\bar{b}_*^{(\nu)}<\kappa\bar{b}_*$, the latter condition can be valid only if

$$\kappa\bar{b}_*^{(\nu)}\geq 1,\quad \nu=1,...,N\,, \quad (9.28)$$

when $G'=const.$ Clearly, Eq. (9.28) implies that $\bar{b}_*^{(\nu)}>1$. This case was considered in the previous section, and it was proved that $\Gamma>0$. The latter result is in conflict with the assumption that equality in Eq. (9.27) can be attained. Consequently,

$$\Gamma(\bar{b}^{(1)},\bar{b}^{(2)},...,\bar{b}^{(N)})>0 \quad (9.29)$$

for arbitrary set of $\bar{b}^{(1)},\bar{b}^{(2)},...,\bar{b}^{(N)}$.

Therefore, a spaced shield with large air gaps performs better than a monolithic shield with the same total thickness.

9.3 Optimization of spaced shields

9.3.1 Solution for a given number of plates

Consider the following problem: how can a monolithic shield be split into plates with large air gaps so as to achieve maximum BLV of the spaced shield. We assume here that the number of plates is given. Equation (9.8) shows that the problem reduces to maximization of the following function:

$$\tilde{\Lambda}(\overline{b}^{(1)},\overline{b}^{(2)},...,\overline{b}^{(N)})=\sum_{j=1}^{N} G(\overline{b}^{(j)})\,. \tag{9.30}$$

Minimization is performed under the constraints:

$$\sum_{j=1}^{N} \overline{b}^{(j)}=\overline{b}_{sum}, \quad \overline{b}^{(i)}>0, \quad i=1,...,N\,, \tag{9.31}$$

where the total thickness of the plates $\overline{b}_{sum}$ is known. In the following exposition, the equivalent formulation of this problem is used:

$$\Lambda(\overline{b}^{(1)},...,\overline{b}^{(N-1)})=\sum_{j=1}^{N-1} G(\overline{b}^{(j)})+G\left(\overline{b}_{sum}-\sum_{j=1}^{N-1}\overline{b}^{(j)}\right)\to max \tag{9.32}$$

subjected to the constraint:

$$\sum_{j=1}^{N-1} \overline{b}^{(j)}<\overline{b}_{sum}, \quad \overline{b}^{(i)}>0, \ i=1,...,N-1\,. \tag{9.33}$$

Let us investigate the behavior of the function Λ. Since the first and second derivatives of G exist, Λ is a sufficiently smooth function. The equations for the first and second derivatives of Λ read:

$$H^{(\nu)}\equiv\frac{\partial\Lambda}{\partial\overline{b}^{(\nu)}}=G'(\overline{b}^{(\nu)})-G'\left(\overline{b}_{sum}-\sum_{j=1}^{N-1}\overline{b}^{(j)}\right), \tag{9.34}$$

$$H^{(\nu,\mu)}\equiv\frac{\partial^2\Lambda}{\partial\overline{b}^{(\nu)}\partial\overline{b}^{(\mu)}}=\delta_{\nu\mu}G''(\overline{b}^{(\nu)})+G''\left(\overline{b}_{sum}-\sum_{j=1}^{N-1}\overline{b}^{(j)}\right), \tag{9.35}$$

where $\nu = 1,\ldots,N-1$, $\mu = 1,\ldots,N-1$ and

$$\delta_{\nu\mu} = \begin{cases} 1 & if \quad \nu = \mu \\ 0 & if \quad \nu \neq \mu \end{cases}. \tag{9.36}$$

The *Hessian* (the matrix with the elements $H^{(\nu\mu)}$) plays the role of the second derivative of Λ, considered as a function of many variables (Korn and Korn, 1968). Properties of the Hessian are associated with the convexity of function Λ, and they depend on the signs of the roots of the characteristic equation:

$$\begin{vmatrix} H^{(1,1)}-\eta & H^{(1,2)} & \ldots & H^{(1,N-1)} \\ H^{(2,1)} & H^{(2,2)}-\eta & \ldots & H^{(2,N-1)} \\ \ldots & \ldots & \ldots & \ldots \\ H^{(N-1,1)} & H^{(N-1,2)} & \ldots & H^{(N-1,N-1)}-\eta \end{vmatrix} = 0. \tag{9.37}$$

Since the last term in Eq. (9.35) is the same for all $H^{(\nu,\mu)}$, this term can be dropped out in the determinant in Eq. (9.37). Nonzero elements remain only at the diagonal, and the characteristic equation reads:

$$\prod_{j=1}^{N-1}\left[G''(\bar{b}^{(\nu)})-\eta\right]=0. \tag{9.38}$$

The domain determined by linear inequalities in Eq. (9.33) is convex. Since $G''(z) \leq 0$ in this domain, all roots of Eq. (9.38) are nonpositive, and hence the Hessian is negative semidefinite. In classical mathematical analysis such functions are known as non-concave functions. Note that in mathematical programming such functions are defined as "concave functions".

Let us prove that the considered problem always has the unique solution if:

$$\bar{b}_{sum} \leq N. \tag{9.39}$$

Using Eq. (9.34), we rewrite the conditions $H^{(\nu)} = 0\ (\nu = 1,\ldots,N-1)$ in the following form:

$$G'(\bar{b}^{(\nu)}) = G'(\bar{b}^{(1)}), \quad \nu = 2,\ldots,N-1, \tag{9.40}$$

$$G'\left(\bar{b}_{sum} - \sum_{j=1}^{N-1} \bar{b}^{(j)}\right) = G'(\bar{b}^{(1)}) . \tag{9.41}$$

Assume that there exists a solution of Eqs. (9.40) and (9.41 with:

$$\bar{b}^{(1)} \geq 1 . \tag{9.42}$$

Since $G'(z)$ decreases in the interval $0 < z < 1$ and is constant for $z \geq 1$, Eqs. (9.40) and (9.41) yield, respectively:

$$\bar{b}^{(\nu)} > 1, \quad \nu = 2,...,N-1, \tag{9.43}$$

$$\bar{b}_{sum} - \sum_{j=1}^{N-1} \bar{b}^{(j)} > 1 . \tag{9.44}$$

Summation of Eqs. (9.42)-(9.44) yields:

$$\bar{b}_{sum} > N , \tag{9.45}$$

which is in conflict with Eq. (9.39).

The alternative assumption, $\bar{b}^{(1)} < 1$, implies the unique solution of Eqs. (9.40) and (9.41):

$$\bar{b}^{(\nu)} = \bar{b}_*^{(\nu)} = \bar{b}_{sum} / N , \quad \nu = 1 \div N . \tag{9.46}$$

Since $\bar{b}_*^{(\nu)} < 1$, then $G''(\bar{b}_*^{(\nu)}) < 0$, Eq. (9.38) has only negative roots and the Hessian is negative definite. Therefore, Λ is strictly convex at the point $\bar{b}^{(\nu)} = \bar{b}_*^{(\nu)}$, and this point is the point of a local maximum. In convex programming (the considered problem belongs to the class of problems considered in convex programming), a local maximum (if it does exist) is also a global maximum (see, e.g., Rockafellar, 1970). Therefore, it can be concluded that the global maximum of function Λ is attained at the point determined by Eq. (9.46), i.e., when the thicknesses of the plates are the same. The equation for the maximum value Λ_* of function Λ reads:

$$\Lambda_* = N\left[\Psi(\lambda_1) - \Psi\left(\lambda_1(1 - \frac{\bar{b}_{sum}}{N})\right)\right] . \tag{9.47}$$

Let us consider the case when

$$\bar{b}_{sum} > N\,. \tag{9.48}$$

Then there exists a non-empty sub-domain of the domain determined by Eq. (9.33) for which Eqs. (9.42)-(9.44) are valid. To prove the latter claim, it is sufficient to find such a sub-domain. It can be verified directly that the following sub-domain

$$1 < \bar{b}^{(\nu)} \le \bar{b}_{sum} / N\,, \quad \nu = 1,\dots,N-1 \tag{9.49}$$

satisfies Eqs. (9.33) and Eqs. (9.42)-(9.44). The structure of the sub-domain (denoted by E) determined by these equations is shown in *Figure 10-15a-b* for $N = 2$ and $N = 3$.

Since $G(z) = \Psi(\lambda_1) + \lambda_1(z-1)$ for $z > 1$, the magnitude of function Λ at the points that belong to E is given by the following formula:

$$\Lambda_* = N[\Psi(\lambda_1) - \lambda_1] + \lambda_1 \bar{b}_{sum}\,, \tag{9.50}$$

i.e., Λ_* is constant.

The sub-domain E is a connected convex domain because it is determined by linear inequalities. Clearly, if the non-concave function Λ takes the constant value in E, then this value is the maximum of this function.

Therefore, if $\bar{b}_{sum} > N$, then there exists a sub-domain in the space of variables $\bar{b}^{(\nu)} > 1$ that is determined by Eq. (9.33). In this sub-domain the BLV attains its maximum value Λ_* that is determined by Eq. (9.50). In particular, the variables determined by Eq. (9.46) belong to this domain.

9.3.2 Influence of the number of plates on the optimal solution

The solution given by Eqs. (9.47) and (9.50) can be represented in the following form:

$$\Lambda_* = \begin{cases} N[\Psi(\lambda_1) - \lambda_1] + \lambda_1 \bar{b}_{sum} & if \quad N < \bar{b}_{sum} \\ N\left[\Psi(\lambda_1) - \Psi\left(\lambda_1 (1 - \dfrac{\bar{b}_{sum}}{N})\right)\right] & if \quad N \ge \bar{b}_{sum} \end{cases}. \tag{9.51}$$

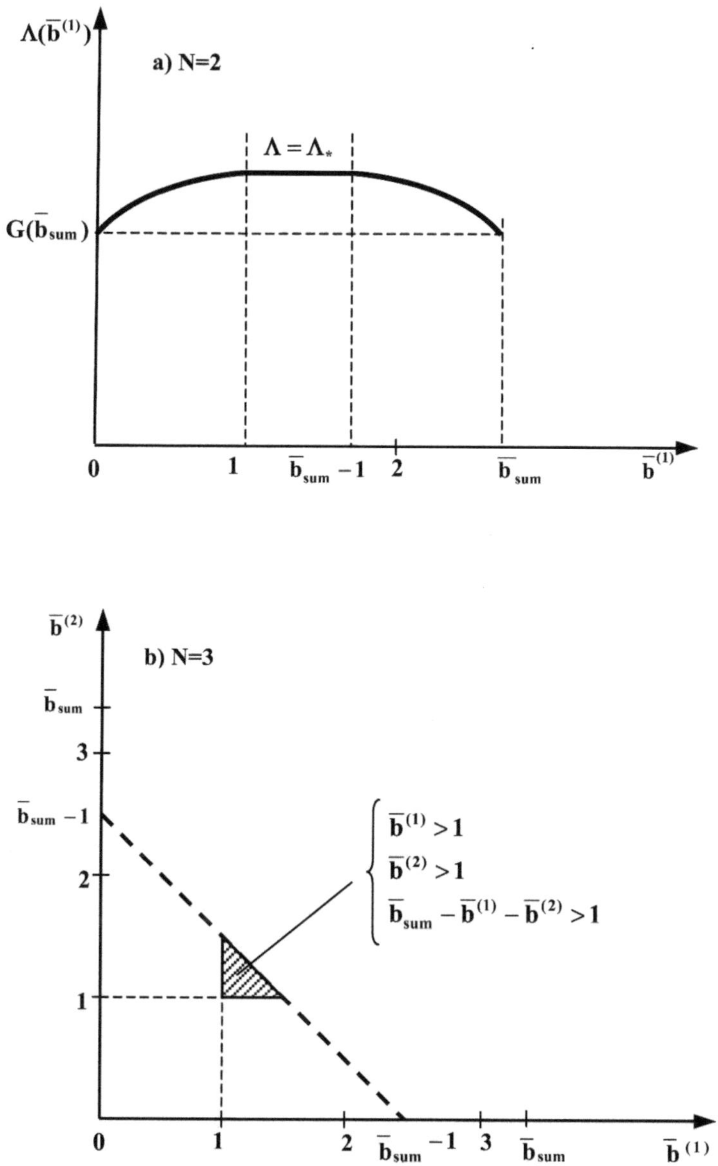

Figure 10-15. Sub-domain determined by Eqs. (9.33) and (9.42)-(9.44).

Assuming that N varies continuously, let us study the behavior of Λ_* as a function of N. Let us calculate the first derivative:

$$\frac{d\Lambda_*}{dN} = \begin{cases} \Psi(\lambda_1) - \lambda_1 & \text{if} \quad N < \bar{b}_{sum} \\ \Psi(\lambda_1) - \Psi\left(\lambda_1 - \frac{\lambda_1 \bar{b}_{sum}}{N}\right) - \frac{\lambda_1 \bar{b}_{sum}}{N} \Psi'\left(\lambda_1 - \frac{\lambda_1 \bar{b}_{sum}}{N}\right) & \text{if} \quad N \geq \bar{b}_{sum} \end{cases} \tag{9.52}$$

Since

$$\Psi(\lambda_1) - \lambda_1 = \int_0^{\lambda_1} \frac{z^3 dz}{1 - z^3} > 0, \tag{9.53}$$

$d\Lambda_*/dN > 0$ for $N < \bar{b}_{sum}$. The expression $d\Lambda_*/dN$ for $N \geq \bar{b}_{sum}$ can be transformed by using Lagrange theorem (Korn and Korn, 1968) twice. In the first step we obtain:

$$\Psi(\lambda_1) - \Psi\left(\lambda_1 - \frac{\lambda_1 \bar{b}_{sum}}{N}\right) = \frac{\lambda_1 \bar{b}_{sum}}{N} \Psi'\left(\lambda_1 - \frac{\lambda_1 \bar{b}_{sum}}{N_*}\right), \quad N_* > N, \tag{9.54}$$

and finally:

$$\begin{aligned} \frac{d\Lambda_*}{dN} &= \frac{\lambda_1 \bar{b}_{sum}}{N} \left[\Psi'\left(\lambda_1 - \frac{\lambda_1 \bar{b}_{sum}}{N_*}\right) - \Psi'\left(\lambda_1 - \frac{\lambda_1 \bar{b}_{sum}}{N}\right) \right] \\ &= \left(\frac{\lambda_1 \bar{b}_{sum}}{N}\right)^2 \frac{N_* - N}{N_*} \Psi''\left(\lambda_1 - \frac{\lambda_1 \bar{b}_{sum}}{N_{**}}\right), \quad N < N_{**} < N_*. \end{aligned} \tag{9.55}$$

Since $\Psi''(z) = 3z^2/(1 - z^3)^2 > 0$, $d\Lambda_*/dy < 0$ for $N \geq \bar{b}_{sum}$. Therefore, Λ_* is increasing function of variable N.

Using Eqs. (9.51) and (9.54), we can calculate the limit:

$$\begin{aligned} \lim_{N \to \infty} \Lambda_* &= \lim_{N \to \infty} N \left[\Psi(\lambda_1) - \Psi\left(\lambda_1 - \frac{\bar{b}_{sum} \lambda_1}{N}\right) \right] \\ &= \bar{b}_{sum} \lambda_1 \lim_{N_* \to \infty} \Psi'\left(\lambda_1 - \frac{\lambda_1 \bar{b}_{sum}}{N_*}\right) = \bar{b}_{sum} \lambda_1 \Psi'(\lambda_1) = \frac{\lambda_1 \bar{b}_{sum}}{1 - \lambda_1^3}. \end{aligned} \tag{9.56}$$

Hence, the BLV increases if the number of plates with the same thicknesses in the shield increases and tends to a finite value when the number of plates approaches infinity.

Let us investigate the behavior of the BLV for large N. The BLV v_{bl} is determined by the following equation:

$$\frac{1}{\lambda_0} v_{bl}^2 = exp\left(\frac{3a_2\lambda_1^2}{a_1}\Lambda_*\right) - 1, \qquad (9.57)$$

where Λ_* is determined by Eq. (9.47). Differentiating both the sides of this equation over N, we obtain:

$$\frac{2}{\lambda_0} v_{bl} \frac{dv_{bl}}{dN} = \frac{3a_2\lambda_1^2}{a_1} exp\left(\frac{3a_2\lambda_1^2}{a_1}\Lambda_*\right)\frac{d\Lambda_*}{dN}. \qquad (9.58)$$

Equation (9.55) implies that:

$$\lim_{N\to\infty} \frac{d\Lambda_*}{dN} = 0. \qquad (9.59)$$

Using Eq. (9.58), we find that $dv_{bl}/dN \to 0$ when $N \to \infty$. Therefore, it can be expected that the BLV depends only weakly on the number of plates for large N.

9.4 Model for a shield with arbitrary air gaps

In the general case, we can construct a model similar to that considered in *Section 10-1*. For a sharp impactor having the shape of a body of revolution, Eq. (3-3.31) can be rewritten as:

$$\sigma_{n\ plast} = \varepsilon(h-x)[(a_2\Phi'^2 + a_1\Phi\Phi'')\dot{h}^2 + a_1\Phi\Phi'\ddot{h} + a_0], \qquad (9.60)$$

where function $\varepsilon(\xi)$ is determined by Eq. (4.2) and the notations are shown in *Figure 10-5*. Since the sequential numbers of all layers (plates and gaps) in the shield correspond to their location in the shield, plates have odd numbers while gaps have even numbers. Parameter μ in *Figure 10-5* denotes the sequential number of pair that comprises the plate with the sequential number $2\mu - 1$ and the gap with number 2μ (except for the last plate).

Substituting $\sigma_{n\ plast}$ from Eq. (9.60) into Eq. (3-3.28), we obtain Eq. (3-3.19) with

$$g_0(h) = 2\pi a_0 \int_{\theta(h)}^{\Theta(h)} \varepsilon(h-x)\Phi\Phi' dx\,, \tag{9.61}$$

$$g_1(h) = 2\pi\left[a_2 \int_{\theta(h)}^{\Theta(h)} \varepsilon(h-x)\Phi\Phi'^3 dx + a_1 \int_{\theta(h)}^{\Theta(h)} \varepsilon(h-x)\Phi\Phi'\Phi'' dx\right], \tag{9.62}$$

$$g_2(h) = \pi a_1 \int_{\theta(h)}^{\Theta(h)} \varepsilon(h-x)\Phi^2\Phi'^2 dx\,. \tag{9.63}$$

The BLV is determined by Eq. (3-3.25), where $\alpha(h)$ and $\beta(h)$ are determined by Eq. (3-3.36) and must be calculated using Eqs. (9.61)-(9.63).

In the case of a *sharp cone of revolution* when $\Phi(x) = kx$, Eqs. (9.61)-(9.63) can be written as follows:

$$g_0(h) = 2\pi a_0 k^2 \hat{J}_1, \quad g_1(h) = 2\pi a_2 k^4 \hat{J}_1, \quad g_2(h) = \pi a_1 k^4 \hat{J}_2\,, \tag{9.64}$$

where

$$\hat{J}_\nu = \int_{\theta(h)}^{\Theta(h)} x^\nu \varepsilon(h-x) dx, \quad \nu = 1,2\,. \tag{9.65}$$

To present $\hat{J}_\nu$ in a more convenient form, we use the identity given by Eq. (2.15). Replacing $a_i(h-x)$ by $\varepsilon(h-x)$, we can rewrite this identity for a spaced shield (see *Figure 10-5*):

$$\begin{aligned}&\int_{\theta(h)}^{\Theta(h)} \Xi(x) g(h-x) dx \\ &= \sum_{\mu=1}^{\mu_{max}} \varepsilon^{(2\mu-1)} \int_{x_*\left(h-\xi^{(2\mu-1)}\right)}^{x_*\left(h-\xi^{(2\mu-2)}\right)} \Xi(x) dx + \sum_{\mu=1}^{\mu_{max}-1} \varepsilon^{(2\mu)} \int_{x_*\left(h-\xi^{(2\mu)}\right)}^{x_*\left(h-\xi^{(2\mu-1)}\right)} \Xi(x) dx\end{aligned} \tag{9.66}$$

Taking into account Eq. (4.2) for $\Xi(x) = x^\nu$, we obtain:

$$\hat{\bar{J}}_\nu = \frac{\hat{J}_\nu}{L^{\nu+1}} = \frac{1}{L^{\nu+1}} \sum_{\mu=1}^{\mu_{max}} \int_{x_*\left(h-\xi^{(2\mu-1)}\right)}^{x_*\left(h-\xi^{(2\mu-2)}\right)} x^\nu \, dx$$
$$= \frac{1}{\nu+1} \sum_{\mu=1}^{\mu_{max}} \left[X_\Diamond^{\nu+1}\left(\bar{h} - \bar{\xi}^{(2\mu-2)}\right) - X_\Diamond^{\nu+1}\left(\bar{h} - \bar{\xi}^{(2\mu-1)}\right) \right], \tag{9.67}$$

where $\bar{\xi}^{(\nu)} = \xi^{(\nu)}/L, \quad \bar{h} = h/L$.

Using transformations similar to those described in *Section 8-2.3.2*, we obtain an equation for the BLV that coincides with Eqs. (3-4.19), (3-4.21) and (3-4.30), where

$$\bar{I} = 2 \int_0^{\bar{b}+1} \frac{\hat{\bar{J}}_1 d\bar{h}}{\bar{m}_{imp} + 3\hat{\bar{J}}_2} \, . \tag{9.68}$$

It is easy to show that when $b^{(2\mu)} = \xi^{(2\mu)} - \xi^{(2\mu-1)} \geq L$ for $\mu = 1 \div \mu_{max} - 1$, Eq. (9.68) yields the following expression:

$$\bar{I} = \lambda_1^2 \sum_{\mu=1}^{\mu_{max}} G(\bar{b}^{(2\mu-1)}) \, . \tag{9.69}$$

Let us show that the spacing has only a minor effect on the BLV for large $\bar{m}_{imp}$. Eq. (9.65) yields the following estimate:

$$\hat{\bar{J}}_2(\bar{h}) = \frac{\hat{J}_2}{L^3} \leq \int_{\theta(h)}^{\Theta(h)} x^2 dx = \frac{1}{3L^3}\left[\Theta^3(h) - \theta^3(h)\right] \leq \frac{1}{3} \, . \tag{9.70}$$

Equations (9.68) and (9.70) show that when

$$\bar{m}_{imp} >> 1 \tag{9.71}$$

then $\bar{m}_{imp} >> 3\hat{\bar{J}}_2$ and:

$$\bar{I} \approx \frac{1}{\bar{m}_{imp}} \int_0^{\bar{b}+1} 2\bar{J}_1 d\bar{h} = \frac{1}{\bar{m}_{imp}} \sum_{\mu=1}^{\mu_{max}} \int_0^1 d\bar{x}\bar{x} \int_{\bar{x}+\bar{\xi}^{(2\mu-2)}}^{\bar{x}+\bar{\xi}^{(2\mu-1)}} d\bar{h} = \frac{\bar{b}_{sum}}{2\bar{m}_{imp}} \, , \tag{9.72}$$

i.e., $\hat{T}$ in Eq. (3-4.30) becomes:

$$\hat{T} \approx exp\left(\frac{\pi a_2 b_{sum} k^4 L^2}{m}\right), \tag{9.73}$$

where $\overline{b}_{sum}$ is the sum of the thicknesses of the plates normalized by L. Thus, we have an equation for the BLV of a monolithic shield with the same total thickness.

It is often convenient to use the following alternative notations (see *Figure 10-16*):

$$b^{(2\mu-1)} \rightarrow b^{(\mu)}, \mu = 1 \div N, \quad b^{(2\mu)} \rightarrow \Delta^{(\mu)}, \mu = 1 \div N-1, \tag{9.74}$$

$$\xi^{(2\mu-1)} \rightarrow \xi^{(\mu)}, \quad \xi^{(2\mu)} \rightarrow \xi^{(\mu+1)} - b^{(\mu+1)}, \quad \mu = 1 \div N, \tag{9.75}$$

where N is the number of plates in the shield. Then Eqs. (9.67) and (9.69) can be rewritten as:

$$\overline{I} = \lambda_1^2 \sum_{j=1}^{N} G(\overline{b}^{(j)}), \tag{9.76}$$

$$\hat{\overline{J}}_\nu = \frac{1}{\nu+1} \sum_{j=1}^{N} \left[X_\Diamond^{\nu+1}\left(\overline{h} - \overline{\xi}^{(j)} + \overline{b}^{(j)}\right) - X_\Diamond^{\nu+1}\left(\overline{h} - \overline{\xi}^{(j)}\right) \right]. \tag{9.77}$$

Therefore, the BLV of an impactor with a sharp cone-shaped nose against a spaced shield with arbitrary air gaps consisting of plates manufactured from the same material , $v_{bl} = \sqrt{w_{bl}}$, can be calculated using Eq. (3-4.21), where $\lambda_0, \hat{T}, \overline{I}, \overline{J}_\nu$ and $\overline{m}_{imp}$ are determined by Eqs. (3-4.16), (3-4.30), (9.68), (9.67) and (3-4.23), respectively. In the case of large air gaps (with a width larger than the length of the impactor's nose), $\overline{I}$ can be obtained from Eq. (9.76), where the parameter λ_1 and function G are determined by Eqs. (3-4.29) and (3-4.28), respectively.

9.5 Numerical simulations

We have performed numerical simulations for two shield materials, namely, aluminum alloy AA5083-H116 and soft steel. In the calculations, the following values of the parameters characterizing the mechanical properties of the materials were used: density $\gamma_{sh} = 2660\, kg/m^3$, Young's modulus $\mu_9 = 71\, GPa$, shear strength $\mu_{11} = 190\, MPa$, Poisson's ratio $\mu_{10} = 0.33$ for the aluminum alloy (the data is taken from MatWeb, http://www.matweb.com), and $\gamma_{sh} = 7830\ kg/m^3$, $\mu_9 = 206\ GPa$, $\mu_{11} = 206$ MPa (Recht, 1990) with $\mu_{10} = 0.3$ for the soft steel. Hereafter in

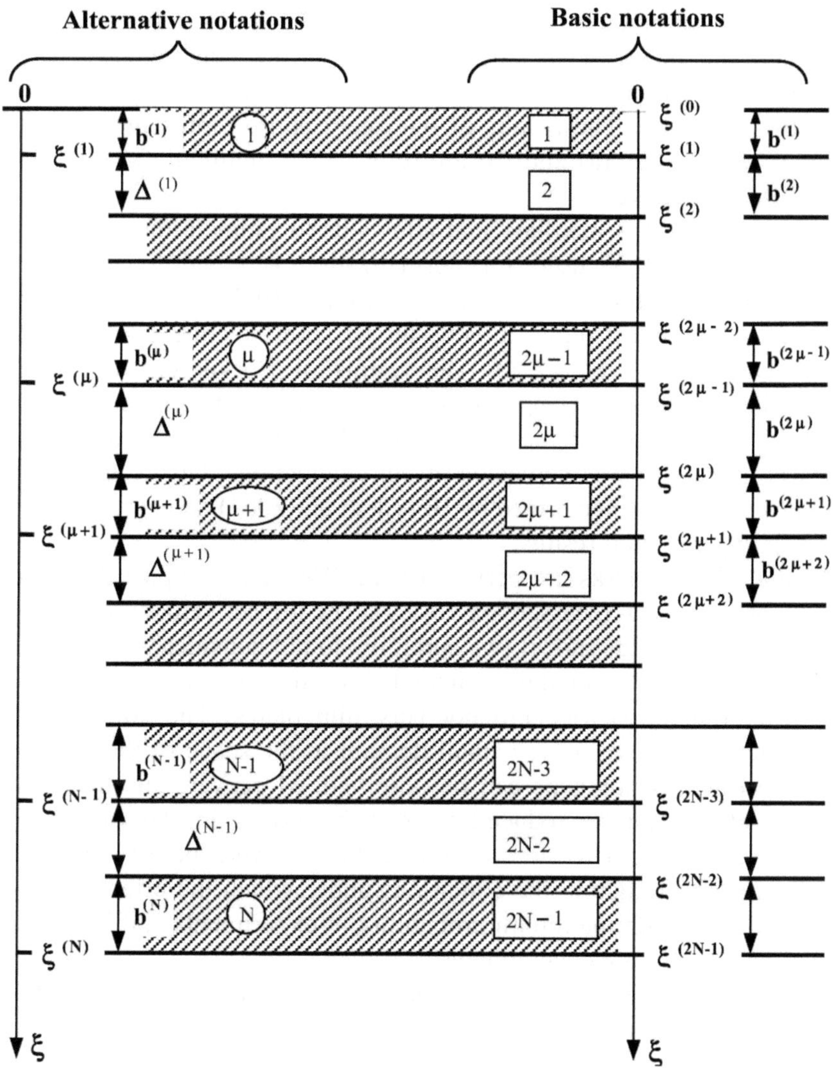

Figure 10-16. Alternative notations.

this section, we use the parameter $v_{bl}/v_{bl\,mon}$ to characterize the difference in the BLV of the spaced shields, v_{bl}, and the monolithic shield with the same total thickness, $v_{bl\,mon}$. This dimensionless parameter depends (for a given dimensionless thicknesses of the plates in a spaced shield) only on two dimensionless parameters, a_2/a_1 and λ_1 (or $\overline{m}_{imp}$). Therefore, we should investigate the behavior of this parameter in the general case, without selecting particular materials of the shield. For purposes of clarity, we did not follow this path. The results of our analysis are represented in several figures where figures (a) and (b) are refer to the aluminum shield and the steel shield, respectively.

Figure 10-17a-b is auxiliary and allows us to estimate the ranges of the parameters. The BLV used for normalization is equal to the BLV of 90^0 conical impactors when $k=1$. *Figure 10-18a-b* is similar to *Figure 10-17a-b*, but it shows directly the BLV of the cone-shaped impactor with the angle $\approx 36°$ ($k \approx 1/3$). The results of Børvik et al. (2004) for numerical simulations (triangles) and experiments (circles) are shown in *Figure 10-18a.* Note that the very small bluntness of the cone's nose in Børvik et al. (2004) was neglected in our calculations. Our results are very close the calculations performed by Børvik et al. (2004) and agree fairly well with their experimental results.

A number of remarks are pertinent here. We cannot expect the effect of spacing for $\overline{m}_{imp} >> 1$ (see *Section 10-9.4*). The impact velocity is limited by the requirements of the model (we certainly do not consider velocities larger than $1000\ m/s$). The model is intended to describe the penetration of sufficiently thin impactors into not very thin plates. The combination of these limitations leaves a very small domain in the space of the parameters where the effect of air gaps can be observed. Moreover, this theoretically predicted effect can be reduced to zero by errors in the model or by experimental errors. Therefore, our calculations were performed, mainly, with the aim of predicting the conditions under which the effect of spacing can be observed with maximum probability in experiments and in numerical simulations using more exact models.

The effect of the width of the air gap on the BLV of a shield consisting of two plates is shown in *Figure 10-19a-b* for a dimensionless air gap $\overline{\Delta}^{(1)} = \Delta^{(1)}/L < 1$ (this case with narrow air gaps was not analyzed in the analytical study). The effect of air gap on the BLV of the spaced shield increases with the increase of the width of the air gap. Inspection of *Figure 10-19a-b* shows that the order of the plates has a weak effect on the BLV.

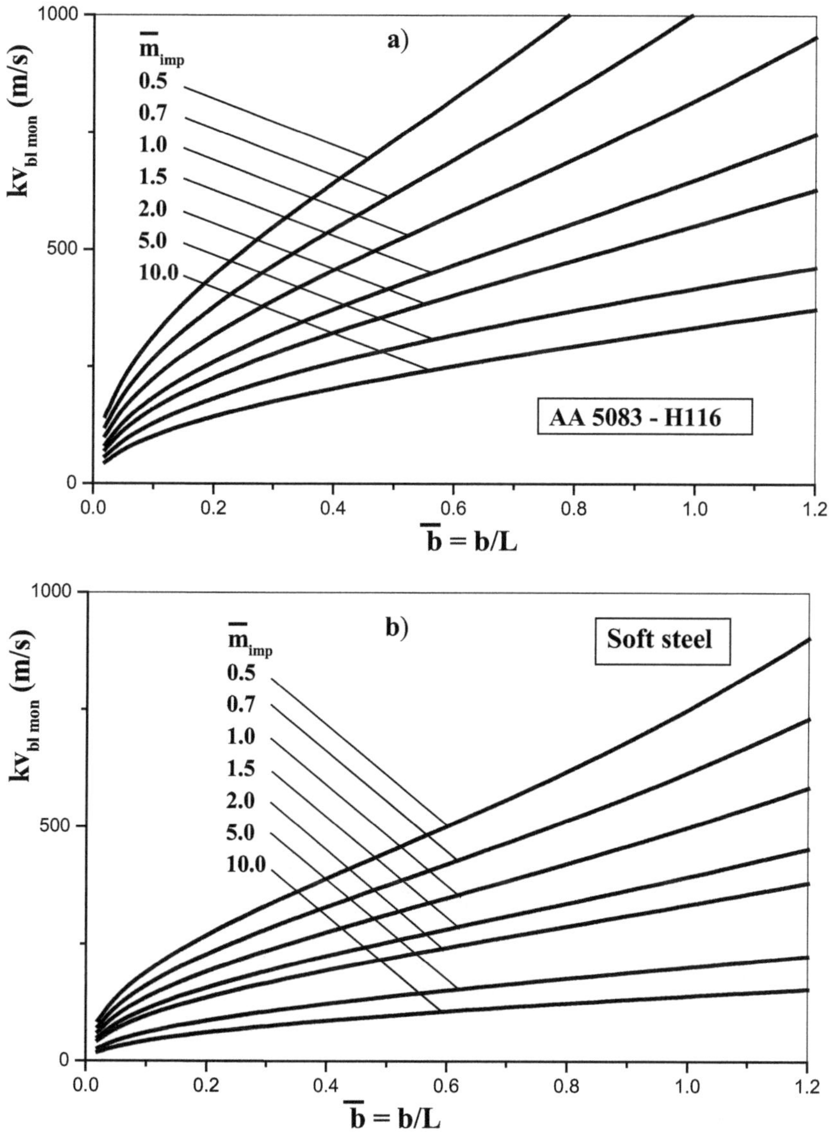

Figure 10-17. Normalized BLV of monolithic shield vs. dimensionless thickness for a conical-nose impactor with the angle $2\tan^{-1}k$ and length L. $\overline{m}_{imp}$ is determined by Eq. (3-4.23).

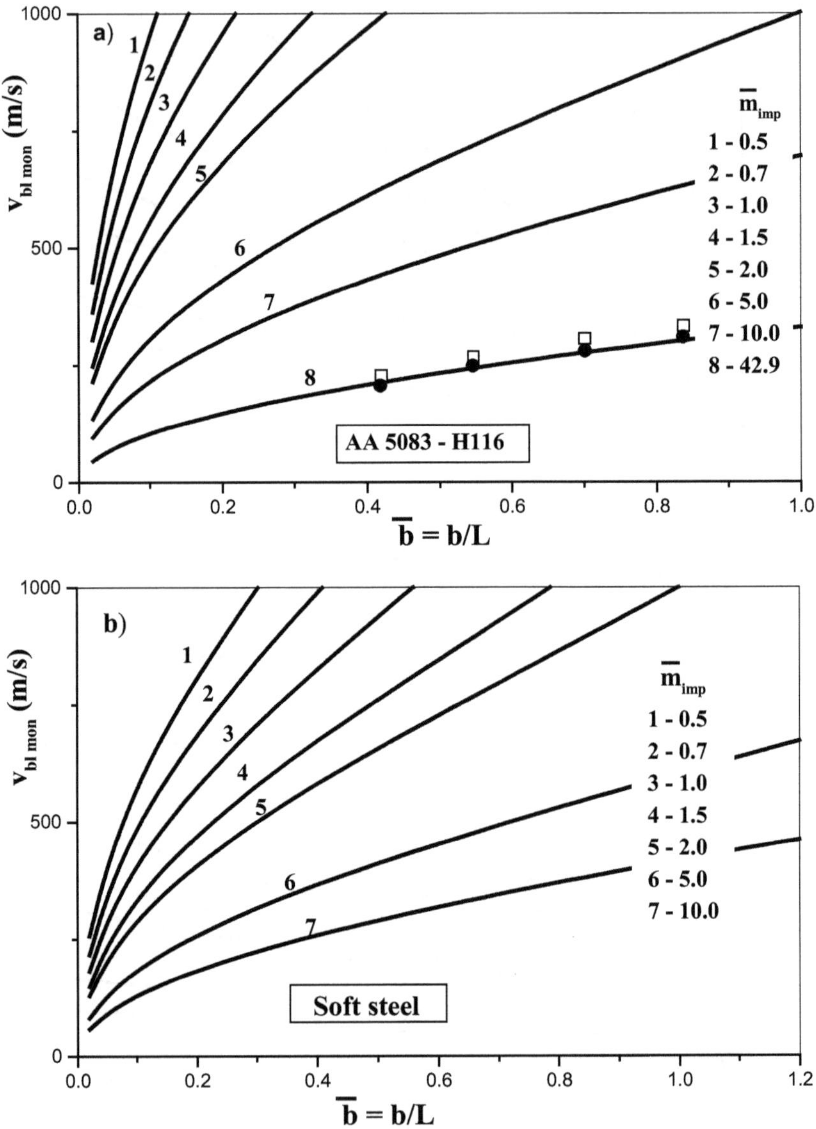

Figure 10-18. BLV of monolithic shield against conical-nose impactor with the angle 36° vs. dimensionless thickness of the shield. Triangles and circles indicate the results of numerical simulation and experiments of Børvik et al. (2004), respectively.

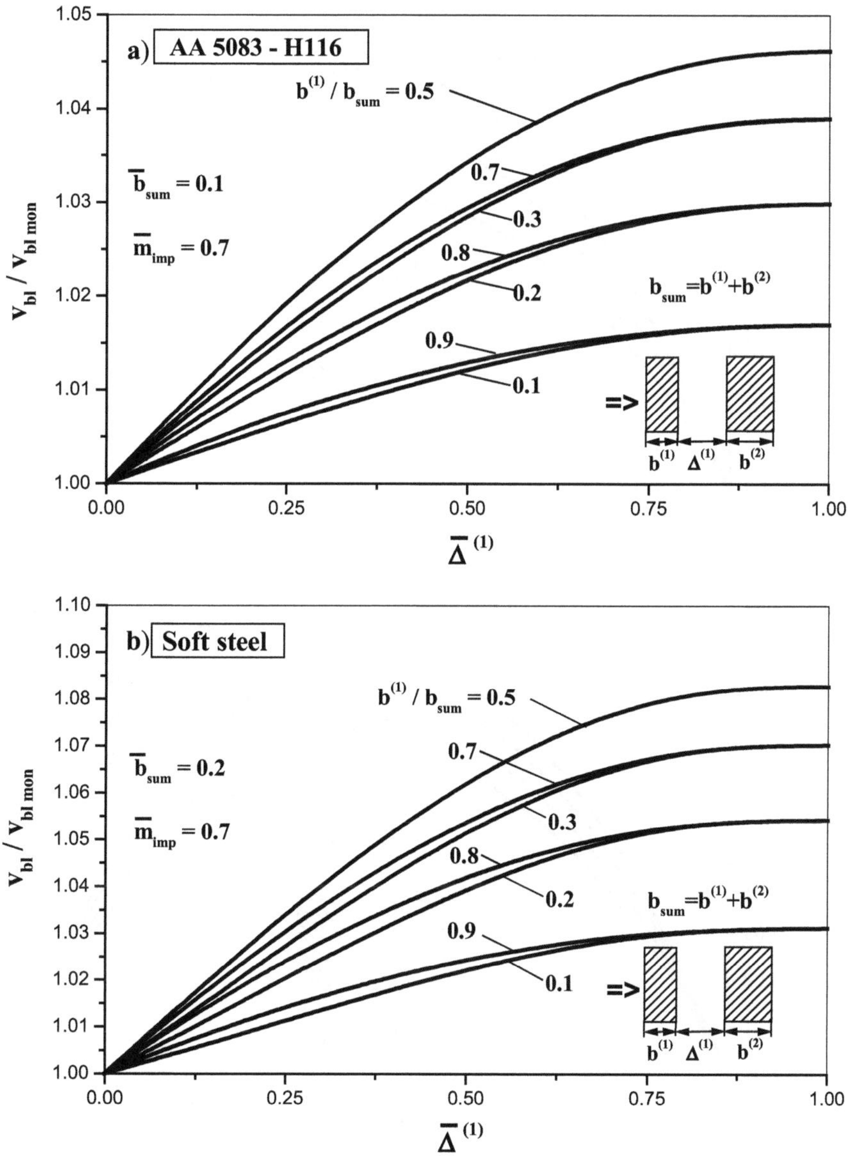

Figure 10-19. Dimensionless BLV of the spaced shield vs. dimensionless width of the air gap.

Figure 10-20a-b shows the influence of the thicknesses of the plates on the BLV when the monolithic shield is separated into two plates with wide air gap. Inspection of *Figure 10-20a-b* shows that in compliance with the theoretical predictions the maximum efficiency of spacing is achieved when the plates have the same thicknesses, and the maximum is well pronounced.

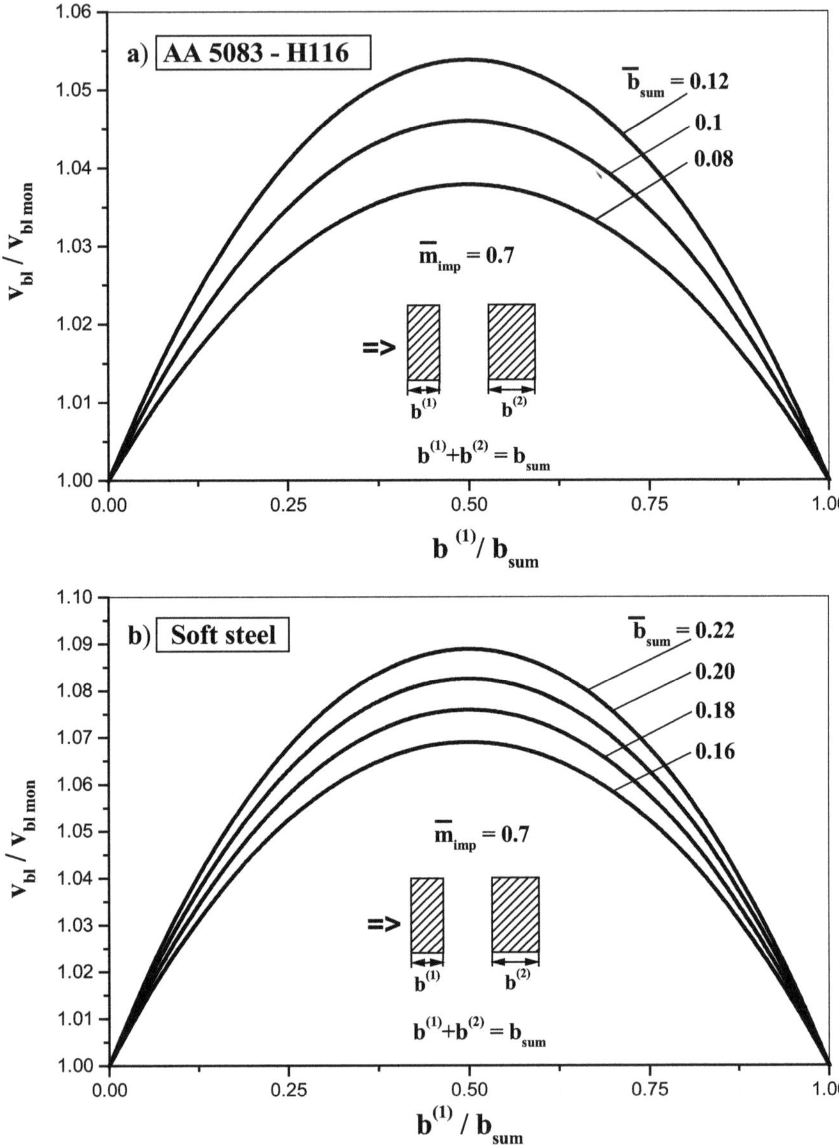

Figure 10-20. Dimensionless BLV of the spaced shield consisting of two plates with a wide air gap vs. dimensionless width of the first plate; b_{sum} is the total thickness of the shield.

The results of a similar analysis for a shield separated into three plates are shown in *Figure 10-21a-b*. The results are pertinent to the case of wide air gaps, and the total thickness of the shield is given. These results confirm the theoretical prediction that the maximum BLV is attained when all plates have the same thicknesses. Every curve in this figure shows the dependence of the BLV on the dimensionless thickness of the second plate when the thickness of the first plate is given. Note that each curve has maximum at the point where the thicknesses of the second and the third plates are equal.

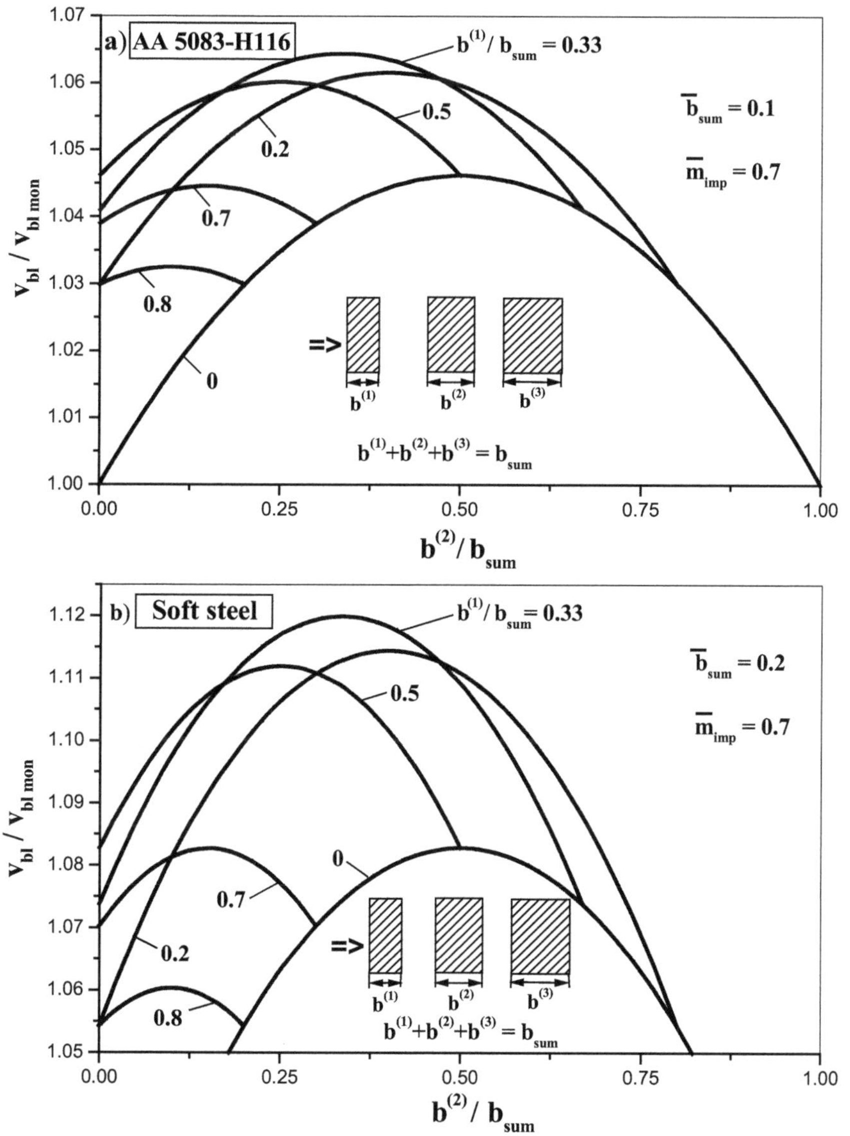

Figure 10-21. Normalized BLV of a spaced shield consisting of three plates with wide air gaps vs. dimensionless width of the second plate; b_{sum} is the total thickness of the shield.

Figure 10-22a-b shows the effect of partitioning the shield into N plates having the same thicknesses and with wide air gaps. Clearly, the normalized BLV of the spaced shield strongly depends on N, especially for relatively small N.

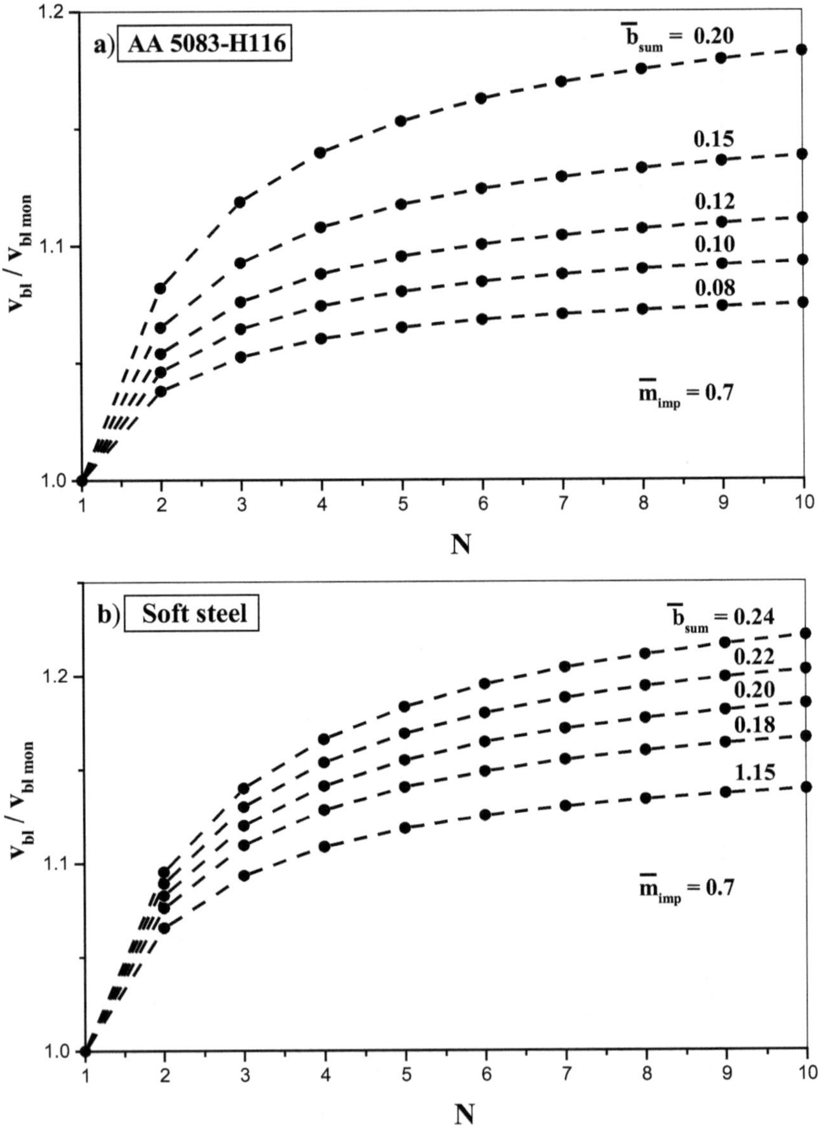

Figure 10-22. Normalized BLV of the spaced shield with large air gaps vs. the number of plates; b_{sum} is the total thickness of the shield.

Numerical simulations show that an increase of the width of the air gaps up to L causes an increase of the BLV. Therefore, we have to search for the shield with the maximum BLV among the spaced shields with large air gaps. We have shown that the shield with the same thicknesses of the plates is optimum for a given N and that the BLV increases with increasing N. Equations (9.56) and (9.57) yield the following estimate:

$$v_{bl} < \frac{1}{k}\sqrt{\frac{a_0}{a_2}\left[exp\left(\frac{2\pi a_2 k^4 b_{sum} L^2}{m}\right) - 1\right]}\,. \tag{9.78}$$

Equation (9.78) is valid for monolithic and arbitrary spaced shields with total thickness b_{sum} against a cone-shaped impactor with angle equal to $2\,tan^{-1}k$, mass m_{imp} and length of the nose L. Note that the limiting value of the BLV does not depend on the parameter of the model a_1.

Chapter 11

OPTIMIZATION OF TWO-COMPONENT CERAMIC SHIELDS

1. PERFORATION MODEL

In this section, we consider a two-component composite armor consisting of a ceramic front plate and a ductile back plate. An analysis of the properties of this armor and a discussion of problems associated with improving efficiency of the armor can be found in Anderson (2002), where the functions of the two components of the armor are summarized as follows: a hard element is needed to erode and decelerate the bullet, while a ductile element is required to capture the remnants of the eroded bullet.

Florence (1969) developed an analytical model for this type of armor impacted normally by a rigid projectile at the ballistic velocity. This model, as re-worked by Hetherington and Rajagopalan, 1991 (see also Smith and Hetherington, 1994), yields the following expression for the BLV:

$$v_{bl}^2 = \frac{\alpha \mu_5^{(2)} \mu_4^{(2)} b^{(2)} s\left[\left(\gamma^{(1)} b^{(1)} + \gamma^{(2)} b^{(2)}\right)s + m\right]}{0.91 m^2}, \; s = \pi(R + 2b^{(1)})^2, \; (1.1)$$

where v_{bl} is the BLV, m is the mass of the projectile, R is the radius of the projectile, $b^{(i)}$ are the thicknesses of the plates, $\mu_4^{(i)}$ are the ultimate tensile strengths, $\mu_5^{(2)}$ is the breaking strain, $\gamma^{(i)}$ are the densities of the materials of the plates, superscripts 1 and 2 refer to a ceramic plate and a back plate, respectively, and $\alpha = 1$.

The model is illustrated in *Figure 11-1*. The impactor was modeled as a short cylindrical rod that strikes the ceramic plate. The ceramic plate breaks progressively into a cone of fractured material. The impact energy is transferred to the back plate which is deformed like a uniform membrane. The simplifying assumptions that allowed an analytical expression to be obtained for the BLV were presented by Hetherington and Rajagopalan (1991) as follows: (i) the diameter of the circular area at the back plate over which the momentum is distributed is equal to the base diameter of the fracture conoid in the ceramic facing, and the angle of the conoid is chosen to be equal to 63°; (ii) the deformation history of the back plate may be modeled by the motion of a membrane clamped around the perimeter of the base of the fractured conoid, and the initial conditions for the membrane's motion are determined by the condition imposed by the projectile's impact and by the conservation of momentum within the projectile/shield system; (iii) failure occurs when the maximum tensile strain in the membrane reaches the ultimate breaking strain of the back plate.

Although other models have been suggested to describe perforation of two-component ceramic shields (Woodward, 1990; Reijer den, 1991; Zaera and Sanchez-Galvez, 1998; Chocron-Benloulo and Sanchez-Galvez, 1998; Zaera et al., 2000; Fellows and Barton, 1999; Zhang et al., 2002; James, 2002, Gonçalves et al., 2004; Ravid et al., 2001; Navarro et al., 1994), Florence's model was found to be the most suitable for solving problems associated with armor optimization. Some numerical results obtained with this approach have been presented by Florence (1969) and by Hetherington and Rajagopalan (1991). Later, Hetherington (1992a) investigated analytically the problem of determining the structure of two-component armor with a given areal density that provides the maximum BLV. He suggested an approximate expression for the optimum value of the ratio of the front plate width to the back plate width. Wang and Lu (1996) investigated a similar problem where the total thickness of the armor rather than the areal density was a given. Lee and Yoo (2001) conducted a comprehensive numerical and experimental study that supported the results of the armor optimization based on the Florence's model. Navarro et al. (1993), Ko et al. (1996), and Vaidya et al. (2000) also used Florence's model in their studies. Hetherington and Lemieux (1994) and Sadanandan and Hetherington (1997) generalized Florence's model to the case of an oblique impact. Based on the experimental results, Hohler et al. (2001) suggested a phenomenological formula for estimating the optimal ratio between the thicknesses of the ceramic and metal plates for a general case of an oblique impact. In a very useful study by Holmquist et al. (1999), the authors collected and systemized experimental data on properties of ceramic armor materials including data on perforation of two-component ceramic shields.

The development of numerical methods should also be mentioned (for example, Espinosa et al., 1998; Roeder and Sun, 2001; Fawaz et al., 2004)

This chapter is based on the investigations of Ben-Dor et al. (2000b, 2005). In these studies, using appropriate dimensionless variables, the solutions of the optimization problems for an arbitrary two-component composite armor have been determined in an analytical form.

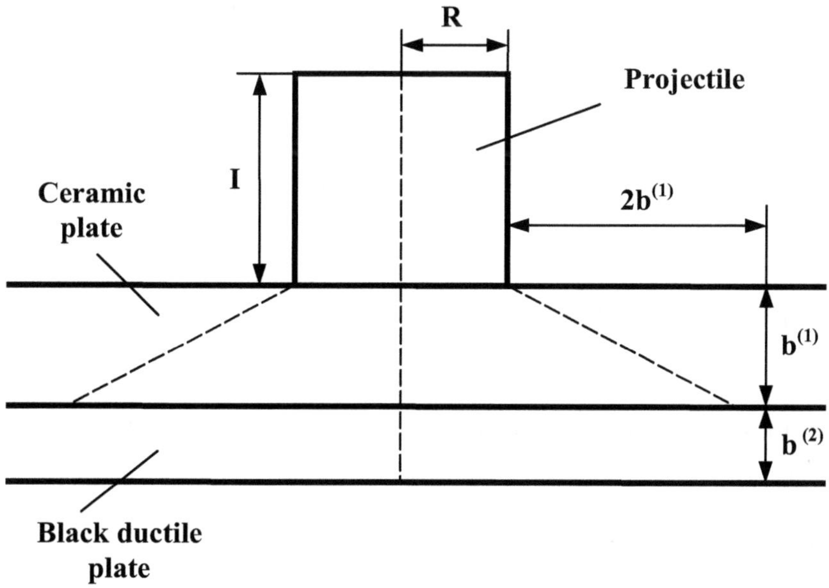

Figure 11-1. Schematic representation of two-component armor.

Here, we employ a slightly generalized model determined by Eq. (1.1), with coefficient $\alpha \neq 1$, which can be determined using the available experimental data to increase the accuracy of the predictions. The availability of experimental data on the BLV v_{bl} allows us to determine the fitting parameter α as a regression coefficient in Eq. (1.1). For the data presented in Hetherington and Rajagopalan (1991), we found that $\alpha = 0.90$.

2. SHIELD WITH THE MINIMUM AREAL DENSITY

2.1 Formulation of the problem. Dimensionless variables

The goal of this section is to determine the thicknesses of the plates $b^{(1)}, b^{(2)}$ that provide the minimum areal density of the armor:

$$A = \gamma^{(1)} b^{(1)} + \gamma^{(2)} b^{(2)} \tag{2.1}$$

for a given BLV, v_{bl}.

Let us introduce the following dimensionless variables $\overline{b}^{(1)}, \overline{b}^{(2)}, \overline{\gamma}^{(1)}, \overline{\gamma}^{(2)}, \overline{w}, \overline{A}$:

$$b^{(i)} = \overline{b}^{(i)} R,\ \gamma^{(i)} = \frac{m}{\pi R^3}\overline{\gamma}^{(i)},\ i = 1,2,\ v_{bl} = \overline{v}_{bl}\sqrt{\frac{\alpha\mu_5^{(2)}\mu_4^{(2)}}{0.91\gamma^{(2)}}},\ A = \frac{m}{\pi R^2}\overline{A} \tag{2.2}$$

Using these dimensionless variables, we can rewrite Eqs. (1.1) and (2.1) as follows:

$$\overline{v}_{bl}^2 = \overline{\gamma}^{(2)}\overline{b}^{(2)}\,\overline{s}\left[\left(\overline{\gamma}^{(1)}\overline{b}^{(1)} + \overline{\gamma}^{(2)}\overline{b}^{(2)}\right)\overline{s} + 1\right], \tag{2.3}$$

$$\overline{A} = \overline{\gamma}^{(1)}\overline{b}^{(1)} + \overline{\gamma}^{(2)}\overline{b}^{(2)}, \tag{2.4}$$

where

$$\overline{s} = \frac{s}{\pi R^2} = (1 + 2\overline{b}^{(1)})^2. \tag{2.5}$$

Equations (2.3) and (2.4) allow us to obtain an explicit formula for the criterion of optimization. Equation (2.3) is a quadratic equation with respect to $\overline{A}^{(2)} = \overline{\gamma}^{(2)}\overline{b}^{(2)}$:

$$\overline{A}^{(2)2} + (\overline{\gamma}^{(1)}\overline{b}^{(1)} + \overline{s}^{-1})\overline{A}^{(2)} - (\overline{v}_{bl}/\overline{s})^2 = 0, \tag{2.6}$$

and Eq. (2.6) has only one positive root:

$$\bar{\gamma}^{(2)}\bar{b}^{(2)} = \bar{A}^{(2)}$$
$$= 0.5\left[\sqrt{(\bar{\gamma}^{(1)}\bar{b}^{(1)} + \bar{s}^{-1})^2 + 4(\bar{v}_{bl}/\bar{s})^2} - (\bar{\gamma}^{(1)}\bar{b}^{(1)} + \bar{s}^{-1})\right] \tag{2.7}$$

Substitution of Eq. (2.7) into Eq. (2.4) yields:

$$\bar{A}(\bar{b}^{(1)}, \bar{\gamma}^{(1)}, \bar{v}_{bl}) = \frac{\bar{\gamma}^{(1)}\bar{b}^{(1)}\bar{s} - 1 + \sqrt{(\bar{\gamma}^{(1)}\bar{b}^{(1)}\bar{s} + 1)^2 + 4\bar{v}_{bl}^2}}{2\bar{s}} \tag{2.8}$$

Thus, the problem is reduced to finding a positive $\bar{b}^{(1)}$ that provides the minimum $\bar{A}$. Then $\bar{b}^{(2)}$ can be found from Eq. (2.4).

It is important to emphasize that the dimensionless areal density $\bar{A}$ is a function of one variable $\bar{b}^{(1)}$ and it depends only on two parameters, $\bar{\gamma}^{(1)}$ and $\bar{v}_{bl}$. Assume that

$$\bar{b}^{(1)}_{opt} = \varphi_1(\bar{\gamma}^{(1)}, \bar{v}_{bl}) \tag{2.9}$$

yields the minimum $\bar{A}$. Then, the dimensionless minimum areal density $\bar{A}_{opt}$ and the optimal ratio of the areal density of the second plate to the areal density of the first plate, $A^{(2)}_{opt} / A^{(1)}_{opt}$, are also some functions of $\bar{\gamma}^{(1)}$ and $\bar{v}_{bl}$:

$$\bar{A}_{opt} = \bar{A}\left(\varphi_1(\bar{\gamma}^{(1)}, \bar{v}_{bl}), \bar{\gamma}^{(1)}, \bar{v}_{bl}\right) = \varphi_2(\bar{\gamma}^{(1)}, \bar{v}_{bl}), \tag{2.10}$$

$$\frac{\bar{A}^{(2)}_{opt}}{\bar{A}^{(1)}_{opt}} = \frac{\bar{A}^{opt}}{\bar{\gamma}^{(1)}\bar{b}^{(1)}_{opt}} - 1 = \frac{\varphi_2(\bar{\gamma}^{(1)}, \bar{v}_{bl})}{\bar{\gamma}^{(1)}\varphi_1(\bar{\gamma}^{(1)}, \bar{v}_{bl})} - 1 = \varphi_3(\bar{\gamma}^{(1)}, \bar{v}_{bl}), \tag{2.11}$$

where the subscript *opt* indicates the corresponding optimum variables. It can be easily shown that:

$$\bar{\gamma}^{(2)}\bar{b}^{(2)}_{opt} = \bar{A}_{opt} - \bar{\gamma}^{(1)}\bar{b}^{(1)}_{opt} = \varphi_2(\bar{\gamma}^{(1)}, \bar{v}_{bl}) - \bar{\gamma}^{(1)}\varphi_1(\bar{\gamma}^{(1)}, \bar{v}_{bl})$$
$$= \varphi_4(\bar{\gamma}^{(1)}, \bar{v}_{bl}) \tag{2.12}$$

Therefore, the solution of the optimization problem depends on two dimensionless parameters. This allows us to investigate the problem completely and for a general case, namely, for an arbitrary combination of materials of the plates. The results for a particular combination of materials can be determined by applying the solution obtained in the dimensionless form:

$$b_{opt}^{(1)} = R\varphi_1(\overline{\gamma}^{(1)}, \overline{v}_{bl}), \tag{2.13}$$

$$b_{opt}^{(2)} = \frac{m}{\pi R^2 \gamma^{(2)}} \varphi_4(\overline{\gamma}^{(1)}, \overline{v}_{bl}), \tag{2.14}$$

$$A^{opt} = \frac{m}{\pi R^2} \varphi_2(\overline{\gamma}^{(1)}, \overline{v}_{bl}), \tag{2.15}$$

where $\overline{\gamma}^{(1)}$ and $\overline{v}_{bl}$ are determined by Eq. (2.2).

To elucidate the analysis based on the dimensionless variables, we will show in the following exposition (where possible) the results obtained for a reference armor, which we call a "basic armor" (BA). As a BA we selected a ceramic/glass fiber reinforced plastic armor and used the experimental data of Hetherington and Rajagopalan (1991) on perforation of the armor with different thicknesses of the plates by a 0.50-inch projectile. For BA a change from dimensionless variables to dimensional variables for areal density, $A(kg/m^2)$, the widths of the plates, $b^{(i)}(mm)$, and the BLV $v_{bl}(m/s)$ is performed as follows: $A = 370\overline{A}$, $b^{(i)} = 6.35\overline{b}^{(i)}$, $v_{bl} = 133\overline{v}_{bl}$ ($\overline{\gamma}^{(1)} = 0.060$ corresponds to $\gamma^{(1)} = 3{,}499\, kg/m^3$).

2.2 Properties of the function $\overline{A}(\overline{b}^{(1)})$

Consider the dependence of $\overline{A}$ vs. $\overline{b}^{(1)}$. Since

$$\lim_{\overline{b}^{(1)} \to +\infty} \frac{\overline{A}(\overline{b}^{(1)})}{\overline{b}^{(1)}} = \overline{\gamma}^{(1)}, \quad \lim_{\overline{b}^{(1)} \to +\infty} [\overline{A}(\overline{b}^{(1)}) - \overline{\gamma}^{(1)}\overline{b}^{(1)}] = 0, \tag{2.16}$$

then

$$\overline{A} = \overline{\gamma}^{(1)}\overline{b}^{(1)} \tag{2.17}$$

is the asymptote of the curve determined by the function $\overline{A} = \overline{A}(\overline{b}^{(1)})$. Eq. (2.8) implies that:

$$\overline{A}(\overline{b}^{(1)}) - \overline{\gamma}^{(1)}\overline{b}^{(1)} = \frac{\sqrt{(\overline{\gamma}^{(1)}\overline{b}^{(1)}\overline{s}+1)^2 + 4\overline{v}_{bl}^2} - (\overline{\gamma}^{(1)}\overline{b}^{(1)}\overline{s}+1)}{2\overline{s}} > 0, \tag{2.18}$$

i.e., the function $\overline{A} = \overline{A}(\overline{b}^{(1)})$ is located above the asymptote. Expression for the derivative:

$$\overline{A}' = \frac{d\overline{A}}{d\overline{b}^{(1)}} = \frac{\overline{\gamma}^{(1)}}{2} + \frac{2}{\overline{s}\sqrt{\overline{s}}} + \frac{(\overline{\gamma}^{(1)}\overline{b}^{(1)}\overline{s}+1)(\overline{\gamma}^{(1)}\overline{s}\sqrt{\overline{s}}-4)\sqrt{\overline{s}} - 16\overline{v}_{bl}^2}{2\overline{s}^2\sqrt{(\overline{\gamma}^{(1)}\overline{b}^{(1)}\overline{s}+1)^2 + 4\overline{v}_{bl}^2}} \tag{2.19}$$

and Eq. (2.17) show that $\overline{A}(\overline{b}^{(1)})$ is an increasing function for large $\overline{b}^{(1)}$.

Consider now the behavior of this function for small $\overline{b}^{(1)}$. Equations (2.5), (2.8) and (2.19) yield:

$$\overline{A}(0) = (q-1)/2 > 0 , \tag{2.20}$$

$$\overline{A}'(0) = \frac{-4q^2 + (\overline{\gamma}^{(1)} + 4)q + \overline{\gamma}^{(1)}}{2q} , \tag{2.21}$$

where

$$q = \sqrt{4\overline{v}_{bl}^2 + 1} > 1 . \tag{2.22}$$

Therefore, the sign of the derivative $\overline{A}'(0)$ is determined by the sign of the expression $-4q^2 + (\overline{\gamma}^{(1)} + 4)q + \overline{\gamma}^{(1)}$. In particular,

$$\overline{A}'(0) > 0 , \tag{2.23}$$

when

$$-4q^2 + (\overline{\gamma}^{(1)} + 4)q + \overline{\gamma}^{(1)} > 0 . \tag{2.24}$$

The latter inequality is equivalent to the following inequality:

$$\overline{\gamma}^{(1)} > \varphi(q) , \tag{2.25}$$

where

$$\varphi(q) = \frac{4q(q-1)}{q+1}. \tag{2.26}$$

Since for $q > 1$,

$$\frac{d\varphi}{dq} = \frac{4(q^2+2q-1)}{(q+1)^2} > 0, \quad \frac{d^2\varphi}{dq^2} = \frac{8}{(q+1)^3} > 0, \tag{2.27}$$

$$\frac{dq}{d\overline{v}_{bl}} = \frac{4\overline{v}_{bl}}{\sqrt{4\overline{v}_{bl}^2+1}} > 0, \quad \frac{d^2q}{d\overline{v}_{bl}^2} = \frac{4}{(4\overline{v}_{bl}^2+1)^{3/2}} > 0, \tag{2.28}$$

then

$$\frac{d\varphi}{d\overline{v}_{bl}} = \frac{d\varphi}{dq}\frac{dq}{d\overline{v}_{bl}} > 0, \tag{2.29}$$

$$\frac{d^2\varphi}{d\overline{v}_{bl}^2} = \frac{d^2\varphi}{dq^2}\left(\frac{dq}{d\overline{v}_{bl}}\right)^2 + \frac{d\varphi}{dq}\frac{d^2q}{d\overline{v}_{bl}^2} > 0. \tag{2.30}$$

Therefore, the domain where $\overline{A}'(0) > 0$ and the domain where $\overline{A}'(0) < 0$ on the plane $(\overline{\gamma}^{(1)}, \overline{v}_{bl})$ are separated by a curve determined by the concave increasing function $\overline{\gamma}^{(1)} = \varphi(q(\overline{v}_{bl}))$ (see *Figure 11-2*). Clearly,

$$\overline{A}'(0) < 0 \tag{2.31}$$

when

$$\overline{\gamma}^{(1)} < \varphi(q). \tag{2.32}$$

Let us consider both cases, i.e., the case when $\overline{A}'(0) > 0$ and the case when $\overline{A}'(0) < 0$.

Assume that Eqs. (2.23) and (2.25) are valid. Numerical simulations have shown that function $\overline{A}(\overline{b}^{(1)})$ increases for all $\overline{b}^{(1)} > 0$. The typical behavior of this function is shown in *Figure 11-3*. It should be noted that function $\overline{A}(\overline{b}^{(1)})$ attains its maximum at the boundary, $\overline{b}^{(1)} = 0$, when the used physical model is not valid. However, the latter case is of no practical significance. The mass of the cylindrical impactor can be expressed through its density γ_{imp}, length L and radius of the base R:

$$m = \pi R^2 L \gamma_{imp} . \tag{2.33}$$

Substituting Eq. (2.33) into the equation for $\gamma^{(1)}$ in Eqs. (2.2), we obtain:

$$\bar{\gamma}^{(1)} = \frac{R}{L}\frac{\gamma^{(1)}}{\gamma_{imp}} . \tag{2.34}$$

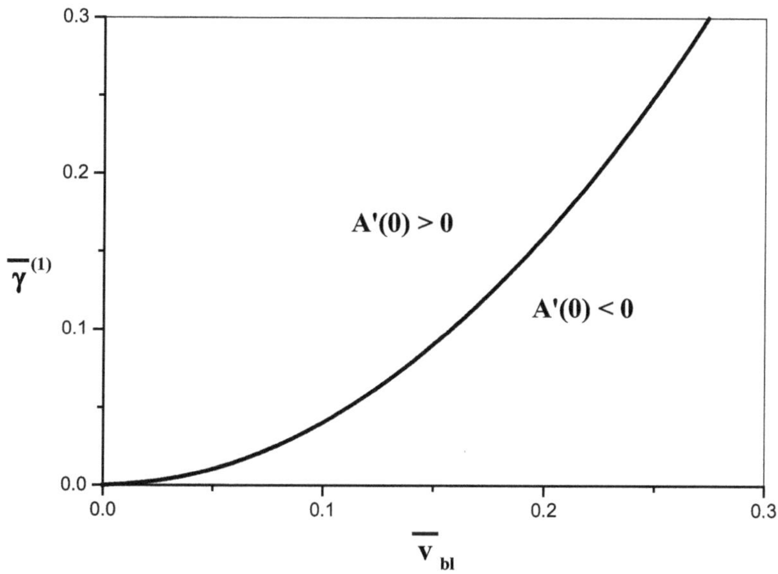

Figure 11-2. Domains with positive and negative values of $\bar{A}'(0)$ on the plane of parameters $\bar{v}_{bl}$ and $\gamma^{(1)}$.

Equation (2.34) shows that $\bar{\gamma}^{(1)}$ is much smaller than 1. Therefore (see *Figure 11-2*), Eq. (2.23) is valid only for very small values of the BLV $\bar{v}_{bl}$, that do not correspond to ballistic impact conditions. Thus, for example, for the BA, $\bar{\gamma}^{(1)} = 0.06$, and the case described by Eq. (2.23) occurs when $v_{bl} < 17\, m/s$.

Let us consider the case when Eqs. (2.31) and (2.32) are valid. Since $\bar{A}'(0) < 0$ and $\bar{A}'(\bar{b}^{(1)}) > 0$ for large $\bar{b}^{(1)}$, there is at least one point where $\bar{A}' = 0$ and, consequently, function $\bar{A}(\bar{b}^{(1)})$ attains its minimum. Numerical simulations show that function $\bar{A}(\bar{b}^{(1)})$ has only one minimum.

The typical behavior of function $\overline{A}(\overline{b}^{(1)})$ is shown in *Figure 11-4* for $\overline{\gamma}^{(1)} = 0.06$ and different $\overline{v}_{bl}$.

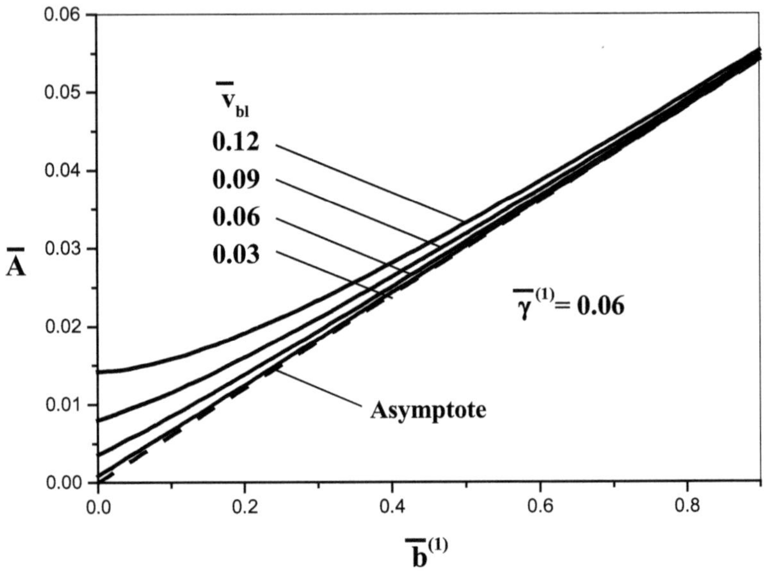

Figure 11-3. Areal density of armor vs. thickness of the ceramic plate for the case $\overline{A}'(0) > 0$.

The minimum of function $\overline{A}(\overline{b}^{(1)})$ provides the solution of the optimization problem. However, the variation of this function in the neighborhood of the minimum is quite small (see also Hetherington and Rajagopalan, 1991). Therefore, the thickness of the ceramic plate can be changed in the vicinity of the optimal value without a considerable loss in areal density.

Note that qualitatively the curves $\overline{A}(\overline{b}^{(1)})$ with $v_{bl} = const.$ plotted using experimental data and theoretical curves have the same behavior. The latter conclusion is supported by the results of calculations shown in *Figure 11-5a-b* without additional smoothing. In *Figure 11-5a*, the experimental data of Hetherington (1992a) on 7.62 AP ammunition against alumina/aluminum armor are used while in *Figure 11-5b* the experimental data of Wilkins (1978) for shields manufactured of AD85 backed by 6061-T6 aluminum struck by a 30-caliber bullet are plotted.

Since experimental data in ballistic experiments are presented as values of BLV for different thicknesses of the plates, in the following expositions we describe a procedure that allowed us to obtain the dependencies of the areal density vs. the thickness of the ceramic plate for a given BLV (Ben-Dor et al., 1999a).

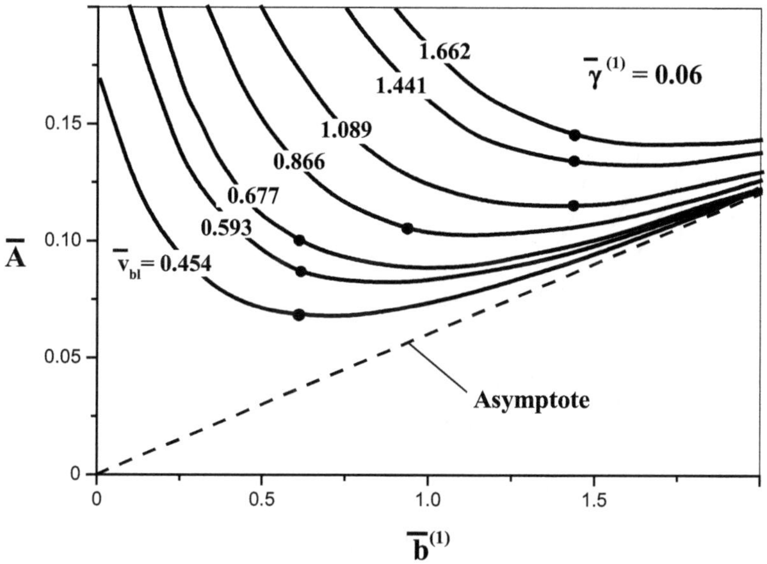

Figure 11-4. Areal density of armor vs. thickness of the ceramic plate for the case $\overline{A}'(0) > 0$. Solid circles-experimental data of Hetherington and Rajagopalan (1991).

Let us consider the BLV v_{bl} as a function of the thicknesses of the plates $b^{(1)}$ and $b^{(2)}$, $v_{bl} = \Psi(\overline{b}^{(1)}, \overline{b}^{(2)})$. It is clear on physical grounds that Ψ is an increasing function of both arguments. Then, there exists a decreasing inverse function $b^{(2)} = \xi(b^{(1)}, v_{bl})$, where v_{bl} is a parameter. The areal density can be considered as a function of $b^{(1)}$ for a given BLV, $A = \gamma^{(1)} b^{(1)} + \gamma^{(2)} \xi(b^{(1)}, v_{bl})$. The values of BLV are known from experiments for all versions of the armor, i.e., v_{bl} are known for pairs of the thicknesses, $b^{(1)}$ and $b^{(2)}$. Then the following procedure can be used for determining the dependence $A = A(b^{(1)})$.

Step 1. Approximating function Ψ as a function of two variables on a non-uniform grid and determining the domain on plane $(b^{(1)}, b^{(2)})$, where a relatively good approximation can be constructed. In our calculations, we used spline approximation.

Step 2. Constructing functions $A = A(b^{(1)})$ for given values of the BLV using the approximation obtained in Step 1 and determining the range of $b^{(1)}$ where this function is defined.

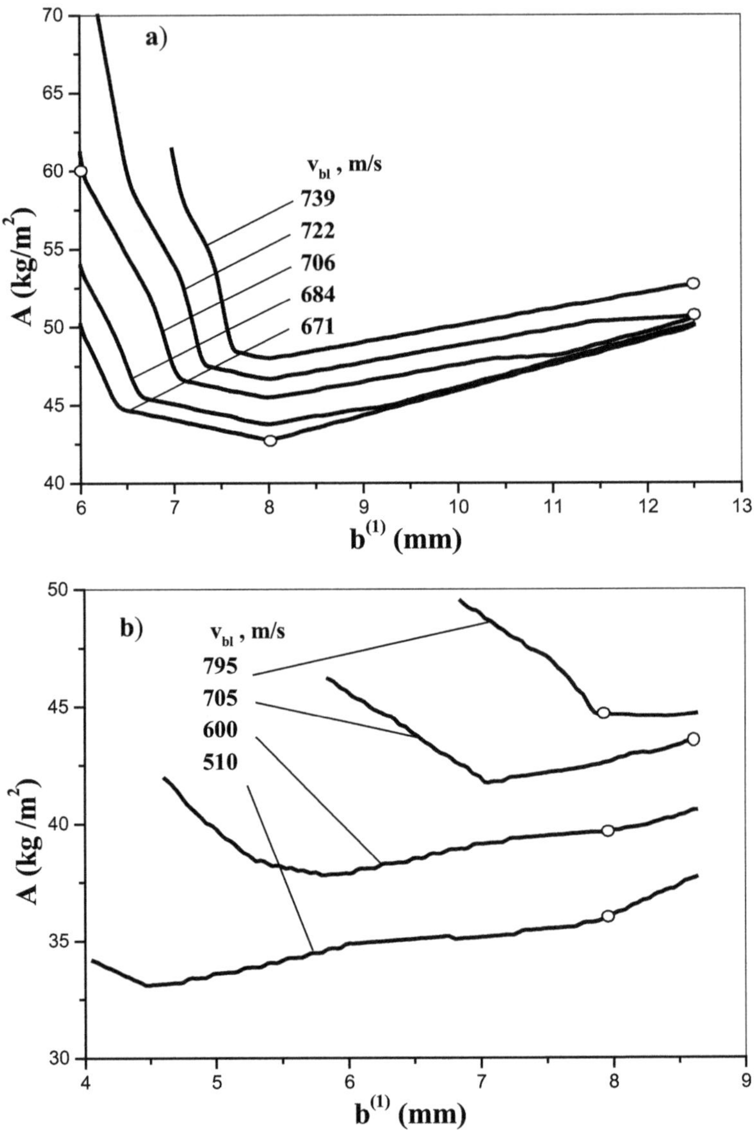

Figure 11-5. Areal density of the armor vs. thickness of the ceramic plate. Processing of the experimental data of a) Hetherington, 1992a, and b) Wilkins, 1978. Circles – processed results.

2.3 Optimal shield

Areal density of the optimal shield and optimal thickness of the ceramic plate vs. given BLV are shown in *Figure 11-6* and *Figure 11-7*, respectively.

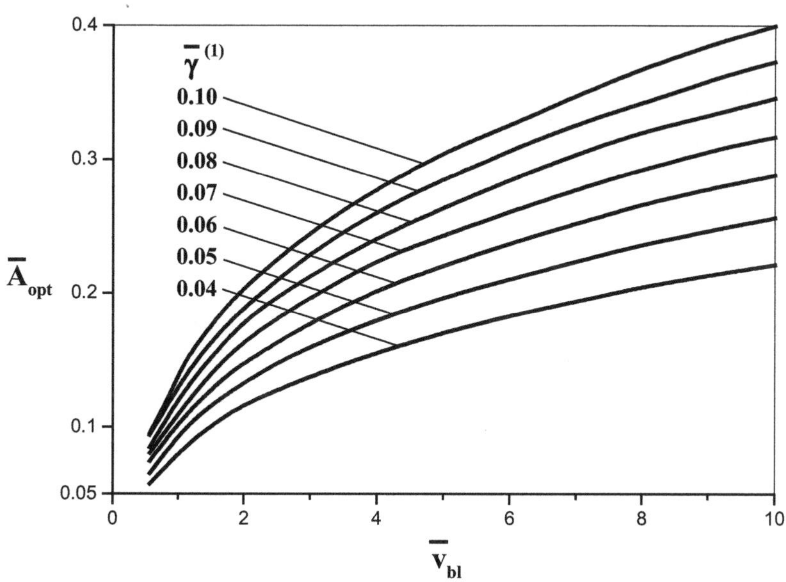

Figure 11-6. Areal density of the optimal armor vs. BLV.

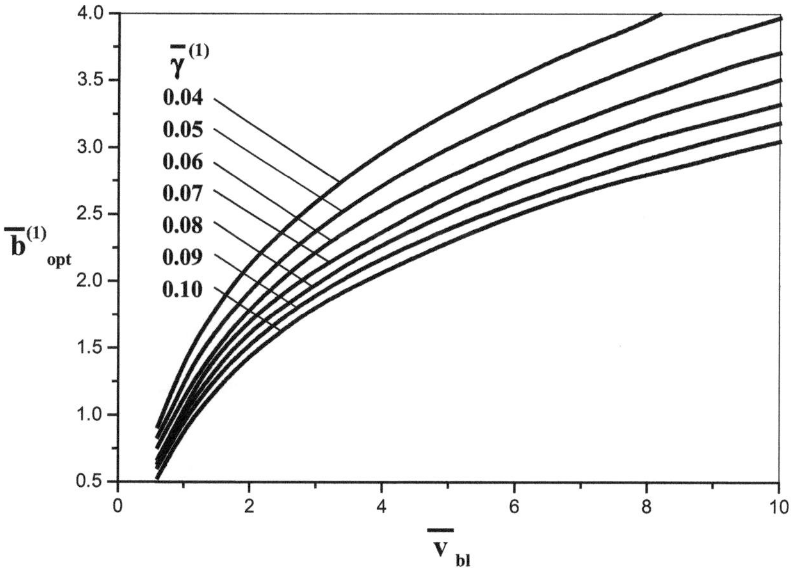

Figure 11-7. Optimal thickness of the ceramic plate vs. BLV.

Inspection of *Figure 11-8* allows us to arrive at some useful conclusions about the properties of optimal armor. Indeed, *Figure 11-8* shows that even for a relatively small $\bar{v}_{bl}$, the ratio of the areal densities of the plates in the optimal armor, $A^{(2)}_{opt} / A^{(1)}_{opt}$, is close to a constant value ≈ 0.30. The latter value corresponds to $v_{bl} \approx 400\, m/s$ for the BA. Therefore, for $\bar{v}_{bl} > 4$ the ratio of the thicknesses of the plates in the optimal armor is inversely proportional to the ratio of their densities:

$$\frac{b^{(2)}_{opt}}{b^{(1)}_{opt}} \approx 0.3 \frac{\gamma^{(1)}}{\gamma^{(2)}} \cdot \tag{2.35}$$

The families of curves plotted in *Figure 11-6* and *Figure 11-8* can be approximated with an average accuracy of 3% in the range $0.04 \le \bar{\gamma}^{(1)} \le 0.1,\ 1 \le \bar{v}_{bl} \le 10$ as follows:

$$\bar{A}_{opt} = \hat{\varphi}_2\left(\bar{\gamma}^{(1)}, \bar{v}_{bl}\right) = \left(0.04 + 1.12\bar{\gamma}^{(1)}\right)\left(\bar{v}_{bl}\right)^{0.425}, \tag{2.36}$$

$$\frac{\bar{A}^{(2)}_{opt}}{\bar{A}^{(1)}_{opt}} = \frac{\bar{\gamma}^{(2)}\bar{b}^{(2)}_{opt}}{\bar{\gamma}^{(1)}\bar{b}^{(1)}_{opt}} = \hat{\varphi}_3\left(\bar{\gamma}^{(1)}, \bar{v}_{bl}\right) = 0.29 + \left(0.1 + \bar{\gamma}^{(1)}\right)\left(\bar{v}_{bl}\right)^{-1.47}. \tag{2.37}$$

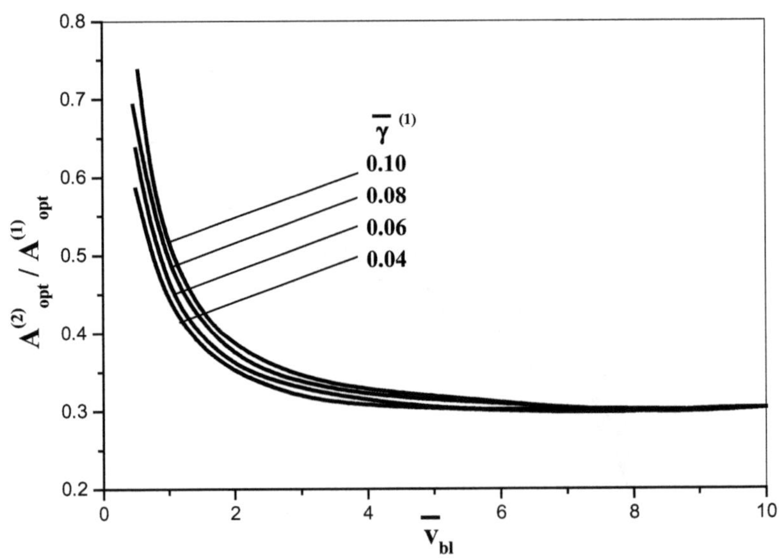

Figure 11-8. Optimal ratio of the areal densities of the plates vs. BLV.

Using Eqs. (2.10)-(2.12), we can express $\bar{b}^{(1)}_{opt}$ and $\bar{b}^{(2)}_{opt}$ through functions $\hat{\varphi}_2$ and $\hat{\varphi}_3$:

$$\overline{b}_{opt}^{(1)} = \hat{\varphi}_1(\overline{\gamma}^{(1)}, \overline{v}_{bl}) = \frac{\hat{\varphi}_2(\overline{\gamma}^{(1)}, \overline{v}_{bl})}{\overline{\gamma}^{(1)}\left[\hat{\varphi}_3(\overline{\gamma}^{(1)}, \overline{v}_{bl}) + 1\right]} = \frac{(0.04 + 1.12\overline{\gamma}^{(1)})(\overline{v}_{bl})^{1.895}}{\overline{\gamma}^{(1)}(\overline{\gamma}^{(1)} + 1.29\overline{v}_{bl}^{1.47} + 0.1)}, \tag{2.38}$$

$$\overline{b}_{opt}^{(2)} = \frac{1}{\overline{\gamma}^{(2)}} \hat{\varphi}_4(\overline{\gamma}^{(1)}, \overline{v}_{bl}) = \frac{\hat{\varphi}_2(\overline{\gamma}^{(1)}, \overline{v}_{bl}) \hat{\varphi}_3(\overline{\gamma}^{(1)}, \overline{v}_{bl})}{\overline{\gamma}^{(2)}\left[\hat{\varphi}_3(\overline{\gamma}^{(1)}, \overline{v}_{bl}) + 1\right]}. \tag{2.39}$$

Together with Eqs. (2.13)-(2.15), the solution given by Eqs. (2.36)-(2.39) provides the characteristics of the optimal armor in terms of the parameters determining the material properties of the armor's components, its cross-section area, the mass of the impactor and the expected impact velocity.

3. SHIELD WITH THE MAXIMUM BALLISTIC LIMIT VELOCITY

3.1 Formulation of the problem

Let us determine the thicknesses of the plates, $\overline{b}^{(1)}$ and $\overline{b}^{(2)}$, that provide the maximum BLV for a given areal density of the armor. In some sense, this problem is the inverse to the problem that was investigated in *Section 11-2*. Here, we use the same model to describe perforation.

Substituting $\overline{\gamma}^{(2)}\overline{b}^{(2)}$ from Eq. (2.4) and $\overline{s}$ from Eq. (2.5) into Eq. (2.3), we obtain:

$$\overline{v}_{bl}^2 = \overline{A}\,\psi(\overline{A}, \beta, z), \tag{3.1}$$

where

$$\psi(\overline{A}, \beta, z) = (1 - z)(\beta z + 1)[\overline{A}(\beta z + 1)^2 + 1], \tag{3.2}$$

$$z = \frac{\overline{A}^{(1)}}{\overline{A}} = \frac{\overline{A}^{(1)}}{\overline{A}^{(1)} + \overline{A}^{(2)}}, \quad \beta = \frac{2\overline{A}}{\overline{\gamma}^{(1)}}. \tag{3.3}$$

Therefore, the problem is reduced to the problem of determining z,

$$0 \le z \le 1, \tag{3.4}$$

that provides the minimum of ψ considered as a function of z.

The solution of this problem depends only on two parameters, $\overline{A}$ and $\gamma^{(1)}$. Suppose that

$$z_{opt} = \varphi_0(\overline{A}, \overline{\gamma}_1) \tag{3.5}$$

provides the minimum ψ. Then, the principal dimensionless parameters associated with the optimal solution (the thickness of the ceramic plate $\overline{b}_{opt}^{(1)}$, the areal densities of the plates $\overline{A}_{opt}^{(1)}$ and $\overline{A}_{opt}^{(2)}$, their ratio, and BLV, $\overline{v}_{bl}^{opt}$) are also some functions of $\overline{A}$ and $\overline{\gamma}^{(1)}$:

$$\overline{b}_{opt}^{(1)} = \frac{\overline{A}}{\overline{\gamma}^{(1)}}\varphi_0(\overline{A}, \overline{\gamma}^{(1)}) = \varphi_1(\overline{A}, \overline{\gamma}^{(1)}), \tag{3.6}$$

$$\overline{A}_{opt}^{(1)} = \overline{A}\,\varphi_0(\overline{A}, \overline{\gamma}^{(1)}), \tag{3.7}$$

$$\overline{A}_{opt}^{(2)} = \overline{\gamma}^{(2)}\overline{b}_{opt}^{(2)} = \overline{A}[1 - \varphi_0(\overline{A}, \overline{\gamma}^{(1)})] = \varphi_2(\overline{A}, \overline{\gamma}^{(1)}), \tag{3.8}$$

$$\frac{\overline{A}_{opt}^{(2)}}{\overline{A}_{opt}^{(1)}} = \frac{A_{opt}^{(2)}}{A_{opt}^{(1)}} = \frac{1}{\varphi_0(\overline{A}, \overline{\gamma}^{(1)})} - 1 = \varphi_3(\overline{A}, \overline{\gamma}^{(1)}), \tag{3.9}$$

$$\overline{v}_{bl}^{opt} = \sqrt{\overline{A}\,\psi\left(\overline{A}, \frac{2\overline{A}}{\overline{\gamma}^{(1)}}, \varphi_0(\overline{A}, \overline{\gamma}^{(1)})\right)} = \varphi_4(\overline{A}, \overline{\gamma}^{(1)}). \tag{3.10}$$

3.2 Investigation of the function $\psi\,(\overline{A}, \beta, z)$

Let us calculate the derivative:

$$\psi_z(\overline{A}, \beta, z) = \frac{\partial \psi}{\partial z} = (\beta z + 1)\, f(\overline{A}, \beta, z), \tag{3.11}$$

where

$$f(\overline{A}, \beta, x) = c_3 z^3 + c_2 z^2 + c_1 z + c_0, \tag{3.12}$$

$$c_0 = \overline{A}(4\beta - 1) + 2\beta - 1, \quad c_1 = \beta[\overline{A}(8\beta - 7) - 3],$$
$$c_2 = \overline{A}\beta^2(4\beta - 11), \quad c_3 = -5\overline{A}\beta^3. \tag{3.13}$$

Hereafter, the parameters $\overline{A}$ and β are not listed as arguments of the functions.

Consider the behavior of the function at the interval determined by Eq. (3.04). Let us calculate the values of the functions ψ and f at the boundaries of the interval:

$$\psi(0) = \overline{A} + 1 > 0, \tag{3.14}$$

$$\psi(1) = 0, \tag{3.15}$$

$$f(0) = c_0 = \frac{\overline{A} + 1}{\overline{\gamma}^{(1)}} g(\overline{A}, \overline{\gamma}^{(1)}), \tag{3.16}$$

$$f(1) = -[\overline{A}\beta^3 + 3\overline{A}\beta^2 + (3\overline{A} + 1)\beta + \overline{A} + 1] < 0, \tag{3.17}$$

where

$$g(\overline{A}, \overline{\gamma}^{(1)}) = \frac{4\overline{A}(2\overline{A} + 1)}{\overline{A} + 1} - \overline{\gamma}^{(1)}. \tag{3.18}$$

The curve $g(\overline{A}, \overline{\gamma}^{(1)}) = 0$ separates the domain, $\overline{A} \geq 0$, $\overline{\gamma}^{(1)} \geq 0$, into two sub-domains determined by the conditions $g < 0$ and $g > 0$, respectively (see *Figure 11-9*). Consider now two these cases in more detail, taking into account that the third-order polynomial $f(z)$ can have one or three real roots.

Assume that

$$g(\overline{A}, \overline{\gamma}^{(1)}) < 0. \tag{3.19}$$

Since $f(z) \to +\infty$ when $z \to -\infty$, and $f(0) < 0$, then the equation $f(z) = 0$ has a root in the interval $-\infty < z < 0$ and, consequently, 0 or 2 roots in the interval $0 < z < 1$.

Consider the case when

$$g(\overline{A}, \overline{\gamma}^{(1)}) > 0. \tag{3.20}$$

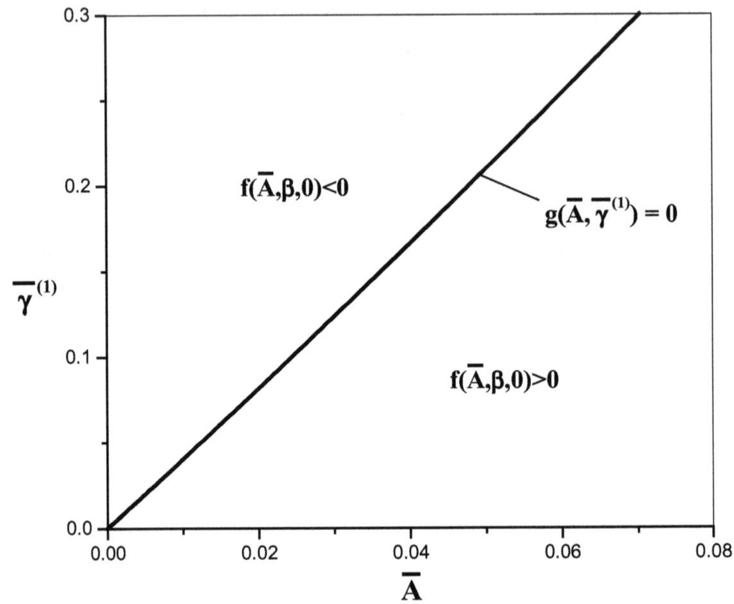

Figure 11-9. Sign of $f(\overline{A},\beta,0)$ depending on the parameters $\overline{A}$ and $\overline{\gamma}^{(1)}$.

Taking into account Eq. (3.17) we can conclude that equation $f(z)=0$ has one or three roots in the interval $0<z<1$. Since $\psi_z(0)>0$, then there exists an arbitrary small ς such that $\psi(\varsigma)>\psi(0)>\psi(1)$ and, consequently, a maximum $\psi(z)$ can be attained not at the boundaries of the segment $[0,1]$ but inside its interior point where $\psi_z(z)=f(z)=0$.

Numerical simulations showed that the roots of the equation $f(z)=0$ in the interval $[0,1]$ were not found when Eq. (3.19) was valid, and there was only one root in the opposite case (Eq. 3.20). Therefore, the maximum BLV is attained at the point $z=0$ (the first case) and at the point where $f(z)=0$ (the second case). The behavior of the function $\overline{v}_{bl}$ is shown in *Figure 11-10a-b*, where *Figure 11-10a* illustrates the transition from case 1, $\overline{A}<\overline{A}^*$, to case 2, $\overline{A}>\overline{A}^*$, where $\overline{A}=\overline{A}^*$ is the solution of the equation $g(\overline{A},\overline{\gamma}^{(1)})=0$.

Let us consider in more detail the case determined by Eq. (3.19) and show that this case is of no practical significance. Eq. (3.19) can be solved with respect to positive $\overline{A}$:

$$\overline{A}<\Theta(\overline{\gamma}^{(1)}), \tag{3.21}$$

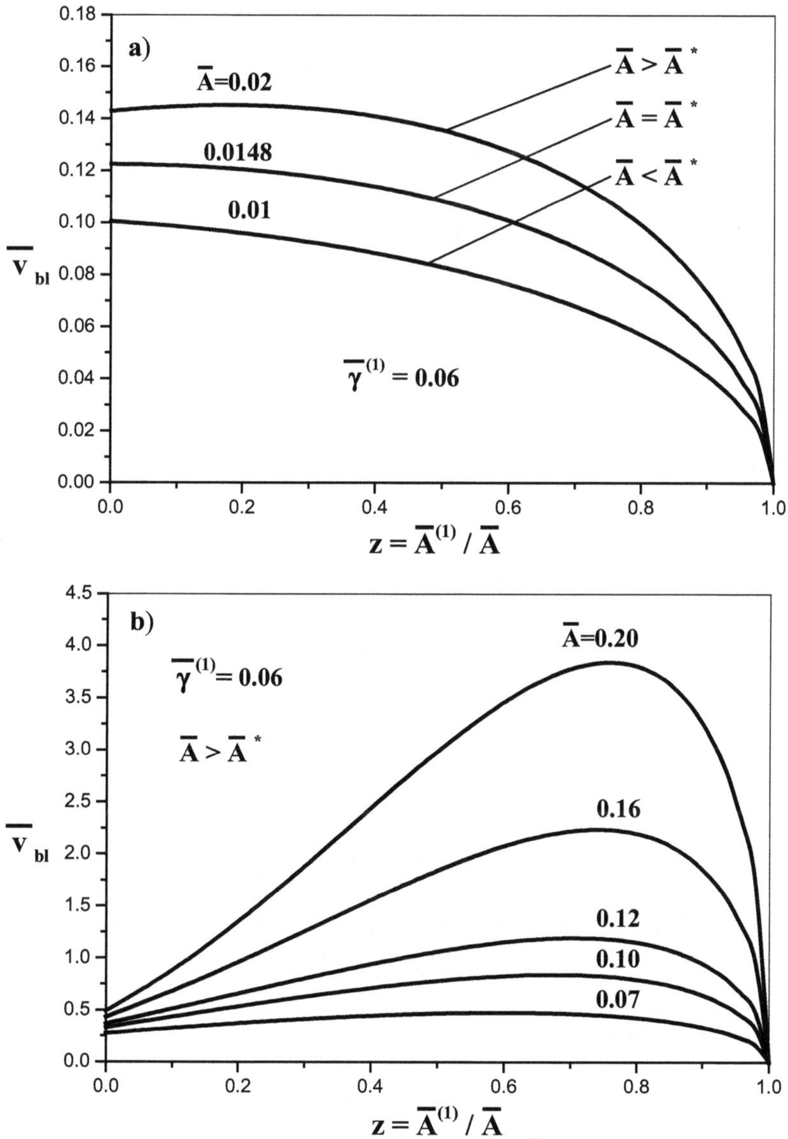

Figure 11-10. Function $\overline{v}_{bl}(z)$ for different $\overline{A}$.

where

$$\Theta(\bar{\gamma}^{(1)})=(1/16)[\bar{\gamma}^{(1)}-4+\sqrt{\bar{\gamma}^{(1)2}+24\bar{\gamma}^{(1)}+16}\,]\,. \tag{3.22}$$

On the other hand, since $\psi(x)$ is a decreasing function, then:

$$\bar{v}_{bl}\leq\bar{v}_{bl}(0)=\sqrt{\bar{A}\,\psi(0)}=\sqrt{\bar{A}(\bar{A}+1)}\,. \tag{3.23}$$

Equations (3.21) and (3.23) yield:

$$\bar{v}_{bl}<\theta(\bar{\gamma}^{(1)})\,, \tag{3.24}$$

where

$$\theta(\bar{\gamma}^{(1)})=\sqrt{\Theta(\bar{\gamma}^{(1)})[\Theta(\bar{\gamma}^{(1)})+1]}\,. \tag{3.25}$$

It was shown in the previous section that $\bar{\gamma}^{(1)}$ is much smaller than 1. Using a Taylor series expansion for small $\bar{\gamma}^{(1)}$, we find that:

$$\Theta(\bar{\gamma}^{(1)})=0.25\bar{\gamma}^{(1)}-(0.25\bar{\gamma}^{(1)})^{2}+O(\bar{\gamma}^{(1)}), \tag{3.26}$$

$$\theta(\bar{\gamma}^{(1)})=0.5\sqrt{\bar{\gamma}^{(1)}}+O(\bar{\gamma}^{(1)5/2}). \tag{3.27}$$

Taking into account Eq. (3.27), we can conclude that Eq. (3.24) is valid only for very small values of $\bar{v}_{bl}$. The latter do not correspond to ballistic impact conditions. For instance, for the BA (see *Section 11-2-1*) Eq. (3.24) implies that $v_{bl}<17\,m/s$. Note also that equation $\bar{A}-0.25\bar{\gamma}^{(1)}=0$ is a good approximation for the curve shown in *Figure 11-9*.

3.3 Optimal shield

In the case when Eq. (3.20) is valid, the equation $f(z)=0$ has only one real root, which is the point of the maximum of function $\psi(z)$. This root can be determined using Cardano's formulas (Korn and Korn, 1968):

$$z_{opt}=\sqrt[3]{\sqrt{C_0}-0.5C_2}-\sqrt[3]{\sqrt{C_0}+0.5C_2}-(1/3)c_2/c_3\,, \tag{3.28}$$

where

$$C_0 = (C_1/3)^3 + (C_2/2)^2, \qquad (3.29)$$

$$C_1 = \frac{3c_1c_3 - c_2^2}{3c_3^2}, \quad C_2 = \frac{2c_2^3}{27c_3^3} - \frac{c_1c_2}{3c_3^2} + \frac{c_0}{c_3}. \qquad (3.30)$$

Areal density (*Figure 11-11*) and thickness (*Figure 11-12*) of the ceramic plate, ratio of densities of the plates (*Figure 11-13*) and BLV (*Figure 11-14*) for the optimum armor vs. given areal density represent the solution of the optimization problem. Although Eq. (3.28) represents the solution of the considered problem in the closed form, *Figure 11-12* shows that $\bar{b}_{opt}^{(1)}$ vs. $\bar{A}$ is practically a linear function for every $\bar{\gamma}^{(1)}$. The family of curves plotted in *Figure 11-12* can be more simply approximated with an average accuracy of 3% in the range $0.04 \le \bar{\gamma}^{(1)} \le 0.1,\ 0.05 \le \bar{A} \le 0.35$ as follows:

$$\bar{b}_{opt}^{(1)} = \varphi_1(\bar{A}, \bar{\gamma}^{(1)}) = (588.5\bar{\gamma}^{(1)} - 407.2\sqrt{\bar{\gamma}^{(1)}} + 78.2)\bar{A} - 0.25. \qquad (3.31)$$

After substituting

$$\varphi_0(\bar{A}, \bar{\gamma}^{(1)}) = \frac{\bar{\gamma}_1}{\bar{A}}\varphi_1(\bar{A}, \bar{\gamma}^{(1)}), \qquad (3.32)$$

Eqs. (3.7)-(3.10) can also be rewritten using function φ_1.

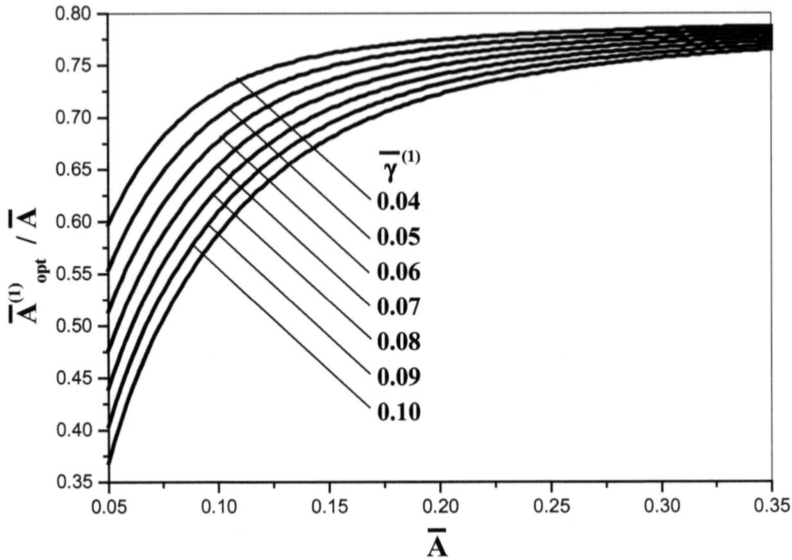

Figure 11-11. Relative area density of the ceramic plate of the optimal armor vs. given areal density of the armor.

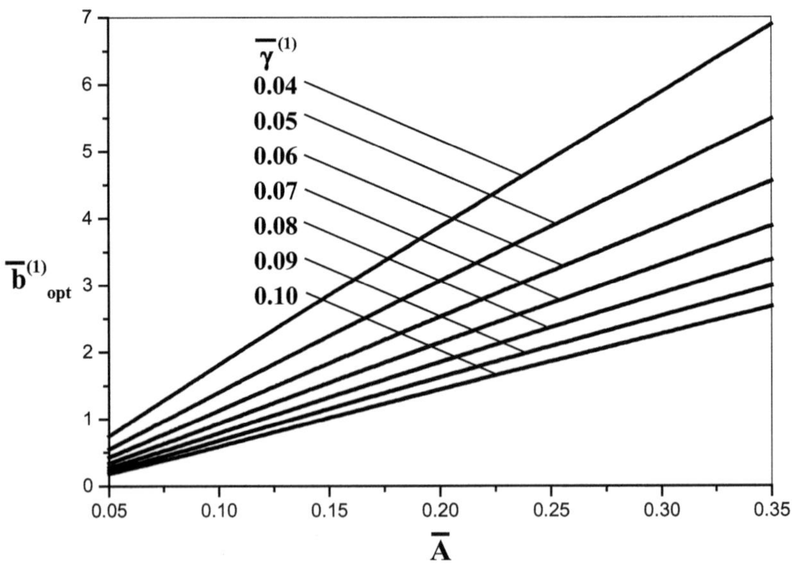

Figure 11-12. Optimal thickness of the ceramic plate vs. given areal density of the armor.

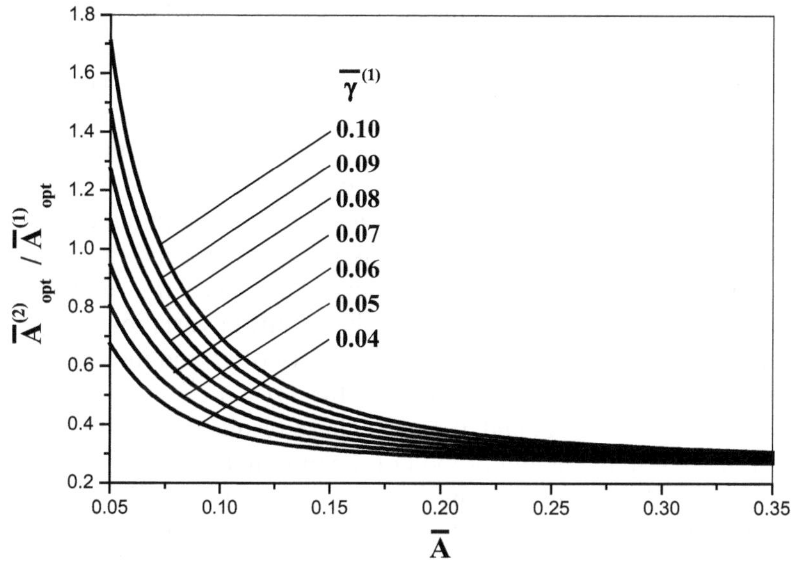

Figure 11-13. Optimal ratio of the areal density of the back plate, $\overline{A}^{(2)}$, to the areal density of the ceramic plate, $\overline{A}^{(1)}$, vs. given areal density of the armor.

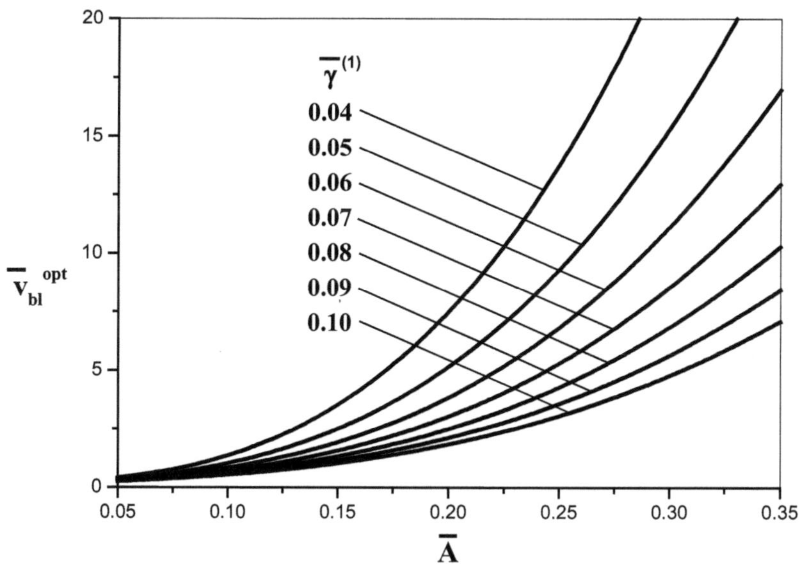

Figure 11-14. Maximum BLV vs. given areal density of the armor.

It was noted in literature (Hetherington and Rajagopalan, 1991; Lee and Yoo, 2001) that variation of BLV in the neighborhood of the minimum is quite small, i.e., the thicknesses of the plates may be changed in the vicinity of the optimal values for the given areal density of the armor, without considerable loss in the BLV. The characteristic results of the analysis of this feature of the solution are shown in *Figure 11-15* for $\bar{\gamma}^{(1)} = 0.06$.

Let us investigate how the BLV $\bar{v}_{bl}$ varies vs. some parameter η (e.g., the thickness of the ceramic plate), taking into account the constraints imposed in the formulation of the problem. The lower η_{inf} and the upper boundaries of variation of parameter η that result in variation of $\bar{v}_{bl}$ inside the interval $[(1-\varepsilon)v_{bl}^{opt}, v_{bl}^{opt}]$ for several values of ε and the optimal value η_{opt} are shown in *Figure 11-15a-c* for $\eta = \bar{A}^{(2)} / \bar{A}^{(1)} = A^{(2)} / A^{(1)}$, $\eta = z = \bar{A}^{(1)} / \bar{A} = A^{(1)} / A$, and $\eta = \bar{b}^{(1)}$ These results support the claim that variation of BLV in the neighborhood of the minimum is quite small.

Therefore, there are many possibilities to select lightweight two-component armor among different versions having almost the same ballistic properties.

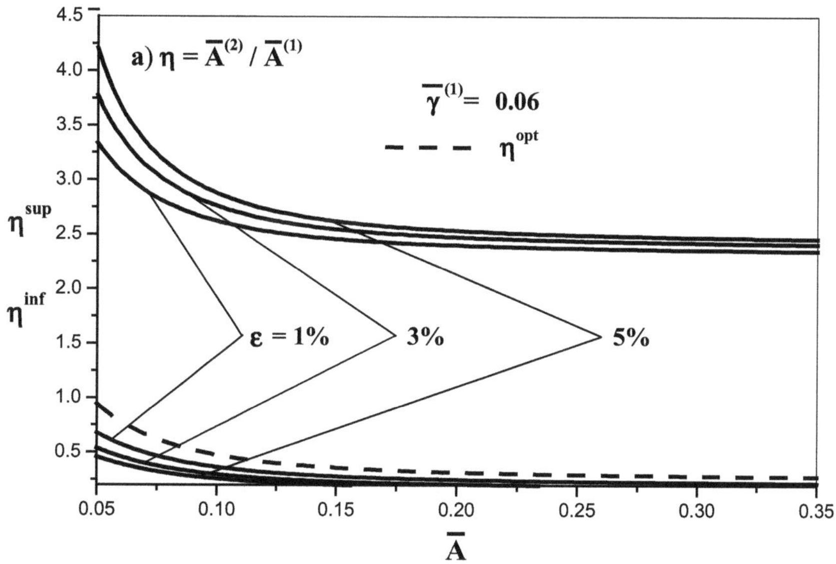
a) η = A(2) / A(1)
γ(1) = 0.06
η opt
η sup
η inf
ε = 1%
3%
5%
A

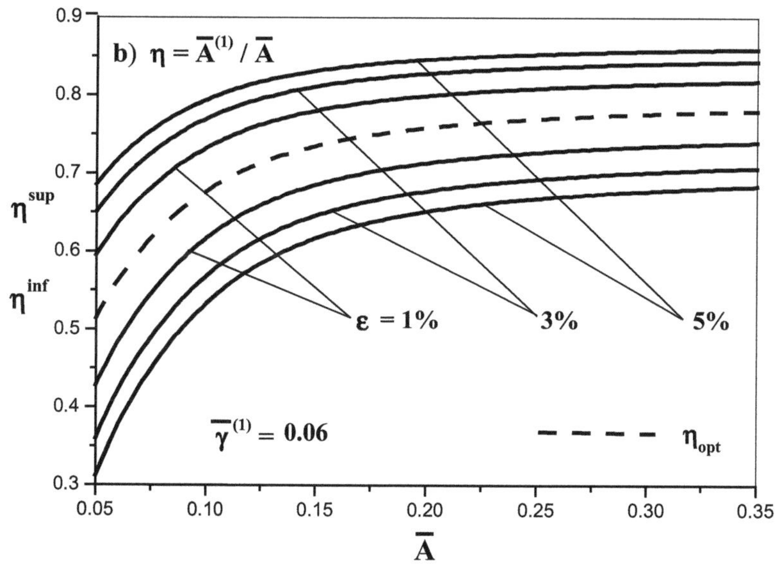
b) η = A(1) / A
η sup
η inf
ε = 1%
3%
5%
γ(1) = 0.06
η opt
A

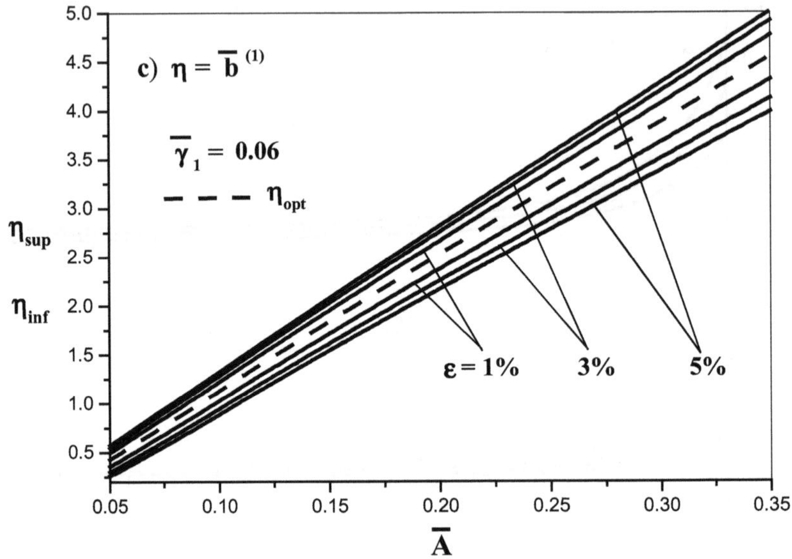

Figure 11-15. Boundaries of the interval $[\eta_{inf}, \eta_{sup}]$ and the optimal value η whereby the BLV, $\bar{v}_{bl}$, varies inside the interval $[(1-\varepsilon)\bar{v}_{bl}^{opt}, \bar{v}_{bl}^{opt}]$ vs. given areal density, $\bar{A}$.

Appendix

Appendix A1

NOTATIONS

All notations are explained in the text. For purposes of convenience, some of notations are summarized in *Table A1-1*, abbreviations are given in *Table A1-2*. Two types of notations are used in the monograph. "Local" notations (most of the notations are of this type) are used in each chapter and they may have different meanings in other chapters. "Global" notations have the same meaning through out the whole monograph; these are presented mainly in *Table A1-1*. The same abbreviations are used throughout the monograph. Local notation is considered valid in a case of conflict between the "local" and "global" notations.

Notations like μ_i и λ_i (i is number) are used only to denote parameters that characterize material properties of the shield (μ_i) and parameters that depend upon material properties of the shield and characteristics of the striker (λ_i). All these notations are presented in *Table A1-1*.

Bar above the variable always denotes a dimensionless variable, although some dimensionless variables do not have bars. Throughout the monograph (except for *Chapter 11*) all linear dimensions are normalized by L, i.e.,

$$\bar{z} = \frac{z}{L}, \quad z = x, \Phi, h, \xi, r, \xi^{(j)}, b^{(j)}, b, b_{sum}, \Delta^{(j)} \tag{A1.1}$$

In *Chapter 2* we describe a number of different models. These models are determined by different functions, since these functions depend on the model and on the shape of the impactor. Specific situations are explained in tables in *Chapter 2*. *Table A1-1* contains only general explanations and refers to the general case (commonly to a three-dimensional impactor and general local interaction model).

Table A1-1. Basic notations

Notation	Meaning	Equation or figure that explains or determines parameters or function	Comment
A	Areal density of the shield	Eqs. (10-6.1) and (11-2.1)	
$A^{(j)}$	Areal density of the j-th plate in non-homogeneous shield		
$a_0, a_1, \ldots$	Parameters of impactor-barrier interaction model (homogeneous shield)		See *Section 2-1*
$a_i^{(j)}$	Parameters of impactor-barrier interaction model (non-homogeneous shield); i and j are the number of parameter and of the layer, correspondingly	Eq. (10-2.12)	
b	Thickness of the shield including possible air gaps		Except for *Sections 5-2* and *5-3*
b_{sum}	Total thickness of the plates in the shield		
$b^{(j)}$	Thickness of the j-th layer of the shield		
$\tilde{c}$	Parameter of the two-stage model for concrete shield	Eq. (3-1.9)	*Chapters 3* and *7*
D	Drag force	Eq. (2-1.4)	
G	Function	Eq. (3-4.28)	*Chapters 3* and *10*
H	Depth of penetration (DOP)	Eq. (2-2.26)	
h	Coordinate, current location of the impactor's nose	*Figure 2-2*	
k_0, k	Coefficients in equation of conical longitudinal contour of impactor	Eq. (2-4.3)	
k_1	Parameter	Eq. (8-2.4)	
L	Length of the nose of the impactor	*Figure 2-2*	
L_0	Length of the impactor's cylindrical part	*Figure 2-2*	
M	Function determined by an integral	*Appendix 2*	
m	Constant mass of impactor		Except for *Chapter 4*

Notation	Meaning	Equation or figure that explains or determines parameters or function	Comment
m	Mass of impactor (function)		*Chapter 4*
$\overline{m}_{imp}$	Dimensionless mass of impactor	Eq. (3-4.23)	
$\vec{n}^0$	Inner normal vectors at a given location at the projectile's surface	*Figure 2-1*	
p	Stress at the boundary of a cavity in CEA		See *Sections 3-1.1* and *3-1.2*
R	Shank radius of the impactor of revolution		
r	Radius of the impactor's flat bluntness for impactor of revolution		Except for *Chapter 9*
r	Function determining the shape of the reference impactor		*Chapter 9*
t	(1) Time; (2) Parameter in parametric description of the generator of the optimum impactor		
t_0	$=\Phi'(0)$, impactor of revolution		*Chapters 6-8*
t_1	$=\Phi'(L)$, impactor of revolution		*Chapters 6-8*
t_0^*, t_1^*	Value t_0, t_1 for optimal impactor, respectively		*Chapters 6-8*
$U, u, \widetilde{u}$	Function describing impactor-shield interaction	*Tables 2-1* and *2-2*	
u_0, u_1	Ibid	Eq. (2-2.12)	
V	Function determining the solution of the equation of motion of impactor	*Table 2-1*	
V_{imp}	Volume of the impactor	Eq. (2-4.1)	
$\overline{V}_{imp}$	$=V_{imp}/L^3$		
v	Impactor's velocity		
$\vec{v}^0$	(1) Unit vector of the local velocity at the projectile's surface; (2) Unit vector of the projectile's velocity		
v_{imp}	Impact velocity		
v_{bl}	Ballistic limit velocity (BLV)	Eq. (2-2.25)	Except for *Chapters 11*
v_{bl}	Ballistic limit velocity (BLV)	Eq. (11-1.1)	*Chapters 11*
v_{res}	Residual velocity		

Notation	Meaning	Equation or figure that explains or determines parameters or function	Comment
$v_0, \overline{v}_0$	Velocity and dimensionless velocity threshold whereby the type of the model for a concrete shield is changed	Eqs. (7-1.13) and (7-3.3)	*Chapter 7*
w, w_{bl}, w_{res}	$= v^2, v_{bl}^2, v_{res}^2$		*Chapters 3* and *9*
w, w_{bl}, w_{res}	$= v^{\beta}, v_{bl}^{\beta}, v_{res}^{\beta}$		Except for *Chapters 3* and *9*
$X_{\diamond}$	Function used for description of the impactor-shield contact surface	Eq. (2-2.10)	
x	Coordinate associated with the impactor	*Figure 2-2*	Except *for Chapter 5*
x_*	Function used for description of the impactor-shield contact surface	Eq. (2-2.10)	
y	Current radius of the cavity in CEA model		See *Section 3-1.1* Except for *Chapter 5*
α	Velocity power in some LIMs		Except for *Chapters 3* and *11*
β	$= 2 - \alpha$		Except for *Chapters 3* and *11*
γ_{imp}	Average density of the impactor	Eq. (2-4.2)	
γ_{sh}	Density of a homogeneous shield		
$\gamma^{(j)}$	Density of the j-th plate in a non-homogeneous shield		
$\Delta^{(\mu)}$	Thickness of the air gap with number μ in the spaced shield (alternative notations)	*Figure 10-16*; Eqs. (10-9.74) and (10-9.75)	*Chapter 10*
δ	Function used in the description of the impactor's nose resistance	Eqs. (2-2.15) and (2-2.16)	
ε	Function determining the location of the air gaps in the shield	Eq. (10-4.2)	*Chapter 10*
$\varepsilon^{(j)}$	Parameters defining the function ε	Eq. (10-4.2)	*Chapter 10*
η	Function determining the shape of the impactor's cross section	Eq. (2-4.3)	Except for *Chapter 11*
θ, Θ	Functions describing area of the impactor-shield interaction	*Table 2-1*	Except for *Chapter 11*

Notation	Meaning	Equation or figure that explains or determines parameters or function	Comment
ϑ	Coordinate associated with the impactor	Figure 2-2	
κ	Parameter in a model	Eq. (3-1.9)	*Chapters 3* and *6*
λ_0	A parameter	Eq. (3-4.16)	
λ_1	Ibid	Eq. (3-4.29)	
λ_2, λ_3	Ibid	Eq. (6-1.3)	
λ_4	Ibid	Eq. (6-2.3)	
λ_5	Ibid	Eq. (6-3.12)	
λ_6	Ibid	Eq. (7-1.11)	
λ_7, λ_8	Ibid	Eq. (7-2.3)	
λ_9, λ_{10}	Ibid	Eq. (7-3.2)	
λ_{11}	Ibid	Eq. (8-1.16)	
λ_{12}	Ibid	Eq. (8-1.25)	
$\lambda_{13}, \lambda_{14}$	Ibid	Eq. (8-3.6)	
λ_{15}	Ibid	Eq. (8-3.13)	
λ_{16}	Ibid	Eq. (9-2.40)	
λ_{17}	Ibid	Eq. (10-3.31)	
λ_{18}	Ibid	Eq. (10-9.9)	
λ_{19}	Ibid	Eq. (3-4.36)	
λ	Parameter, similar to Lagrange multiplier, in non-classical variational problem	Eq. (6-4.6) Eq. (7-3.13)	*Chapters 6-7*
μ_{fr}	Impactor-shield friction coefficient	Eq. (2-3.1)	
μ_1	Parameter	Eq. (3-2.7)	
μ_2	Ibid	Eq. (3-2.7)	
μ_3	Ibid	Eq. (6-3.1)	
μ_4	Ultimate tensile strength		
μ_5	Breaking strain		
μ_6	Quasi-static linear elastic limit		
μ_7	Uni-axial compressive strength		
μ_8	Parameter describing mechanical properties of concrete		
μ_9	Young's modulus		
μ_{10}	Poisson's ratio		

Notation	Meaning	Equation or figure that explains or determines parameters or function	Comment
μ_{11}	Shear strength		
ξ	Coordinate associated with the shield	*Figure 3-1*	*Chapters 3* and *10*
$\xi^{(j)}$	Total thicknesses of the first j layers, basic notations	Eq. (10-2.13)	*Chapter 10*
$\xi^{(j)}$	Total thicknesses of the first j layers. The alternative notations	*Figure 10-1*; Eqs. (10-9.74) and (10-9.75)	*Chapter 10*
ρ	Coordinate associated with impactor	*Figure 2-2*	
σ	Function determining the area of cross-section of the impactor	*Tables 2-1* and *2-2*	
σ_n, σ_τ	Normal and tangential stress at the impactor's surface, respectively		See *Sections 3-1.1* and *3-1.2*
$\vec{\tau}^0$	Inner tangent vector at a given location of the projectile's surface	Eq. (2-1.2); *Figure 2-1*	
τ	$= R/L$		
Φ	Function determining the shape of the impactor.	*Figure 2-2*; *Table 2-2*	
χ	$= a_0/a_2$	Eq. (3-4.16)	*Chapter 3*
$\chi^{(j)}$	$= a_0^{(j)}/a_2^{(j)}$	Eq. (10-3.18)	*Chapter 10*
$\hat{\upsilon}$	Angle between vector $\vec{n}^0$ and vector ($-\vec{v}^0$)	Eq. (2-1.3)	
υ	Half angle of the apex of the cone of revolution		
Ψ	A function	Eq. (3-4.26)	*Chapters 3* and *10*
Ω_n, Ω_τ	Functions determining the impactor-shield interaction model	Eq. (2-1.1)	
Ω_0	Function depending on the LIM	Eq. (2-1.6)	

Table A1-2. Abbreviations

Abbreviation	Meaning
AR	Area rule
BA	Basic armor
BLV	Ballistic limit velocity
CCEA	Cylindrical cavity expansion approximation/approach
CEA	Cavity expansion approximation/approach
DLIM	Degenerate localized interaction model
DOP	Depth of penetration
FRP	Fiber-reinforced plastic
HSPM	High-speed penetration mechanics
LIA	Localized interaction approach
LIM	Localized interaction model
LIT	Localized interaction theory
ODE	Ordinary differential equation
SCEA	Spherical cavity expansion approximation/approach
SFT	Shield with a finite thickness
SIS	Semi-infinite shield

Appendix A2

PROPERTIES OF THE INTEGRAL USED IN PENETRATION DYNAMICS MODELING

In the following, we present explicit formulas for the following integrals:

$$M(i,j,c_0,c_1,c_2;W) = \int_0^W \frac{\zeta^j d\zeta}{(c_2\zeta^2 + c_1\zeta + c_0)^i}. \tag{A2.1}$$

These integrals were evaluated through elementary functions for the particular cases which are used in the monograph.

The following notations are used hereafter:

$$\Psi = c_2W^2 + c_1W + c_0, \ \Delta = 4c_0c_2 - c_1^2, \tag{A2.2}$$

$$\eta = 2c_2W + c_1, \varepsilon = \sqrt{c_0 / c_2^2}. \tag{A2.3}$$

Case 1:

$$c_0 > 0, c_1 > 0, c_2 > 0, \Delta \neq 0. \tag{A2.4}$$

In this case:

$$M(1,0,c_0,c_1,c_2;\,W)=\begin{cases}\dfrac{1}{\sqrt{-\Delta}}\ln\left[\dfrac{(\eta-\sqrt{-\Delta})(c_1+\sqrt{-\Delta})}{(\eta+\sqrt{-\Delta})(c_1-\sqrt{-\Delta})}\right] & \text{if } \Delta<0\\ \dfrac{2}{\sqrt{\Delta}}\left[\tan^{-1}\left(\dfrac{\eta}{\sqrt{\Delta}}\right)-\tan^{-1}\left(\dfrac{c_1}{\sqrt{\Delta}}\right)\right] & \text{if } \Delta>0\end{cases} \tag{A2.5}$$

$$M(1,1,c_0,c_1,c_2;\,W)=\frac{1}{2c_2}\left[\ln\left(\frac{\Psi}{c_0}\right)-c_1M(1,0,c_0,c_1,c_2;\,W)\right], \tag{A2.6}$$

$$M(2,2,c_0,c_1,c_2;\,W)=\frac{1}{\Delta}\left[2c_0M(1,0,c_0,c_1,c_2;W)-\frac{(2c_0+c_1W)W}{\Psi}\right] \tag{A2.7}$$

$$M(2,3,c_0,c_1,c_2;\,W)=\frac{1}{2c_2^2}\left[\ln\left(\frac{\Psi}{c_0}\right)+\frac{2c_2W[c_0c_1+(c_1^2-2c_0c_2)W]}{\Psi\Delta}\right.$$
$$\left.-\frac{c_1(6c_0c_2-c_1^2)}{\Delta}M(1,0,c_0,c_1,c_2;\,W)\right]. \tag{A2.8}$$

Case 2:

$$c_0>0,\,c_1>0,\,c_2>0,\,\Delta=0\,(c_1=2\sqrt{c_0c_2}\,). \tag{A2.9}$$

In this case:

$$M(1,1,c_0,c_1,c_2;\,W)=\frac{1}{c_2}\left[\ln\left(\frac{W+\varepsilon}{\varepsilon}\right)-\frac{W}{W+\varepsilon}\right] \tag{A2.10}$$

$$M(2,2,c_0,c_1,c_2;\,W)=\frac{1}{3c_2^2}\left[\frac{1}{\varepsilon}-\frac{3W^2+3\varepsilon W+\varepsilon^2}{(W+\varepsilon)^3}\right] \tag{A2.11}$$

$$M(2,3,c_0,c_1,c_2;\,W)$$
$$=\frac{1}{c_2^2}\left[\frac{\varepsilon(36W^2+27\varepsilon W+22\varepsilon^2)}{12(W+\varepsilon)^3}+\ln\left(\frac{W+\varepsilon}{\varepsilon}\right)-\frac{11}{6}\right] \tag{A2.12}$$

Case 3:

$$c_0 > 0, c_1 = 0, c_2 > 0 . \tag{A2.13}$$

In this case:

$$M(1,1,c_0,0,c_2;\, W) = \frac{1}{2c_2} ln\left(1 + \frac{c_2}{c_0} W^2\right). \tag{A2.14}$$

It is easy to prove that the following formulas are valid for any $k \neq 0$:

$$M(i,j,kc_0,kc_1,kc_2;\, W) = \frac{1}{k} M(i,j,c_0,c_1,c_2;\, W), \tag{A2.15}$$

$$M(i,j,c_0,kc_1,k^2c_2;\, W) = \frac{1}{k^{j+1}} M(i,j,c_0,c_1,c_2;\, kW). \tag{A2.16}$$

In the analysis of the behavior of $M(i,j,c_0,c_1,c_2;\, W)$ when $W \to \infty$, we use the following notations:

$$M(i,j,c_0,c_1,c_2;\infty) = \lim_{W \to +\infty} M(i,j,c_0,c_1,c_2;\, W). \tag{A2.17}$$

If $\Delta \neq 0$, then:

$$M(1,0,c_0,c_1,c_2;\infty) = \begin{cases} \dfrac{1}{\sqrt{-\Delta}} ln\left(\dfrac{c_1 + \sqrt{-\Delta}}{c_1 - \sqrt{-\Delta}}\right) & if\ \Delta < 0 \\ \dfrac{2}{\sqrt{\Delta}}\left[\dfrac{\pi}{2} - tan^{-1}\left(\dfrac{c_1}{\sqrt{\Delta}}\right)\right] & if\ \Delta > 0 \end{cases}, \tag{A2.18}$$

$$M(2,2,c_0,c_1,c_2;\infty) = \frac{1}{c_2\Delta}\left[2c_0c_2M(1,0,c_0,c_1,c_2;\,\infty) - c_1\right], \tag{A2.19}$$

$$M(2,3,c_0,c_1,c_2;\infty) = \frac{1}{2c_2^2}\left[\lim_{W \to +\infty} ln\left(\frac{\Psi}{c_0}\right) + \frac{2(c_1^2 - 2c_0c_2)}{\Delta} - \frac{c_1(6c_0c_2 - c_1^2)}{\Delta} M(1,0,c_0,c_1,c_2;\infty)\right] = +\infty. \tag{A2.20}$$

If $\Delta = 0$ then:

$$M(2,2,c_0,c_1,c_2;\,\infty) = \frac{1}{3c_2^2\varepsilon} = \frac{1}{3c_2\sqrt{c_0}}, \tag{A2.21}$$

$$M(2,3,c_0,c_1,c_2;\,\infty) = \frac{1}{c_2^2}\left[\lim_{W\to+\infty} \ln\left(\frac{W+\varepsilon}{\varepsilon}\right) - \frac{11}{6}\right] = +\infty\,. \tag{A2.22}$$

The above-mentioned integrals have been calculated using tables in the book of Gradstein and Ryzhik (1980).

References

Abrate, S., 1994, Impact on laminated composites: Recent advances, *Appl. Mech. Rev.* **47**(11): 517-544.

Abrate, S., 1998, *Impact on Composite Structures,* Cambridge University Press, Cambridge.

Alekseev, V. M., Tikhomirov, V. M., and Fomin, S. V., 1987, *Optimal Control,* Consultants Bureau, New York.

Alekseeva, E. V., and Barantsev, R. G., 1976, *Local Method of Aerodynamic Calculations in Rarefied Gas,* Leningrad State University, Leningrad (in Russian).

Allen, W. A., Mayfield, E. B., and Morrison, H. L., 1957, Dynamics of a projectile penetrating sand, *J. Appl. Phys.* **28**(3): 370-376.

Almohandes, A. A., Abdel-Kader, M. S., and Eleiche, A. M., 1996, Experimental investigation of the ballistic resistance of steel-fiberglass reinforced polyester laminated plates, *Composites. Part B* **27**(5): 447-458.

Anderson, C. E., Jr., and Bodner S. R., 1988, Ballistic impact: the status of analytical and numerical modeling, *Int. J. Impact Eng.* **11**(1): 33-40.

Anderson, C. E., Jr., Hohler, V., Walker, J. D., and Stilp, A. J., 1999, The influence of projectile hardness on ballistic performance, *Int. J. Impact Eng.* **22**(6): 619-632.

Anderson, C. E., Jr., 2002, Development an ultra-lightweight armor concept, *Ceramic Transactions,* **134** (Ceramic Armor Materials by Design, J. McCauley et al., eds): 485-498.

Apshtein, E., and Titow, O., 1996, Area rule and volume rule for integral quantities, *Acta Mechanica* **116**: 45-60.

Aptukov, V. N., 1985, Optimal structure of inhomogeneous plate with continuous distribution of properties over the thickness, *Mech. Solids* **20**(3): 148-151.

Aptukov, V. N., 1991a, Expansion of a spherical cavity in a compressible elasto-plastic medium. I. The influence of mechanical characteristics, free surface, and lamination, *Strength Mat.* **23**(12): 1262-1268.

Aptukov, V. N., 1991b, Expansion of a spherical cavity in a compressible elasto-plastic medium. II. Effect of inertial forces. Temperature effects, *Strength Mat.* **23**(12): 1269-1274.

Aptukov, V. N., Belousov, V. L., and Kanibolotskii, M. A., 1986, Optimization of the structure of a layered slab with the penetration of a rigid striker, *Mech. Composite Mat.* **22**(2): 179-183.

Aptukov, V. N., Murzakaev, A. V., and Fonarev, A. V., 1992, *Applied Theory of Penetration*, Nauka, Moscow (in Russian).

Aptukov, V. N., Petrukhin, G. I., and Pozdeev, A. A., 1985, Optimal deceleration of a rigid body by an inhomogeneous plate for the case of normal impact, *Mech. Solids* **20**(1): 155-160.

Aptukov, V. N., and Pozdeev, A. A., 1982, Some minimax problems of the technology and strengths of constructions, *Eng. Cybernetics.* **20**(1): 39-46.

Avallone, E. A., and Baumeister, T., III, eds., 1996. *Marks' Standard Handbook for Mechanical Engineers*, 10th Edition, McGraw-Hill.

Awerbuch, J., 1970, A mechanical approach to projectile penetration, *Isr. J. Technol.* **8**(4): 375-383.

Awerbuch, J., and Bodner, S. R., 1974, Analysis of the mechanics of perforation of projectiles in metallic plates, *Int. J. Solids Struct.* **10**(6): 671-684.

Backman, M., and Goldsmith, W., 1978, The mechanics of penetration of projectiles into targets, *Int. J. Eng. Sci.* **16**(1): 1-99.

Baranov, A. V., Konanykhin, Y. P., Pasechnik, L. P., and Sugak, S. G., 1991, Investigation of dynamical effect of the impactor on multi-layered shields, Universitet Druzhby Narodov, Preprint, Chernogolovka, Russia (in Russian).

Barantsev, R. G., 1978, Local method in rarefied gas dynamics. *Eng. Trans.* (formerly *Rozprawy Inzynierskie*), **26**(1): 3-9.

Barantsev, R. G., Vasil'ev, L. A., Ivanov, E. V., Kozachek, V. V., Minajchev, F. D., Mikhajlov, L. V., and Murzov, N. V., 1969, Aerodynamic calculations in rarefied gas on the basis of the hypothesis of locality, *Aerodinamika Razrezhennykh Gazov* **4**: 170-174 (in Russian).

Barr, P., compiler, 1990, *Guidelines for the Design and Assessment of Concrete Structures Subjected to Impact,* UK Atomic Energy Authority, Safety and Reliability Directorate, UK.

Bashurov, V. V., Stepanov, V. F., and Skorkin, N. A., 1994, Calculation of resistance of deformable media to solids penetration, Res. Inst. of Tech. Phys. (VNIITF), Preprint 30, Cheliabinsk-70 (in Russian).

Ben-Dor, G., Dubinsky, A., and Elperin, T., 1997a, Shape optimization of high velocity impactors using analytical models, *Int. J. Fract.* **87**(1): L7-L10.

Ben-Dor, G., Dubinsky, A., and Elperin, T., 1997b, Area rules for penetrating bodies, *Theoret. Appl. Fract. Mech.* **26**(3): 193-198.

Ben-Dor, G., Dubinsky, A., and Elperin, T., 1997c, Optimal 3D impactors penetrating into layered targets, *Theoret. Appl. Fract. Mech.* **27**(3): 161-166.

Ben-Dor, G., Dubinsky, A., and Elperin, T., 1998a, On the ballistic resistance of multi-layered targets with air gaps, *Int. J. Solids Struct.* **35**(23): 3097-3103.

Ben-Dor, G., Dubinsky, A., and Elperin, T., 1998b, A model of high speed penetration into ductile targets, *Theoret. Appl. Fract. Mech.* **28**(3): 237-239.

Ben-Dor, G., Dubinsky, A., and Elperin, T., 1998c, New area rule for penetrating impactors, *Int. J. Impact Eng.* **21**(1-2): 51-59.

Ben-Dor, G., Dubinsky, A., and Elperin, T., 1998d, Effect of air gaps on ballistic resistance of targets for conical impactors, *Theoret. Appl. Fract. Mech.* **30**(3): 243-249.

Ben-Dor, G., Dubinsky, A., and Elperin, T., 1998e, Analysis of ballistic properties of layered targets using cavity expansion model, *Int. J. Fract.* **90**(4): L63-L67.

Ben-Dor, G., Dubinsky, A., and Elperin, T., 1998f, Optimization of layered shields with a given areal density, *Int. J. Fract.* **91**(1): L9-L14.
Ben-Dor, G., Dubinsky, A., and Elperin, T., 1999a, Optimization of light weight armor using experimental data, *Theoret. Appl. Fract. Mech.* **100**(4): L29-L33.
Ben-Dor, G., Dubinsky, A., and Elperin, T., 1999b, On the order of plates providing the maximum ballistic limit velocity of a layered armor, *Int. J. Impact Eng.* **22**(8): 741-755.
Ben-Dor, G., Dubinsky, A., and Elperin, T., 1999c, Effect of air gap and order of plates on ballistic resistance of two layered armor, *Theoret. Appl. Fract. Mech.* **31**(3): 233-241.
Ben-Dor, G., Dubinsky, A., and Elperin, T., 1999d, Some ballistic properties of non-homogeneous shields, *Composites. Part A.* **30**(6): 733-736.
Ben-Dor, G., Dubinsky, A., and Elperin, T., 2000a, Optimization of the shape of a penetrator taking into account plug formation, *Int. J. Fract.* **106**(3): L29-L34.
Ben-Dor, G., Dubinsky, A., and Elperin, T., and Frage, N, 2000b, Optimization of two component ceramic armor for a given impact velocity, *Theoret. Appl. Fract. Mech.* **33**(3): 185-190.
Ben-Dor, G., Dubinsky, A., and Elperin, T., 2000c, The optimum arrangement of the plates in a multilayered shield, *Int. J. Solids Struct.* **37**(4): 687-696.
Ben-Dor, G., Dubinsky, A., and Elperin, T., 2000d, Analytical solution for penetration by rigid conical impactors using cavity expansion models, *Mech. Res. Communic.* **27**(2): 185-189.
Ben-Dor, G., Dubinsky, A., and Elperin, T., 2001a, Shape optimization of penetrator nose, *Theoret. Appl. Fract. Mech.* **35**(3): 261-270.
Ben-Dor, G., Dubinsky, A., and Elperin, T., 2001b, A class of models implying the Lambert-Jonas relation, *Int. J. Solids Struct.* **38**(40-41): 7113-7119.
Ben-Dor, G., Dubinsky, A., and Elperin, T., 2002a, Optimal nose geometry of the impactor against FRP laminates, *Compos. Struct.* **55**(1): 73-80.
Ben-Dor, G., Dubinsky, A., and Elperin, T., 2002b, Optimization of the nose shape of an impactor against a semi-infinite FRP laminate, *Compos. Sci. Technol.* **62**(5): 663-667.
Ben-Dor, G., Dubinsky, A., and Elperin, T., 2002c, A model for predicting penetration and perforation of FRP laminates by 3-D impactors, *Compos. Struct.* **56**(3): 243-248.
Ben-Dor, G., Dubinsky, A., and Elperin, T., 2002d, On the Lambert-Jonas approximation for ballistic impact, *Mech. Res. Communic.* **29**(2-3): 137-139.
Ben-Dor, G., Dubinsky, A., and Elperin, T., 2003a, Numerical solution for shape optimization of an impactor penetrating into a semi-infinite target, *Comput. Struct.* **81**(1): 9-14.
Ben-Dor, G., Dubinsky, A., and Elperin, T., 2003b, Shape optimization of an impactor penetrating into a concrete or a limestone target, *Int. J. Solids Struct.* **40**(17): 4487-4500.
Ben-Dor, G., Dubinsky, A., and Elperin, T., 2005, Optimization of two-component armor against ballistic impact, *Compos. Struct.* **69**(1): 89-94.
Berdichevsky, V., 1975, Form of minimum-drag body in hypersonic gas flow, *Moscow Univ. Mech. Bulletin* **30**(3-4): 13-18.
Bernard, R. S., 1978, Depth and Motion Prediction for Earth Penetrators, Army Eng. Waterways Exp. Station, Vicksburg MS, Report No. WES-TR-S-78-4.
Bernard, R. S., and Creighton, D. C., 1979, Projectile penetration in soil and rock: analysis for non-normal impact, Army Eng. Waterways Exp. Station, Vicksburg MS, Report No. WES/TR/SL-79-15.
Beth, R. A., 1946, Concrete penetration, Office of Sci. R&D, Washington DC, Report No. OSRD-6459.
Bishop, R. F., Hill, R., and Mott, N. F., 1945, The theory of indentation and hardness tests, *Proc. Phys. Soc.* **57**, Part 3: 147-155.

Bondarchuk, V. S., Vedernikov, Y. A., Dulov, V. G., and Minin, V. F., 1982, On the optimization of star-shaped impactors, *Izvestija Sibirskogo Otdelenija Akademii Nauk SSSR, Serija Tekhnicheskikh Nauk* **13**(3): 60-65 (in Russian).

Børvik, T., Clausen, A. H., Hopperstad, O. S., and Langseth, M., 2004, Perforation of AA5083-H116 aluminium plates with conical-nose steel projectiles – experimental study. *Int. J. Impact Eng.* **30**(4): 367-384.

Børvik, T., Langseth, M., Hopperstad, O. S., and Malo, K. A., 1998, Empirical equations for ballistic penetration of metal plates, The Norwegian Defence Construction Service, Central Staff - Technical Division, Oslo, Norway, Fortifikatorisk Notat No. 260/98.

Børvik, T., Langseth, M., Hopperstad, O. S., and Malo, K. A., 2002, Perforation of 12mm thick steel plates by 20mm diameter projectiles with flat, hemispherical and conical noses. Part I: Experimental study. *Int. J. Impact Eng.* **27**(1): 19-35.

Brady, C. P., 1938, The minimum of a function of integrals in calculus of variations, Thesis (Ph.D), in: Contributions to the calculus of variations, 1930 – Thesis submitted to the Department of Mathematics of the University of Chicago. **4**, 1938-1941. Chicago, University of Chicago Press, 1942.

Brown, S. J., 1986, Energy release protection for pressurized systems. Part 2. Review of studies into impact/terminal ballistics, *Appl. Mech. Rev.* **39**(2), Part 1: 177-202.

Brown, K. H., Koteras, J. R., Longcope, D. B., and Warren, T. L., 2003, Cavity expansion: A library for cavity expansion algorithms, Version 1.0, Sandia Nat. Lab., Albuquerque, NM, Report No. SAND2003-1048.

Bunimovich, A. I., 1973, Relations between the forces acting on a body moving in a rarefied gas, in a light flux and in a hypersonic Newtonian stream, *Fluid Dynamics* **8**(4): 584-589.

Bunimovich, A. I., and Dubinsky, A., 1973, A variational method for a generalized class of functionals and its application to problems of aeromechanics, *Fluid Dynamics* **8**(1): 92-100.

Bunimovich, A. I., and Dubinsky, A., 1995, *Mathematical Models and Methods of Localized Interaction Theory*, World Scientific, Singapore.

Bunimovich, A. I., and Dubinsky, A., 1996, Development, current state of the art, and applications of local interaction theory. Review, *Fluid Dynamics* **31**(3): 339-349.

Bunimovich A. I., and Yakunina, G. E., 1987a, On the shape of minimum-resistance solids of revolution moving in plastically compressible and elastic-plastic media, *J. Appl. Math. Mech.* **51**(3): 386-392.

Bunimovich, A. I., and Yakunina, G. E., 1987b, The shapes of three-dimensional minimum-resistance bodies moving in compressible plastic and elastic media, *Moscow Univ. Mech. Bull.* **42**(3): 59-62.

Bunimovich, A. I., and Yakunina, G. E., 1989, On the shape of a minimum resistance solid of rotation penetrating into plastically compressible media without detachment, *J. Appl. Math. Mech.* **53**(5): 680-683.

Bush, A. W., 1992, *Perturbation Methods for Engineers and Scientists*, CRC Press, London.

Cheeseman, B. A., and Bogetti, T. A., 2003, Ballistic impact into fabric and compliant composite laminates, *Compos. Struct.* **61**(1-2): 161-173.

Chen, L., and Davies, M. C. R., 1997, Analysis of energy absorption of adiabatic shear plugging in thermoviscoplastic targets, *Int. J. Eng. Sci.* **35**(4): 365-373.

Chen, X. W., and Li, Q. M., 2002, Deep penetration of a non-deformable projectile with different geometrical characteristics, *Int. J. Impact Eng.* **27**(6): 619-637.

Chen, X. W., and Li, Q. M., 2003, Perforation of a thick plate by rigid projectiles, *Int. J. Impact Eng.* **28**(7): 743-759.

Chen, X. W., Fan, S. C., and Li, Q. M., 2004, Oblique and normal perforation of concrete targets by a rigid projectile, *Int. J. Impact Eng.* **30**(6): 617-637.

Cherkaev, A. and Cherkaeva, E., 2003, *Calculus of Variations and Applications. Lecture Notes*. University of Utah, Department of Mathematics.

Chernous'ko, F. L., and Banichuk, N. V., 1973, *Variational Problems of Mechanics and Control*, Nauka, Moscow (in Russian).

Chernyi, G. G., 1969, *Introduction to Hypersonic Flow*, Academic Press, New York (original edition in Russian—1959).

Chocron-Benloulo, I. S., and Sanchez-Galvez, V., 1998, A new analytical model to simulate impact onto ceramic/composite armors, *Int. J. Impact Eng.* **21**(6): 461-471.

Choudhury, M. A., Siddiqui, N. A., and Abbas, H., 2002, Reliability analysis of a buried concrete target under missile impact, *Int. J. Impact Eng.* **27**(8): 791-806.

Corben, H. C., and Stehle, P., 1994, *Classical Mechanics*, Dover, New York.

Corbett, G. G., Reid, S. R., and Johnson, W., 1996, Impact loading of plates and shells by free-flying projectiles: a review, *Int. J. Impact Eng.* **18**(2): 141-230.

Corran, R. S. J., Ruiz, C., and Shadbolt, P. J., 1983a, On the design of containment shield, *Comput. and Struct.* **16**(1-4): 563-572.

Corran, R. S. J., Shadbolt, P. J., and Ruiz, C., 1983b, Impact loading of plates-an experimental investigation, *Int. J. Impact Eng.* **1**(1): 3-22.

Dancygier, A. N., 1997, Effect of reinforcement ratio on the resistance of reinforced concrete to hard projectile impact, *Nucl. Eng. Design.* **172**(1-2): 233-245.

Dancygier, A. N., 2000, Scaling of non-proportional non-deforming projectiles impacting reinforced concrete barriers, *Int. J. Impact Eng.* **24**(1): 33-55.

Davis, R. N., 2003, Modeling of high-speed friction using multi-step incrementation of the coefficient of sliding friction, in: *Proc. of the AIAA 54th Annual Southeastern Regional Student Conference, Kill Devil Hills, North Carolina, 27-28 March 2003*, AIAA-RSC2-2003-U-004, pp. 1-10.

Dehn, J. T., 1979, The Particle Dynamics of Target Penetration, Army Ballistic Res. Lab., Aberdeen Proving Ground, MD, Report No. ARBRL-TR-02188.

Dehn, J. T., 1986, A Unified Theory of Penetration, Army Ballistic Res. Lab., Aberdeen Proving Ground, MD, Report No. BRL-TR-2770.

Dehn, J. T., 1987, A unified theory of penetration, *Int. J. Impact Eng.* **5**(1-4), 1987: 238-248.

Dubinsky, A., 1980, Some classes of optimum 3-D bodies determined on the base of localized interaction models, in: *Dinamika razrezhennogo Gaza i Pogranichnogo Sloja*, Moscow State University, Preprint VINITI-4218-80, pp. 50-68 (in Russian).

Dubinsky, A., and Elperin, T., 1997, Method of basic projectiles for calculating force coefficients, *J. Spacecraft and Rockets* **34**(4): 558-560.

Edwards, C. H., 1997, Newton's Nose-Cone Problem, *The Mathematica J.* **7**(1): 64-71.

Eggers, A. J., Jr., Resnikoff, M. M., and Dennis, D. H., 1957, Bodies of revolutions having minimum drag at high supersonic air speeds, NACA Rep. No. 1306.

Eisler, R. D., Chatterjee, A. K., Burghart, G. H., and Loan, P., 1998, Simulates the Tissue Damage from Small Arms Projectiles and Fragments Penetrating the Musculoskeletal System, Mission Res. Corp., Fountain Valley CA, Final Report.

Elek, P., Jaramaz, S., and Mickovic, D., 2005, Modeling of perforation of plates and multi-layered metallic targets, *Int. J. Solids Struct.* **42**(3-4): 1209–1224.

Espinosa, H. D., Dwivedi, S., Zavattieri, P. D., and Yuan, G., 1998, A numerical investigation of penetration in multilayered material/structure system, *Int. J. Solids Struct.* **35**(22): 2975-3001.

Ewing, G. M., 1985, *Calculus of variations with applications,* Dover, New York.

Fawaz, Z., Zheng, W., and Behdinan, K., 2004, Numerical simulation of normal and oblique ballistic impact on ceramic composite armours, *Compos. Struct.* **63**(3-4): 387-395.

Fellows, N. A., and Barton, P. C., 1999, Development of impact model for ceramic-faced semi-infinite armor, *Int. J. Impact Eng.* **22**(8): 793-811.

Florence, A. L., 1969, Interaction of projectiles and composite armor, Stanford Res. Inst., Menlo Park, California, Report No. AMMRC-CR-69-15, Part 2.

Folsom, E. N., Jr., 1987, Projectile penetration into concrete with an inline hole, Master's Thesis, Lawrence Livermore Nat. Lab., Report No. UCRL-53786, Univ. of California, Livermore, California.

Forrestal, M. J., 1983, Forces on conical-nosed penetrators into target with contact shear strength, *Mech. Mat.* **2**(2): 173-177.

Forrestal, M. J., 1986, Penetration into dry porous rock, *Int. J. Solids Struct.* **22**(12): 1485-1500.

Forrestal, M. J., Altman, B. S., Cargile, J. D., and Hanchak, S. J., 1994, An empirical equation for penetration depth of ogive-nose projectiles into concrete targets, *Int. J. Impact Eng.* **15**(4): 395-405.

Forrestal, M. J., Brar, N. S., and Luk, V. K., 1991, Perforation of strain-hardening targets with rigid spherical-nose rods, *ASME J. Appl. Mech.* **58**(1): 7-10.

Forrestal, M. J., Frew, D. J., Hanchak, S. J., and Brar, N. S., 1996, Penetration of grout and concrete targets with ogive-nose steel projectiles, *Int. J. Impact Eng.* **18**(5): 465-476.

Forrestal, M. J., Frew, D. J., Hickerson, J. P., and Rohwer, T. A., 2003, Penetration of concrete targets with deceleration-time measurements, *Int. J. Impact Eng.* **28**(5): 479-497.

Forrestal, M. J., and Hanchak, S. J., 2002, Penetration limit velocity for ogive-nose projectiles and limestone targets, *ASME J. Appl. Mech.* **69**(6): 853-854.

Forrestal, M. J., Lee, L. M., Jenrette, B. D., 1986., Laboratory-scale penetration experiments into geological targets to impact velocities of 2.1 km/s, *ASME J. Appl. Mech.* **53**(2): 317-320.

Forrestal, M. J., Lee, L. M., Jenrette, B. D., and Setchell, R. E., 1984, Gas-gun experiments determine forces on penetrators into geological targets, *ASME J. Appl. Mech.* **51**(3): 602-607.

Forrestal, M. J., and Longcope, D. B., 1982, Closed-form solutions for forces on conical-nosed penetrators into geological targets with constant shear strength, *Mech. Mat.* **1**(4): 285-295.

Forrestal, M. J. and Longcope, D. B., 1990, Target strength of ceramic materials for high-velocity penetration, *J. Appl. Phys.* **67**(8): 3669-3672.

Forrestal, M. J., Longcope, D. B. and Norwood, F. R., 1981a, A model to estimate forces on conical penetrators into dry porous rock, *ASME J. Appl. Mech.* **48**(1): 25-29.

Forrestal, M. J., and Luk, V. K., 1988, Dynamic spherical cavity-expansion in a compressible elastic-plastic solid, *ASME J. Appl. Mech.* **55**(2): 275-279.

Forrestal, M. J., and Luk, V. K., 1992a, Penetration of 7075-T651 aluminum targets with ogival-nose rods, *Int. J. Solids Struct.* **29**(14/15): 1729-1736.

Forrestal, M. J., and Luk, V. K., 1992b, Penetration into soil targets, *Int. J. Impact Eng.* **12**(3): 427-444.

Forrestal, M. J., Luk, V. K., and Brar, N. S., 1990, Perforation of aluminum armor plates with conical-nose projectiles, *Mech. Mat.* **10**(1-2): 97-105.

Forrestal, M. J., Luk, V. K., and Watts, H. A., 1988a, Penetration of reinforced concrete with ogive-nose penetrators, *Int. J. Solids Struct.* **24**(1): 70-87.

Forrestal, M. J., Norwood, F. R., and Longcope, D. B., 1981b, Penetration into targets described by locked hydrostats and shear strength, *Int. J. Solids Struct.* **17**(9): 915-924.

Forrestal, M. J., Okajima, K. and Luk, V. K., 1988b, Penetration of 6061-T651 aluminum target with rigid long rods, *ASME J. Appl. Mech.* **55**(4): 755-760.

Forrestal, M. J., Rosenberg, Z., Luk, V. K., and Bless, S. J., 1987, Perforation of aluminum plates with conical-nosed rods, *ASME J. Appl. Mech.* **54**(1): 230-232.

Forrestal, M. J., and Tzou, D. Y., 1997, A spherical cavity-expansion penetration model for concrete targets, *Int. J. Solids Struct.* **34**(31-32): 4127-4146.

Forrestal, M. J., Tzou, D. Y., Askar, E., and Longcope, D. B., 1995, Penetration into ductile metal targets with rigid spherical-nose rods, *Int. J. Impact Eng.* **16**(5/6): 699-710.

Frew, D. J., Hanchak, S. J., Green, M. L., and Forrestal, M. J., 1998, Penetration of concrete targets with ogive-nose steel rods, *Int. J. Impact Eng.* **21**(6): 489-497.

Frew, D. J., Forrestal, M. J., and Hanchak, S. J., 2000, Penetration experiments with limestone targets and ogive-nose steel projectiles, *ASME J. Appl. Mech.* **67**(4): 841-845.

Galkin, V. S., Erofeev, A. I., and Tolstykh, A. I., 1977, An approximate method of calculating the aerodynamic characteristics of bodies in hypersonic rarefied gas flow. *Trudy TsAGI*, **1833**: 6-10 (in Russian).

Gelfand, I. M., and Fomin, S. V., 1963, *Calculus of Variations*, Prentice-Hall, Englewood Cliffs, N.J.

Gendugov, V. M., Romanova, C. V., and Romodanova, T. V., 1984, Body of revolution with minimum resistance moving in elastic-plastic and plastically compressible media, in: *Problems of Dynamics of Deformable Media*, Armenian Academy of Science, Yerevan, pp. 116-119 (in Russian).

Giere, A. C., 1964, Some energy and momentum considerations in the perforation of plates, *AIAA J.* **2**(8): 1471-1472.

Goldsmith, W., 1960, *Impact*, Arnold, London.

Goldsmith, W., 1999, Non-ideal projectile impact on targets, *Int. J. Impact Eng.* **22**(2-3): 95-395.

Goldsmith, W., and Finnegan, S. A., 1971, Penetration and perforation processes in metal targets at and above ballistic velocities, *Int. J. of Mech. Sci.* **13**(10): 843-866.

Golubev, V. K., and Medvedkin, V. A., 2001, Penetration of a rigid rod into a thick steel plate at elevated velocities, *Strength Mat.* **33**(4): 400-405.

Gomez, J. T., and Shukla, A., 2001, Multiple impact penetration of semi-infinite concrete, *Int. J. Impact Eng.* **25**(10): 965-979.

Gonçalves, D. P., de Melo, F. C. L., Klein, A. N., and Al-Qureshi, H. A., 2004, Analysis and investigation of ballistic impact on ceramic/metal composite armour, *Int. J. Machine Tools & Manufacture* **44**(2-3): 307-316.

Gradstein, I., and Ryzhik, I., 1980, *Table of Integrals, Series, and Products*, Academic Press, New York.

Gupta, N. K., and Madhu, V., 1997, An experimental study of normal and oblique impact of hard-core projectile on single and layered plates, *Int. J. Impact Eng.* **19**(5-6): 395-414.

Hayes, W. D., and Probstein, R. F., 1959, *Hypersonic Flow Theory*, Academic Press, New York & London.

Heimdahl, O. E. R., and Schulz, J. C., 1986, A note on the obtainment of instantaneous penetration information from final penetration data, *ASME J. Appl. Mech.* **53**(1): 226-227.

Hetherington, J. G., 1992a, Optimization of two component composite armours, *Int. J. Impact Eng.* **12**(3): 409-414.

Hetherington, J. G., 1992b, Correspondence on "An investigation into the energy absorbed during ballistic perforation of composite armors", *Int. J. Impact Eng.* **12**(2): 325-327.

Hetherington, J. G., 1996, Energy and momentum changes during ballistic perforation, *Int. J. Impact Eng.* **18**(3): 319-337.

Hetherington, J. G., and Lemieux, P. F., 1994, Effect of obliquity on the ballistic performance of two component composite armors, *Int. J. Impact Eng.* **15**(2): 131-137.

Hetherington, J. G., and Rajagopalan, B. P., 1991, An investigation into the energy absorbed during ballistic perforation of composite armors, *Int. J. Impact Eng.* **11**(1): 33-40.

Heuzé, F. E., 1989, An overview of projectile penetration into geological materials, with emphasis on rocks, Lawrence Livermore Nat. Lab, Report No. UCRL-101559.

Hohler, V., Weber, K., Tham, R., James, B., Barker, A., and Pickup, I., 2001, Comparative analysis of oblique impact on ceramic composite systems, *Int. J. Impact Eng.* **26**(1-10): 333-334.

Holmquist, T. J., Rajendran, A. M., Templeton, D. W., and Bishnoi, K. D., 1999, A Ceramic Armor Material Database, Tacom Res. Dev. and Eng. Center, Warren MI, Report No. TARDEC-TR-13754.

Holt, W. H., Mock, W., Jr., Soper, W. G., Coffey, C. S., Ramachandran, V., and Armstrong, R. W., 1993, Reverse-ballistic impact study of shear plug formation and displacement in Ti_6Al_4V alloy, *J. Appl. Phys.* **73**(8): 3753-3759.

Honda, K., Takamae, G., and Watanabe, T., 1930, On the measurement of the resistance of shield plates to penetration by a rifle bullet, *Tohoku Imperial Univ., 1st Series*, **19**: 703-725.

Hopkins, H. G., 1960, Dynamic expansion of spherical cavities in metals, in: *Progress in Solid Mechanics*, **1**, R. Hill and I. N. Sneddon, eds., Pergamon Press, Oxford, pp. 84-164.

Hurlich, A., 1950, Spaced Armor, Watertown Arsenal Lab., MA, Report No. WAL-710/930-1.

Ipson, T. W., and Recht, R. F. 1975, Ballistic penetration resistance and its measurement, *Exp. Mech.* **15**(7): 249-257.

Isbell, W. M., Anderson, C. E, Jr., Asay, J. R., Bless, S. J., Grady, D. E., and Sternberg, J., 1992, Penetration mechanics research in the former Soviet Union, Sci. Applications Int. Corp., San Diego, CA, Tech. Assessment Report.

Ivanov, A. G., 1998, Software for the course of calculus of variations, in: *Proc. of the Int. Workshop on Nonsmooth and Discontinuous Problems of Control and Optimization*, Chelyabinsk State Univ., Chelyabinsk, pp. 88-90.

James, B., 2002, Practical issues in ceramic armor design, in: *Proc. Ceramic Armor Materials by Design Symp.*, J. W. McCauley et al., eds, Am. Ceramic Soc., *Ceramic Trans.* **134**, pp. 33-44.

Jonas, G. H., and Zukas, J. A., 1978, Mechanics of penetration: analysis and experiment, *Int. J. Eng. Sci.* **16**(1): 879-903.

Jones, S. E., Hughes, M. L., Tonnes, O. A., and Davis, R. N., 2003, A one-dimensional analysis of rigid-body penetration with high-speed friction, *Proc. of the Institution of Mechanical Engineers, Part C, J. of Mech. Eng. Sci.* **217**(C4): 411-422.

Jones, S. E., and Rule, W. K., 2000, On the optimal nose geometry for a rigid penetrator, including the effects of pressure-dependent friction, *Int. J. Impact Eng.* **24**(4): 403-415.

Jones, S. E., Rule, W. K., Jerome, D. M., and Klug, R. T., 1998, On the optimal nose geometry for a rigid penetrator, *Comput. Mech.* **22**(5): 413-417.

Kamke, E., 1959, *Differentialgleichungen; Lösungsmethoden und Lösungen. Gewönlishe Differentialgleichungen*. Akademische Verlagsgesellschaft Geest & Portig, Leipzig.

Kartuzov, V. V., Galanov, B. A., and Ivanov, S. M., 1999, Concept of ultimate fracture velocity in the analysis of spherical cavity expansion in brittle materials: application to penetration problems, *Int. J. Impact Eng.* **23**(1): 431-442.

Kartuzov, V. V., Galanov, B. A., and Ivanov, S. M., 2002, Concept of ultimate fracture-front velocity in cylindrical cavity expansion in a brittle material, *Strength Mat.* **34**(3): 280-286.

Kasano, H., 1999, Recent advances in high-velocity impact perforation of fiber composite laminates, *JSME Int. J., Ser. A.* **42**(2): 147-157.

Kennedy, R. P., 1976, A review of procedures for the analysis and design of concrete structures to resist missile impact effects, *Nucl. Eng. Design* **37**(2): 183-203.

Klepaczko, I. V., 2001, Surface layer thermodynamics of steel penetrator at high and very high sliding velocities, Florida Univ., Shalimar Graduate Engineering and Research Center, Report No. A709014.

Klepaczko, I. V., and Hughes, M. L., 2005, Scaling of wear in kinetic energy penetrators, *Int. J. Impact Eng.* **31**(4): 435-459.

Ko, F., Yu, J. Z., and Song, J. W., 1996, Characterization of multifunctional composite armor, In: Proc. 11th Techn. Conf. of the Am. Soc. for Composites, Oct 7-9, 1996, Atlanta, GA, Technomic Publ Co., pp. 947-957.

Korn, G. A., and Korn, T. M., 1968, *Mathematical Handbook for Scientists and Engineers*, McGraw-Hill Book Comp., New York.

Kraft, J. M., 1955, Surface friction in ballistic penetration, *J. Appl. Phys.* **26**(10): 1248-1253.

Kravchenko, V. P., Skorkin, N. A., and Sapozhnikov, A. A., 1994, Penetration of a solid body of revolution into rock and non-rock soils, Res. Inst. of Tech. Phys. (VNIITF), Cheliabinsk-70, Preprint No. 18 (in Russian).

Kucher, V., 1967, Penetration with optimal work, Ballistic Res. Lab., Aberdeen Proving Ground, MD, Report BRL-R-1384.

Kumar, K. S., and Bhat, T. B., 1998, Response of composite laminates on impact of high velocity projectiles, *Key Eng. Mater.* **141-143**: 337-348.

Lambert, J. P., 1978, A residual velocity predictive model for long rod penetrators, Ballistic Res. Lab., Aberdeen, MD, Report No. ARBRL-MR-02828.

Lambert, J. P., and Jonas, G. H., 1976, Towards standardization of in terminal ballistic testing: velocity representation, Ballistic Res. Lab., Aberdeen, MD, Report No. BRL-R-1852.

Landgrov, I. F., and Sarkisyan, O. A., 1984, Piercing plastic-material barriers with a rigid punch, *J. Appl. Mech. Tech. Phys.* No. 5: 771-773.

Lee, M., and Yoo, Y. H., 2001, Analysis of ceramic/metal armour systems, *Int. J. Impact Eng.* **25**(9): 819-829.

Li, Q. M., and Chen, X. W., 2002, Penetration into concrete targets by a hard projectile, in: *7th Int. Conf. on Structures under Shock and Impact*, N. Jones et al., eds., May 27-29, Montreal, 2002, WIT Press, Southampton, pp. 91-100.

Li, Q. M., and Chen, X. W., 2003, Dimensionless formulae for penetration depth of concrete target impacted by a non-deformable projectile, *Int. J. Impact Eng.* **28**(1): 93-116.

Li, Q. M., and Tong, D. J., 2003, Perforation thickness and ballistic limit of concrete target subjected to rigid projectile impact, *ASCE J. Eng. Mech.*. **129**(9): 1083-1091.

Li, Q. M., Weng, H. J., and Chen, X. W., 2004, A modified model for the penetration into moderately thick plates by a rigid, sharp-nosed projectile, *Int. J. Impact Eng.* **30**(2): 193-204.

Liang, C.-C., Yang, M.-F., Wu, P.-W., and Teng, T.-L., 2005, Resistant performance of perforation of multi-layered targets using an estimation procedure with marine application, *Ocean Eng.* **32**(3-4): 441-468.

Littlefield, D. L., Anderson, C. E., Jr., Partom, Y., and Bless, S. J., 1997, The penetration of steel targets finite in radial extent, *Int. J. Impact Eng.* **19**(1): 49-62.

Longcope, D. B., and Forrestal, M. J., 1981, Closed form approximation for forces on conical penetrators into dry porous rock, *ASME J. Appl. Mech.* **48**(4): 971-972.

Longcope, D. B., and Forrestal, M. J., 1983, Penetration of target described by a Mohr-Coulomb failure criterion with a tension cutoff, *ASME J. Appl. Mech.* **50**(2): 327-333.

Longcope, D. B., Jr., Tabbara, M. R., and Jung, J., 1999, Modeling of oblique penetration into geologic targets using cavity expansion penetrator loading with target free-surface effects, Sandia Nat. Lab., Albuquerque, NM, Report No. SAND99-1104.

Luk, V. K., and Amos, D. E., 1991, Dynamic cylindrical cavity expansion of compressible strain-hardering materials, *ASME J. Appl. Mech.* **58**(2): 334-340.

Luk, V. K., and Forrestal, M. J., 1987, Penetration into semi-infinite reinforced-concrete targets with spherical and ogival nose projectiles, *Int. J. Impact Eng.* **6**(4): 291-301.

Luk, V. K., and Forrestal, M. J., 1989, Comment on "Penetration into semi-infinite reinforced-concrete targets with spherical and ogival nose projectiles (*Int. J. Impact Eng.* **6**, 291-301, 1987), *Int. J. Impact Eng.* **8**(1): 83-84.

Luk, V. K., Forrestal, M. J., and Amos, D. E., 1991, Dynamic spherical cavity expansion of strain-hardening materials, *ASME J. Appl. Mech.* **58**(1): 1-6.

Macek, R. W., and Duffey, T. A., 2000, Finite cavity expansion method for near-surface effects and layering during earth penetration, *Int. J. Impact Eng.* **24**(3): 239-258.

Madhu, V., Bhat, T. B., and Gupta, N. K., 2003, Normal and oblique impacts of hard projectiles on single and layered plates – an experimental study, *Defence Sci. J.* **53**(2): 147-156.

Marom, I., and Bodner, S. R., 1979, Projectile perforation of multi-layered beams, *Int. J. Mech. Sci.* **21**(8): 489-504.

Mastilovic, S., and Krajcinovic, D., 1999a, High-velocity expansion of a cavity within a brittle material, *J. Mech. Phys. Solids* **47**(3): 577-610.

Mastilovic, S., and Krajcinovic, D., 1999b, Penetration of rigid projectiles through quasi-brittle materials, *ASME J. Appl. Mech.* **66**(3): 585-592.

Miele, A., 1962, A study of the slender body of revolution of minimum drag using the Newton-Busemann pressure coefficient law, Boeing Sci. Res. Lab., Flight Sci. Lab., Tech. Report No. 62.

Miele, A., ed., 1965, *Theory of Optimum Aerodynamic Shapes,* Academic Press, New York.

Mileiko, S. T., 1997, *Metal and Ceramic Based Composites*, Elsevier Publ., Amsterdam.

Mileiko, S. T., and Sarkisyan, O. A., 1981, Phenomenological model of punch-through, *J. Appl. Mech. Tech. Phys.* No. 5: 711-713.

Mileiko, S. T., Sarkisyan, O. A., and Kondakov, S. F., 1994, Ballistic limits of Al-6% Mg allow laminated by diffusion bonding, *Theoret. Appl. Fract. Mech.* **21**(1): 9-16.

Mines, R. A. W., Roach, A. M., and Jones, N., 1999, High velocity perforation behaviour of polymer composite laminates, *Int. J. Impact Eng.* **22**(6): 561-588.

Miroshin, R. N., and Khalidov, U. A., 1991, *Locaized Interaction Theory*, Leningrad University, Leningrad (in Russian).

Miroshin, R. N., and Khalidov, U. A., 2002, *Local Methods in Continuum Mechanics*, Saint Petersburg University, S. Petersburg (in Russian).

Mostert, F. J., 1999, Penetration of steel penetrators into concrete targets with pre-drilled cavities of different diameters, in: *Proc. 18th Int. Symp. On Ballistics, San Antonio, Texas, 15-19 Nov. 1999.*

Murphy, M. J., 1984, Performance analysis of two-stage munitions, in: *Proc. 8th Int. Symp. on Ballistics, Orlando, Florida*; Lawrence Livermore Nat. Lab., Report No. UCRL-91301.

Naik, N. K., and Shrirao, P., 2004, Composite structures under ballistic impact, *Compos. Struct.* **66**(1-4): 579-590.

Navarro, C., Martinez, M. A., Cortes, R., and Sanchez-Galvez, V., 1993, Some observations on the normal impact on ceramic faced armours backed by composite plates, *Int. J. Impact Eng.* **13**(1): 145-156.

Navarro, C., Zaera, R., Cortes, R., Martinez, M. A., and Sanchez-Galvez, V., 1994, The response of ceramic faced lightweight armours under projectile impact in: *Proc. 3rd Int.*

Conf. on Structures under Shock and Impact, P. S. Bulson, ed., June 1-3, Madrid, 1994, Comput. Mech. Publ., Southampton, pp. 323-330.

Neilson, A. J., 1985, Empirical equations for the perforation of mikd steel plates, *Int. J. Impact Eng.* **3**(2): 137-142.

Newton, I., 1687, *Philosophiae Naturalis Principia Mathematica.* (Mathematical Principles of Natural Philosophy. Translated into English by A. Motte, revised by F. Cajori. University of California Press, 1966).

Nishiwaki, J., 1951, Resistance to the penetration of a bullet through an aluminium plate, *J. Phys. Soc. Japan* **5**: 374-378.

Nixdorff, K., 1983, Some remarks on the penetration theory of J. Awerbuch and S.R. Bodner, *Trans. CSME* **7**(3): 148-153.

Nixdorff, K., 1984a, Application of the penetration theory of J. Awerbuch and S. R. Bodner on multilayered targets, *Zeitschrift Angewandte Math. Mech.* **64**(4): T147-T149.

Nixdorff, K., 1984b, Some applications of the impact theory of J. Awerbuch and S. R. Bodner, *Trans. CSME* **8**(1): 16-20.

Nixdorff, K., 1987a, Discussion of two theories on the penetration of multilayer metallic targets, *Trans. CSME* **11**(3): 161-178.

Nixdorff, K., 1987b, On the efficiency of different head shapes to perforate thin targets, *Trans. CSME* **11**(2): 109-112.

Norwood, F. R., and Sears, M. P., 1982, A nonlinear model for the dynamics of penetration into geological targets, *ASME J. Appl. Mech.* **49**(1): 26-30.

Ostapenko, N. A., 1997, *Optimum Shapes of Bodies Moving in Dense Media*, Vladar, Moscow (in Russian).

Ostapenko, N. A., Romanchenko, V. I., and Yakunina, G. E., 1994, Optimum forms of three-dimensional bodies for penetration of dense media, *J. Appl. Mech. Tech. Phys. No.* **4**: 515-521.

Ostapenko, N. A., and Yakunina, G. E., 1992, Least-drag bodies moving in media subject to locality hypothesis, *Fluid Dynamics* **27**(1): 71-80.

Ostapenko, N. A., and Yakunina, G. E., 1999, The shape of slender three-dimensional bodies with maximum depth of penetration into dense media, *J. Appl. Math. Mech.* **63**(6): 953-967.

Park, M., Yoo, J., and Chung, D.-T., 2005, An optimization of a multi-layered plate under ballistic impact, *Int. J. Solids Struct.* **42**(1): 123-137.

Partom, Y., 1996, Static cavity expansion model for partially confined targets, The Univ. of Texas at Austin, Inst. Adv. Technol., Report No. IAT.R-0092.

Piekutowski, A. J., Forrestal, M. J., Poormon K. L., and Warren, T. L., 1996, Perforation of aluminum plates with ogive-nose steel rods at normal and oblique impacts, *Int. J. Impact Eng.* **18**(7-8): 877-887.

Project THOR, 1961, The Resistance of Various Metallic Materials to Perforation by Steel Fragments; Empirical Relationships for Fragment Residual Velocity and Residual Weigh, The Johns Hopkins Univ. Ballistic Analysis Lab. Tech. Report No. 47.

Qian, L., Yang, Y., and Liu, T., 2000, A semi-analytical model for truncated-ogive-nose projectiles penetration into semi-infinite concrete targets, *Int. J. Impact Eng.* **24**(9): 947-955.

Radin, J., and Goldsmith, W., 1988, Normal projectile penetration and perforation of layered targets, *Int. J. Impact Eng.* **7**(2): 229-259.

Rakhmatulin, Kh. A., Sagomonyan, A. Ya., and Alekseev, N. A., 1964, *Soils Dynamics*, Moscow State University, Moscow (in Russian).

Ravid, M., Bodner, S. R., and Chocron, I. S., 2001, Penetration analysis of ceramic armor with composite material backing, in: *Proc. 19th Int. Symp. on Ballistics,* **3**, I. R. Crewther, ed., Interlaken, Switzerland, TB 45, pp. 1401-1407.

Recht, R. F., 1990, High velocity impact dynamics: Analytical modeling of plate penetration dynamics, in: *High Velocity Impact Dynamics*, J. A Zukas, ed., John Wiley and Sons, New York, pp. 443-513.

Recht, R. F., and Ipson, T. W., 1963, Ballistic perforation dynamics, *ASME J. Appl. Mech.* **30**(3): 384-390.

Reddy, T. Y., Wen, H. M., Reid, S. R., and Soden, P. D., 1998, Penetration and perforation of composite sandwich panels by hemispherical and conical projectiles, *Trans ASME. J. Pres. Ves. Techn.* **120**(2): 186-194.

Reid, S. R., Reddy, T. Y, Ho, H. M., Crouch, I. G., and Greaves, L. J., 1995, Dynamic indentation of thick fibre-reinforced composites, in: *High Rate Effects on Polymer, Metal and Ceramic Matrix Composites and Other Advanced Materials,* Rajapakse Y. D. S, and Vinson J. R., eds., ASME, AD-vol. **48**, pp. 71-79.

Reid, S. R., and Wen, H. M., 2000, Perforation of FRP laminates and sandwich panels subjected to missile impact, in: *Impact Behaviour of Fibre-Reinforced Composite Materials and Structures*, S. R. Reid and G. Zhou, eds., Woodhead Publ., Cambridge.

Reid, S. R., Wen, H. M., Soden, P. D., and Reddy, T. Y., 1999, Response of single skin laminates and sandwich panels to projectile impact, in: *Composite Materials for Offshore Operation-2*, S. S. Wang, J. J. Williams, and K. H. Lo, eds., Amer. Bur. Shipp., pp. 593-617.

Reid, S. R., and Zhou, G., eds., 2000, *Impact Behaviour of Fibre-Reinforced Composite Materials and Structures*, Woodhead Publ., Cambridge.

Reijer den, P. S., 1991, Impact on ceramic faced armours, Ph.D. Thesis, Delft University of Technology, Netherlands.

Reyes-Villanueva, G., and Cantwell, W. J., 2004, The high velocity impact response of composite and FML-reinforced sandwich structures, *Compos. Sci. Technol.* **64**(1): 35-54.

Rockafellar, R. T., 1970, *Convex Analysis*, Princeton Univ. Press, New Jersey.

Roeder, B. A., and Sun, C. T., 2001, Dynamic penetration of alumina/aluminum laminates: experiments and modeling, *Int. J. Impact Eng.* **25**(2): 169-185.

Rosenberg, Z., and Forrestal, M. J., 1988, Perforation of aluminum plates with conical-nosed rods-additional data and discussion, *ASME J. Appl. Mech.* **55**(1): 236-238.

Sadanandan, S., and Hetherington, J. G., 1997, Characterization of ceramic/steel and ceramic/aluminum armours subjected to oblique impact, *Int. J. Impact Eng.* **19**(9-10): 811-819.

Sagomonyan, A. Ya., 1960, Penetration of sharp bodies of revolution into soils. *Doklady Akademii Nauk SSSR*, **134**(6): 1320-1323 (in Russian).

Sagomonyan, A. Ya., 1974, *Penetration of Solids into Compressible Continuous Media,* Moscow Univ. Publ., Moscow (In Russian).

Sagomonyan, A. Ya., 1975, Plate piercing by a slender solid projectile. *Vestnik Moskovskogo Universiteta, Seria 1, Matematika, Mekhanika,* No. 5: 104-111 (in Russian).

Sagomonyan, A. Ya., 1988, *Dynamics of Barriers Perforation,* Moscow State University, Moscow (in Russian).

Satapathy, S., 1997, Application of cavity expansion analysis to penetration problems, The Univ. of Texas at Austin, Inst. Adv. Technol., Report No. IAT.R-0136.

Satapathy, S., 2001, Dynamic spherical cavity expansion in brittle ceramics, *Int. J. Solids Struct.* **38**(32-33): 5833-5845.

Satapathy, S., and Bless, S., 1996, Calculation of penetration resistance of brittle materials using spherical cavity expansion analysis, *Mech. Mat.* **23**(4): 323-330.

Satapathy, S., and Bless, S., 2000, Cavity expansion resistance of brittle materials obeying a two-curve pressure-shear behavior, *J. Appl. Phys.* **88**(7): 4004-4012.

Seliverstov, V. S., 2001, Estimation and optimization of resistibility of layered barriers to perforation, in: *Proc. VI Zababakhin Scientific Talks (ZST-2001)*, Snezhinsk, Russia (in Russian).

Shirai, T., Kambayashi, A., Ohno T., Taniguchi, H., Ueda, M., and Ishikawa, N., 1997, Experiment and numerical simulation of double-layered RC plates under impact loading, *Nuclear Eng. Design*, **176**(3): 195-205.

Siddiqui, N. A., Choudhury, M. A., and Abbas, H., 2002, Reliability analysis of projectile penetration into geological targets, *Reliability Eng. Syst. Safety* **78**(1): 13-19.

Silva, C. J., and Torres, D. F. M., 2004, On the classical Newton's problem of minimal resistance, in: *III Junior European Meeting on Control, Optimization and Computation*, Univ. of Aveiro, Portugal.

Sjøl, H., and Teland, J. A., 2000, Prediction of concrete penetration using Forrestal's formula, Norwegian Defence Res. Establishment, Report No. FFI/RAPPORT-99/04415.

Sjøl, H., Teland, J. A., and Kaldheim, Ø., 2002, Penetration into concrete – Analysis of small scale experiments with 12 mm projectiles, Norwegian Defence Res. Establishment, Report No. FFI/RAPPORT-2002/04867.

Smith, P. D., and Hetherington, J. G., 1994, Blast and Ballistic Loading of Structures, Butterworth Heinemann, Oxford.

Stone, G. W., 1994, Projectile penetration into representative targets. Sandia Nat. Lab., Albuquerque, NM, Report No. SAND-94-1490.

Szendrei, T., 2000, Resistance of geomaterials to rigid projectile following damage by shaped charge jet penetration, in: *Dynamic Physics Consultants CC*, Johannesburg, South Africa, Febr. 2000.

Teland, J. A., 1998, A review of empirical equations for missile impact effects on concrete, Norwegian Defence Res. Establishment, Report No. FFI/RAPPORT-97/05856.

Teland, J. A., 1999, A review of analytical penetration mechanics, Norwegian Defence Res. Establishment, Report No. FFI/RAPPORT-99/01264.

Teland, J. A., 2001a, A first approach to penetration of tandem charges into concrete, Norwegian Defence Res. Establishment, Report No. FFI/RAPPORT-2001/00624.

Teland, J. A., 2001b, Cavity expansion theory applied to penetration of targets with pre-drilled cavities, in: *Proc. 19th Int. Symp. on Ballistics,* 3, I. R. Crewther, ed., Interlaken, Switzerland, TB 36, pp. 1329-1335.

Teland, J. A., 2002, Multifunctional numerical tool for penetration analysis, Norwegian Defence Res. Establishment, Report No. FFI/RAPPORT-2002/04647.

Teland, J. A., and Moxnes, J. F., 2003, Analytical cavity expansion penetration models compared with numerical simulations, Norwegian Defence Res. Establishment, Report No. FFI/RAPPORT-2003/00934.

Teland, J. A., and Sjøl, H., 2000, Boundary effects in penetration into concrete, Norwegian Defence Res. Establishment, Report No. FFI/RAPPORT-2000/05414.

Teland, J. A., and Sjøl, H., 2004, Penetration into concrete by truncated projectiles, *Int. J. Impact Eng.* **30**(4): 447-464.

Thomson, W. T., 1955, An approximate theory of armor penetration, *J. Appl. Phys.* **26**(1): 80-82.

Ulven, C., Vaidya, U. K., and Hosur, M. V., 2003, Effect of projectile shape during ballistic perforation of VARTM carbon/epoxy composite panels, *Compos. Struct.* **61**(1-2): 143-150.

Vaidya, U. K., Kulkarni, M., Haque, A., Hosur, M. V., and Kulkarni, R., 2000, Ballistic performance of graphite/epoxy and S2-glass/epoxy composites with polycarbonate facing, *Mat. Technol.* **15**(3): 202-214.

Vedernikov, Y. A., Khudiakov, Y. S., and Omelaev, A. I., 1995, *Ballistics: From Arrows to Rockets*, Nauka, Novosibirsk (in Russian).

Vedernikov, Y. A., and Shchepanovsly, V. A., 1995, *Optimization of Reagasdynamic Systems*, Nauka, Novosibirsk (in Russian).

Vitman, F. F., and Ioffe, B. S., 1948, A simple method of determining the dynamical hardness of metals using a double cone, *Zavodskaja Laboratorija* **14**(6): 727-732 (in Russian).

Vitman, F. F., and Stepanov, V. A., 1959, Effect of the strain rate on the resistance of metals to deformation at impact velocities of 100-1000 m/s, in: *Nekotoryje Problemy Prochnosti Tvjordogo Tela.* USSR Acad. of Sci., Moscow-Leningrad, pp. 207-221 (in Russian).

Wang, B., and Lu, G., 1996, On the optimisation of two-component plates against ballistic impact, *J. Mat. Proc. Technol.* **57**(1-2): 141-145.

Warren, T. L., 1999, The effect of strain rate on the dynamic expansion of cylindrical cavities, *ASME J. Appl. Mech.* **66**(3): 818-821.

Warren, T. L., and Forrestal, M. J., 1998, Effect of strain hardening and strain rate sensitivity on penetration of aluminum targets with spherical-nosed rods, *Int. J. Solids Struct.* **35**(28-29): 3737-3753.

Warren, T. L., Hanchak, S. J., and Poormon, K. L., 2001, Penetration of limestone targets by ogive-nosed VAR 4340 steel projectiles at oblique angles: experiments and simulations, *Int. J. Impact Eng.* **30**(10): 1307-1331.

Warren, T. L., and Poormon, K. L., 2001, Penetration of 6061-T6511 aluminum targets by ogive-nosed VAR 4340 steel projectiles at oblique angles: experiments and simulations, *Int. J. Impact Eng.* **25**(10): 993-1022.

Weidemaier, P., Senf, H., Rothenhäusler, H., Filbey, G. L, and Gooch, W. A, 1993, On the ballistic resistance of laminated steel targets: experiments and numerical calculations, in: *14th Int. Symp. On Ballistics, Quebec, Canada, 26-29 Sent. 1993,* pp. 681-690.

Wen, H. M., 2000, Predicting the penetration and perforation of FRP laminates struck normally by projectiles with different nose shapes, *Compos. Struct.* **49**(3): 321-329.

Wen, H. M., 2001, Penetration and perforation of thick FRP laminates, *Compos. Sci. Technol.* **61**(8): 1163-1172.

Wen, H. M., 2002a, Predicting the penetration and perforation of targets struck by projectiles at normal incidence, *Mech. Struct. Machines* **30**(4): 543-577.

Wen, H. M., 2002b, Penetration and perforation of thick metallic targets under impact by missiles, *Chinese J. High Pressure Phys.* **16**(2): 94-104.

Wen, H. M., and Jones, N., 1996, Low-velocity perforation of punch-impact-loaded metal plates, *J. Pressure Vessel Technol.* **118**(2): 181-187.

Wen, H. M., Reddy, T. Y., Reid, S. R., and Soden, P. D., 1998, Indentation penetration and perforation of composite laminates and sandwich panels under quasi-static and projectile loading, *Key Eng Mat.* **141-143**: 501-552.

Wierzbicki, T., 1999, Petalling of plates under explosive and impact loading, *Int. J. Impact Eng.* **22**(9-10): 935-954.

Wilkins, M. L., 1978, Mechanics of penetration and perforation, *Int. J. of Eng. Sci.* **16**(11): 793-807.

Woodward, R. L., 1990, A simple one-dimensional approach to modeling ceramic composite armor defeat, *Int. J. Impact Eng.* **9**(4): 455-474.

Woodward, R. L., and Cimpoeru, S. J., 1998, A study of the perforation of aluminium laminate targets, *Int. J. Impact Eng.* **21**(3): 117-131.

Xu, Y., Keer, L. M., and Luk, V. K., 1997, Elastic-cracked model for penetration into unreinforced concrete targets with ogival nose projectiles, *Int. J. Solids Struct.* **34**(12): 1479-1491.

Yakunina, G. E., 2000a, The construction of three-dimensional shapes within the framework of a model of local interaction, *J. Appl. Math. Mech.* **64**(2): 289-298.

Yakunina, G. E., 2000b, The optimum non-conical and asymmetrical three-dimensional configurations, *J. Appl. Math. Mech.* **64**(4): 583-591.

Yakunina, G. E., 2001, On body shapes providing maximum penetration depth in dense media, *Doklady Physics*, **46**(2): 140-143.

Yakunina, G. E., 2002, Three-Dimensional bodies of minimum drag in hypersonic flow, *J. of Optim. Theory and Appl.* **15**(2): 241-265.

Yankelevsky, D. Z., 1983a, Projectile penetration through a narrow drill in soil, *Int. J. Impact Eng.* **1**(4): 377-391.

Yankelevsky, D. Z., 1983b, Optimal shape of an earth penetrating projectile, *Int. J. Solids Struct.* **19**(1): 25-31.

Yankelevsky, D. Z., 1997, Local response of concrete slabs to low velocity missile impact, *Int. J. Impact Eng.* **19**(4): 331-343.

Yankelevsky, D. Z., and Adin, M. A., 1980, A simplified analytical method for soil penetration analysis, *Int. J. Numer. Anal. Meth. Geomech.* **4**(3): 233-254.

Yankelevsky, D. Z., and Gluck, J., 1980, Nose shape effect on high velocity soil penetration, *Int. J. Mech. Sci.* **22**(5): 297-311.

Yossifon, G., Yarin, A. L., and Rubin, M. B., 2002, Penetration of a rigid projectile into a multi-layered target: theory and numerical computations, *Int. J. Eng. Sci.* **40**(12): 1381-1401.

Young, C. W., 1972, Empirical equations for predicting penetration performation in layered earth materials for complex penetrator configurations, Sandia Nat. Lab., Albuquerque, NM, Report No. SC-DR-72-0523.

Young, C. W., 1997, Penetration equations, Sandia Nat. Lab., Albuquerque, NM, Report No. SAND97-2426.

Yu, H. S., 2000, *Cavity Expansion Methods in Geomechanics*, Kluwer Acad. Publ., Dordrecht.

Zaera, R. and Sanchez-Galvez, V., 1998, Analytical modeling of normal and oblique ballistic impact on ceramic/metal lightweight armors, *Int. J. Impact Eng.* **21**(3): 133-148.

Zaera, R., Sanchez-Saez, S., Perez-Castellanos, J. L. and Navarro, C., 2000, Modelling of the adhesive layer in mixed ceramic/metal armours subjected to impact, *Composites. Part A.* **31**(8): 823-833.

Zaid, A. I. O., El-Kalai, A., and Travis, F. W., 1973, An examination of the perforation of mild steel plate by a flat-ended cylindrical projectile, *Int. J. Mech. Sci.,* **15**(2): 129-143.

Zhang, Z., Shen, J., Zhong, W., and Sun, Z., 2002, A dynamic model of ceramic/fibre-reinforced plastic hybrid composites under projectile striking, *Proc. of the Inst. of Mechanical Eng., Part G,* **216**: 325-331.

Zhu, G., Goldsmith, W., Dharan, C. K. H., 1992, Penetration of laminated Kevlar by projectiles - II. Experimental investigation, *Int. J. Solids Struct.* **29**(4): 399-420.

Zook, J., 1977, An analytical model of kinetic energy projectile/fragment penetration, Army Ballistic Res. Lab., Aberdeen Proving Ground, MD, Report No. BRL-MR-2797.

Zukas, J. A., 1982, Penetration and perforation of solids, in: *Impact Dynamics*, J. A. Zukas, T. Nicholas, H. F. Swift, L. B. Greszczuk, and D. R. Curran, eds., J. Wiley and Sons, New York, pp. 155-214.

Zukas, J. A., ed., 1990, *High Velocity Impact Dynamics*, John Wiley and Sons, New York.

Zukas, J. A., 1996, Effect of lamination and spacing on finite thickness plate perforation, in: *Structures Under Shock and Impact IV*, N. Jones, C. A. Brebbia and J. A. Watson, eds., Comput. Mech. Publ., Southampton, pp. 103-115.

Zukas, J. A., and Scheffler, D. R., 2001, Impact effects in multilayered plates, *Int. J. Solids Struct.* **38**(19): 3321-3328.

Zukas, J. A., and Walters, W. P., 1990, Analytical models for kinetic energy penetration, in: *High Velocity Impact Dynamics*, J. A. Zukas, ed., John Wiley and Sons, New York, pp. 405-441.

Author Index

Mechanics

SOLID MECHANICS AND ITS APPLICATIONS

Series Editor: G.M.L. Gladwell

Aims and Scope of the Series

The fundamental questions arising in mechanics are: *Why?*, *How?*, and *How much?* The aim of this series is to provide lucid accounts written by authoritative researchers giving vision and insight in answering these questions on the subject of mechanics as it relates to solids. The scope of the series covers the entire spectrum of solid mechanics. Thus it includes the foundation of mechanics; variational formulations; computational mechanics; statics, kinematics and dynamics of rigid and elastic bodies; vibrations of solids and structures; dynamical systems and chaos; the theories of elasticity, plasticity and viscoelasticity; composite materials; rods, beams, shells and membranes; structural control and stability; soils, rocks and geomechanics; fracture; tribology; experimental mechanics; biomechanics and machine design.

1. R.T. Haftka, Z. Gürdal and M.P. Kamat: *Elements of Structural Optimization*. 2nd rev.ed., 1990 ISBN 0-7923-0608-2
2. J.J. Kalker: *Three-Dimensional Elastic Bodies in Rolling Contact.* 1990 ISBN 0-7923-0712-7
3. P. Karasudhi: *Foundations of Solid Mechanics.* 1991 ISBN 0-7923-0772-0
4. *Not published*
5. *Not published.*
6. J.F. Doyle: *Static and Dynamic Analysis of Structures.* With an Emphasis on Mechanics and Computer Matrix Methods. 1991 ISBN 0-7923-1124-8; Pb 0-7923-1208-2
7. O.O. Ochoa and J.N. Reddy: *Finite Element Analysis of Composite Laminates.* ISBN 0-7923-1125-6
8. M.H. Aliabadi and D.P. Rooke: *Numerical Fracture Mechanics.* ISBN 0-7923-1175-2
9. J. Angeles and C.S. López-Cajún: *Optimization of Cam Mechanisms.* 1991 ISBN 0-7923-1355-0
10. D.E. Grierson, A. Franchi and P. Riva (eds.): *Progress in Structural Engineering.* 1991 ISBN 0-7923-1396-8
11. R.T. Haftka and Z. Gürdal: *Elements of Structural Optimization.* 3rd rev. and exp. ed. 1992 ISBN 0-7923-1504-9; Pb 0-7923-1505-7
12. J.R. Barber: *Elasticity.* 1992 ISBN 0-7923-1609-6; Pb 0-7923-1610-X
13. H.S. Tzou and G.L. Anderson (eds.): *Intelligent Structural Systems.* 1992 ISBN 0-7923-1920-6
14. E.E. Gdoutos: *Fracture Mechanics.* An Introduction. 1993 ISBN 0-7923-1932-X
15. J.P. Ward: *Solid Mechanics.* An Introduction. 1992 ISBN 0-7923-1949-4
16. M. Farshad: *Design and Analysis of Shell Structures.* 1992 ISBN 0-7923-1950-8
17. H.S. Tzou and T. Fukuda (eds.): *Precision Sensors, Actuators and Systems.* 1992 ISBN 0-7923-2015-8
18. J.R. Vinson: *The Behavior of Shells Composed of Isotropic and Composite Materials.* 1993 ISBN 0-7923-2113-8
19. H.S. Tzou: *Piezoelectric Shells.* Distributed Sensing and Control of Continua. 1993 ISBN 0-7923-2186-3
20. W. Schiehlen (ed.): *Advanced Multibody System Dynamics.* Simulation and Software Tools. 1993 ISBN 0-7923-2192-8
21. C.-W. Lee: *Vibration Analysis of Rotors.* 1993 ISBN 0-7923-2300-9
22. D.R. Smith: *An Introduction to Continuum Mechanics.* 1993 ISBN 0-7923-2454-4
23. G.M.L. Gladwell: *Inverse Problems in Scattering.* An Introduction. 1993 ISBN 0-7923-2478-1

Mechanics

SOLID MECHANICS AND ITS APPLICATIONS

Series Editor: G.M.L. Gladwell

24. G. Prathap: *The Finite Element Method in Structural Mechanics*. 1993 ISBN 0-7923-2492-7
25. J. Herskovits (ed.): *Advances in Structural Optimization*. 1995 ISBN 0-7923-2510-9
26. M.A. González-Palacios and J. Angeles: *Cam Synthesis*. 1993 ISBN 0-7923-2536-2
27. W.S. Hall: *The Boundary Element Method*. 1993 ISBN 0-7923-2580-X
28. J. Angeles, G. Hommel and P. Kovács (eds.): *Computational Kinematics*. 1993 ISBN 0-7923-2585-0
29. A. Curnier: *Computational Methods in Solid Mechanics*. 1994 ISBN 0-7923-2761-6
30. D.A. Hills and D. Nowell: *Mechanics of Fretting Fatigue*. 1994 ISBN 0-7923-2866-3
31. B. Tabarrok and F.P.J. Rimrott: *Variational Methods and Complementary Formulations in Dynamics*. 1994 ISBN 0-7923-2923-6
32. E.H. Dowell (ed.), E.F. Crawley, H.C. Curtiss Jr., D.A. Peters, R. H. Scanlan and F. Sisto: *A Modern Course in Aeroelasticity*. Third Revised and Enlarged Edition. 1995 ISBN 0-7923-2788-8; Pb: 0-7923-2789-6
33. A. Preumont: *Random Vibration and Spectral Analysis*. 1994 ISBN 0-7923-3036-6
34. J.N. Reddy (ed.): *Mechanics of Composite Materials*. Selected works of Nicholas J. Pagano. 1994 ISBN 0-7923-3041-2
35. A.P.S. Selvadurai (ed.): *Mechanics of Poroelastic Media*. 1996 ISBN 0-7923-3329-2
36. Z. Mróz, D. Weichert, S. Dorosz (eds.): *Inelastic Behaviour of Structures under Variable Loads*. 1995 ISBN 0-7923-3397-7
37. R. Pyrz (ed.): *IUTAM Symposium on Microstructure-Property Interactions in Composite Materials*. Proceedings of the IUTAM Symposium held in Aalborg, Denmark. 1995 ISBN 0-7923-3427-2
38. M.I. Friswell and J.E. Mottershead: *Finite Element Model Updating in Structural Dynamics*. 1995 ISBN 0-7923-3431-0
39. D.F. Parker and A.H. England (eds.): *IUTAM Symposium on Anisotropy, Inhomogeneity and Nonlinearity in Solid Mechanics*. Proceedings of the IUTAM Symposium held in Nottingham, U.K. 1995 ISBN 0-7923-3594-5
40. J.-P. Merlet and B. Ravani (eds.): *Computational Kinematics '95*. 1995 ISBN 0-7923-3673-9
41. L.P. Lebedev, I.I. Vorovich and G.M.L. Gladwell: *Functional Analysis*. Applications in Mechanics and Inverse Problems. 1996 ISBN 0-7923-3849-9
42. J. Menčik: *Mechanics of Components with Treated or Coated Surfaces*. 1996 ISBN 0-7923-3700-X
43. D. Bestle and W. Schiehlen (eds.): *IUTAM Symposium on Optimization of Mechanical Systems*. Proceedings of the IUTAM Symposium held in Stuttgart, Germany. 1996 ISBN 0-7923-3830-8
44. D.A. Hills, P.A. Kelly, D.N. Dai and A.M. Korsunsky: *Solution of Crack Problems*. The Distributed Dislocation Technique. 1996 ISBN 0-7923-3848-0
45. V.A. Squire, R.J. Hosking, A.D. Kerr and P.J. Langhorne: *Moving Loads on Ice Plates*. 1996 ISBN 0-7923-3953-3
46. A. Pineau and A. Zaoui (eds.): *IUTAM Symposium on Micromechanics of Plasticity and Damage of Multiphase Materials*. Proceedings of the IUTAM Symposium held in Sèvres, Paris, France. 1996 ISBN 0-7923-4188-0
47. A. Naess and S. Krenk (eds.): *IUTAM Symposium on Advances in Nonlinear Stochastic Mechanics*. Proceedings of the IUTAM Symposium held in Trondheim, Norway. 1996 ISBN 0-7923-4193-7
48. D. Ieşan and A. Scalia: *Thermoelastic Deformations*. 1996 ISBN 0-7923-4230-5

Mechanics

SOLID MECHANICS AND ITS APPLICATIONS

Series Editor: G.M.L. Gladwell

49. J.R. Willis (ed.): *IUTAM Symposium on Nonlinear Analysis of Fracture*. Proceedings of the IUTAM Symposium held in Cambridge, U.K. 1997 ISBN 0-7923-4378-6
50. A. Preumont: *Vibration Control of Active Structures*. An Introduction. 1997 ISBN 0-7923-4392-1
51. G.P. Cherepanov: *Methods of Fracture Mechanics: Solid Matter Physics*. 1997 ISBN 0-7923-4408-1
52. D.H. van Campen (ed.): *IUTAM Symposium on Interaction between Dynamics and Control in Advanced Mechanical Systems*. Proceedings of the IUTAM Symposium held in Eindhoven, The Netherlands. 1997 ISBN 0-7923-4429-4
53. N.A. Fleck and A.C.F. Cocks (eds.): *IUTAM Symposium on Mechanics of Granular and Porous Materials*. Proceedings of the IUTAM Symposium held in Cambridge, U.K. 1997 ISBN 0-7923-4553-3
54. J. Roorda and N.K. Srivastava (eds.): *Trends in Structural Mechanics*. Theory, Practice, Education. 1997 ISBN 0-7923-4603-3
55. Yu.A. Mitropolskii and N. Van Dao: *Applied Asymptotic Methods in Nonlinear Oscillations*. 1997 ISBN 0-7923-4605-X
56. C. Guedes Soares (ed.): *Probabilistic Methods for Structural Design*. 1997 ISBN 0-7923-4670-X
57. D. François, A. Pineau and A. Zaoui: *Mechanical Behaviour of Materials*. Volume I: Elasticity and Plasticity. 1998 ISBN 0-7923-4894-X
58. D. François, A. Pineau and A. Zaoui: *Mechanical Behaviour of Materials*. Volume II: Viscoplasticity, Damage, Fracture and Contact Mechanics. 1998 ISBN 0-7923-4895-8
59. L.T. Tenek and J. Argyris: *Finite Element Analysis for Composite Structures*. 1998 ISBN 0-7923-4899-0
60. Y.A. Bahei-El-Din and G.J. Dvorak (eds.): *IUTAM Symposium on Transformation Problems in Composite and Active Materials*. Proceedings of the IUTAM Symposium held in Cairo, Egypt. 1998 ISBN 0-7923-5122-3
61. I.G. Goryacheva: *Contact Mechanics in Tribology*. 1998 ISBN 0-7923-5257-2
62. O.T. Bruhns and E. Stein (eds.): *IUTAM Symposium on Micro- and Macrostructural Aspects of Thermoplasticity*. Proceedings of the IUTAM Symposium held in Bochum, Germany. 1999 ISBN 0-7923-5265-3
63. F.C. Moon: *IUTAM Symposium on New Applications of Nonlinear and Chaotic Dynamics in Mechanics*. Proceedings of the IUTAM Symposium held in Ithaca, NY, USA. 1998 ISBN 0-7923-5276-9
64. R. Wang: *IUTAM Symposium on Rheology of Bodies with Defects*. Proceedings of the IUTAM Symposium held in Beijing, China. 1999 ISBN 0-7923-5297-1
65. Yu.I. Dimitrienko: *Thermomechanics of Composites under High Temperatures*. 1999 ISBN 0-7923-4899-0
66. P. Argoul, M. Frémond and Q.S. Nguyen (eds.): *IUTAM Symposium on Variations of Domains and Free-Boundary Problems in Solid Mechanics*. Proceedings of the IUTAM Symposium held in Paris, France. 1999 ISBN 0-7923-5450-8
67. F.J. Fahy and W.G. Price (eds.): *IUTAM Symposium on Statistical Energy Analysis*. Proceedings of the IUTAM Symposium held in Southampton, U.K. 1999 ISBN 0-7923-5457-5
68. H.A. Mang and F.G. Rammerstorfer (eds.): *IUTAM Symposium on Discretization Methods in Structural Mechanics*. Proceedings of the IUTAM Symposium held in Vienna, Austria. 1999 ISBN 0-7923-5591-1

Mechanics

SOLID MECHANICS AND ITS APPLICATIONS

Series Editor: G.M.L. Gladwell

69. P. Pedersen and M.P. Bendsøe (eds.): *IUTAM Symposium on Synthesis in Bio Solid Mechanics.* Proceedings of the IUTAM Symposium held in Copenhagen, Denmark. 1999 ISBN 0-7923-5615-2
70. S.K. Agrawal and B.C. Fabien: *Optimization of Dynamic Systems.* 1999 ISBN 0-7923-5681-0
71. A. Carpinteri: *Nonlinear Crack Models for Nonmetallic Materials.* 1999 ISBN 0-7923-5750-7
72. F. Pfeifer (ed.): *IUTAM Symposium on Unilateral Multibody Contacts.* Proceedings of the IUTAM Symposium held in Munich, Germany. 1999 ISBN 0-7923-6030-3
73. E. Lavendelis and M. Zakrzhevsky (eds.): *IUTAM/IFToMM Symposium on Synthesis of Nonlinear Dynamical Systems.* Proceedings of the IUTAM/IFToMM Symposium held in Riga, Latvia. 2000 ISBN 0-7923-6106-7
74. J.-P. Merlet: *Parallel Robots.* 2000 ISBN 0-7923-6308-6
75. J.T. Pindera: *Techniques of Tomographic Isodyne Stress Analysis.* 2000 ISBN 0-7923-6388-4
76. G.A. Maugin, R. Drouot and F. Sidoroff (eds.): *Continuum Thermomechanics.* The Art and Science of Modelling Material Behaviour. 2000 ISBN 0-7923-6407-4
77. N. Van Dao and E.J. Kreuzer (eds.): *IUTAM Symposium on Recent Developments in Non-linear Oscillations of Mechanical Systems.* 2000 ISBN 0-7923-6470-8
78. S.D. Akbarov and A.N. Guz: *Mechanics of Curved Composites.* 2000 ISBN 0-7923-6477-5
79. M.B. Rubin: *Cosserat Theories: Shells, Rods and Points.* 2000 ISBN 0-7923-6489-9
80. S. Pellegrino and S.D. Guest (eds.): *IUTAM-IASS Symposium on Deployable Structures: Theory and Applications.* Proceedings of the IUTAM-IASS Symposium held in Cambridge, U.K., 6–9 September 1998. 2000 ISBN 0-7923-6516-X
81. A.D. Rosato and D.L. Blackmore (eds.): *IUTAM Symposium on Segregation in Granular Flows.* Proceedings of the IUTAM Symposium held in Cape May, NJ, U.S.A., June 5–10, 1999. 2000 ISBN 0-7923-6547-X
82. A. Lagarde (ed.): *IUTAM Symposium on Advanced Optical Methods and Applications in Solid Mechanics.* Proceedings of the IUTAM Symposium held in Futuroscope, Poitiers, France, August 31–September 4, 1998. 2000 ISBN 0-7923-6604-2
83. D. Weichert and G. Maier (eds.): *Inelastic Analysis of Structures under Variable Loads.* Theory and Engineering Applications. 2000 ISBN 0-7923-6645-X
84. T.-J. Chuang and J.W. Rudnicki (eds.): *Multiscale Deformation and Fracture in Materials and Structures.* The James R. Rice 60th Anniversary Volume. 2001 ISBN 0-7923-6718-9
85. S. Narayanan and R.N. Iyengar (eds.): *IUTAM Symposium on Nonlinearity and Stochastic Structural Dynamics.* Proceedings of the IUTAM Symposium held in Madras, Chennai, India, 4–8 January 1999 ISBN 0-7923-6733-2
86. S. Murakami and N. Ohno (eds.): *IUTAM Symposium on Creep in Structures.* Proceedings of the IUTAM Symposium held in Nagoya, Japan, 3-7 April 2000. 2001 ISBN 0-7923-6737-5
87. W. Ehlers (ed.): *IUTAM Symposium on Theoretical and Numerical Methods in Continuum Mechanics of Porous Materials.* Proceedings of the IUTAM Symposium held at the University of Stuttgart, Germany, September 5-10, 1999. 2001 ISBN 0-7923-6766-9
88. D. Durban, D. Givoli and J.G. Simmonds (eds.): *Advances in the Mechanis of Plates and Shells The Avinoam Libai Anniversary Volume.* 2001 ISBN 0-7923-6785-5
89. U. Gabbert and H.-S. Tzou (eds.): *IUTAM Symposium on Smart Structures and Structonic Systems.* Proceedings of the IUTAM Symposium held in Magdeburg, Germany, 26–29 September 2000. 2001 ISBN 0-7923-6968-8

Mechanics

SOLID MECHANICS AND ITS APPLICATIONS

Series Editor: G.M.L. Gladwell

90. Y. Ivanov, V. Cheshkov and M. Natova: *Polymer Composite Materials – Interface Phenomena & Processes*. 2001 ISBN 0-7923-7008-2
91. R.C. McPhedran, L.C. Botten and N.A. Nicorovici (eds.): *IUTAM Symposium on Mechanical and Electromagnetic Waves in Structured Media*. Proceedings of the IUTAM Symposium held in Sydney, NSW, Australia, 18-22 Januari 1999. 2001 ISBN 0-7923-7038-4
92. D.A. Sotiropoulos (ed.): *IUTAM Symposium on Mechanical Waves for Composite Structures Characterization*. Proceedings of the IUTAM Symposium held in Chania, Crete, Greece, June 14-17, 2000. 2001 ISBN 0-7923-7164-X
93. V.M. Alexandrov and D.A. Pozharskii: *Three-Dimensional Contact Problems*. 2001 ISBN 0-7923-7165-8
94. J.P. Dempsey and H.H. Shen (eds.): *IUTAM Symposium on Scaling Laws in Ice Mechanics and Ice Dynamics*. Proceedings of the IUTAM Symposium held in Fairbanks, Alaska, U.S.A., 13-16 June 2000. 2001 ISBN 1-4020-0171-1
95. U. Kirsch: *Design-Oriented Analysis of Structures*. A Unified Approach. 2002 ISBN 1-4020-0443-5
96. A. Preumont: *Vibration Control of Active Structures*. An Introduction (2nd Edition). 2002 ISBN 1-4020-0496-6
97. B.L. Karihaloo (ed.): *IUTAM Symposium on Analytical and Computational Fracture Mechanics of Non-Homogeneous Materials*. Proceedings of the IUTAM Symposium held in Cardiff, U.K., 18-22 June 2001. 2002 ISBN 1-4020-0510-5
98. S.M. Han and H. Benaroya: *Nonlinear and Stochastic Dynamics of Compliant Offshore Structures*. 2002 ISBN 1-4020-0573-3
99. A.M. Linkov: *Boundary Integral Equations in Elasticity Theory*. 2002 ISBN 1-4020-0574-1
100. L.P. Lebedev, I.I. Vorovich and G.M.L. Gladwell: *Functional Analysis*. Applications in Mechanics and Inverse Problems (2nd Edition). 2002 ISBN 1-4020-0667-5; Pb: 1-4020-0756-6
101. Q.P. Sun (ed.): *IUTAM Symposium on Mechanics of Martensitic Phase Transformation in Solids*. Proceedings of the IUTAM Symposium held in Hong Kong, China, 11-15 June 2001. 2002 ISBN 1-4020-0741-8
102. M.L. Munjal (ed.): *IUTAM Symposium on Designing for Quietness*. Proceedings of the IUTAM Symposium held in Bangkok, India, 12-14 December 2000. 2002 ISBN 1-4020-0765-5
103. J.A.C. Martins and M.D.P. Monteiro Marques (eds.): *Contact Mechanics*. Proceedings of the 3rd Contact Mechanics International Symposium, Praia da Consolação, Peniche, Portugal, 17-21 June 2001. 2002 ISBN 1-4020-0811-2
104. H.R. Drew and S. Pellegrino (eds.): *New Approaches to Structural Mechanics, Shells and Biological Structures*. 2002 ISBN 1-4020-0862-7
105. J.R. Vinson and R.L. Sierakowski: *The Behavior of Structures Composed of Composite Materials*. Second Edition. 2002 ISBN 1-4020-0904-6
106. Not yet published.
107. J.R. Barber: *Elasticity*. Second Edition. 2002 ISBN Hb 1-4020-0964-X; Pb 1-4020-0966-6
108. C. Miehe (ed.): *IUTAM Symposium on Computational Mechanics of Solid Materials at Large Strains*. Proceedings of the IUTAM Symposium held in Stuttgart, Germany, 20-24 August 2001. 2003 ISBN 1-4020-1170-9

Mechanics

SOLID MECHANICS AND ITS APPLICATIONS

Series Editor: G.M.L. Gladwell

109. P. Ståhle and K.G. Sundin (eds.): *IUTAM Symposium on Field Analyses for Determination of Material Parameters – Experimental and Numerical Aspects.* Proceedings of the IUTAM Symposium held in Abisko National Park, Kiruna, Sweden, July 31 – August 4, 2000. 2003 ISBN 1-4020-1283-7
110. N. Sri Namachchivaya and Y.K. Lin (eds.): *IUTAM Symposium on Nonlinear Stochastic Dynamics.* Proceedings of the IUTAM Symposium held in Monticello, IL, USA, 26 – 30 August, 2000. 2003 ISBN 1-4020-1471-6
111. H. Sobieckzky (ed.): *IUTAM Symposium Transsonicum IV.* Proceedings of the IUTAM Symposium held in Göttingen, Germany, 2–6 September 2002, 2003 ISBN 1-4020-1608-5
112. J.-C. Samin and P. Fisette: *Symbolic Modeling of Multibody Systems.* 2003 ISBN 1-4020-1629-8
113. A.B. Movchan (ed.): *IUTAM Symposium on Asymptotics, Singularities and Homogenisation in Problems of Mechanics.* Proceedings of the IUTAM Symposium held in Liverpool, United Kingdom, 8-11 July 2002. 2003 ISBN 1-4020-1780-4
114. S. Ahzi, M. Cherkaoui, M.A. Khaleel, H.M. Zbib, M.A. Zikry and B. LaMatina (eds.): *IUTAM Symposium on Multiscale Modeling and Characterization of Elastic-Inelastic Behavior of Engineering Materials.* Proceedings of the IUTAM Symposium held in Marrakech, Morocco, 20-25 October 2002. 2004 ISBN 1-4020-1861-4
115. H. Kitagawa and Y. Shibutani (eds.): *IUTAM Symposium on Mesoscopic Dynamics of Fracture Process and Materials Strength.* Proceedings of the IUTAM Symposium held in Osaka, Japan, 6-11 July 2003. Volume in celebration of Professor Kitagawa's retirement. 2004 ISBN 1-4020-2037-6
116. E.H. Dowell, R.L. Clark, D. Cox, H.C. Curtiss, Jr., K.C. Hall, D.A. Peters, R.H. Scanlan, E. Simiu, F. Sisto and D. Tang: *A Modern Course in Aeroelasticity.* 4th Edition, 2004 ISBN 1-4020-2039-2
117. T. Burczyński and A. Osyczka (eds.): *IUTAM Symposium on Evolutionary Methods in Mechanics.* Proceedings of the IUTAM Symposium held in Cracow, Poland, 24-27 September 2002. 2004 ISBN 1-4020-2266-2
118. D. Ieşan: *Thermoelastic Models of Continua.* 2004 ISBN 1-4020-2309-X
119. G.M.L. Gladwell: *Inverse Problems in Vibration.* Second Edition. 2004 ISBN 1-4020-2670-6
120. J.R. Vinson: *Plate and Panel Structures of Isotropic, Composite and Piezoelectric Materials, Including Sandwich Construction.* 2005 ISBN 1-4020-3110-6
121. *Forthcoming*
122. G. Rega and F. Vestroni (eds.): *IUTAM Symposium on Chaotic Dynamics and Control of Systems and Processes in Mechanics.* Proceedings of the IUTAM Symposium held in Rome, Italy, 8–13 June 2003. 2005 ISBN 1-4020-3267-6
123. E.E. Gdoutos: *Fracture Mechanics. An Introduction. 2nd edition.* 2005 ISBN 1-4020-3267-6
124. M.D. Gilchrist (ed.): *IUTAM Symposium on Impact Biomechanics from Fundamental Insights to Applications.* 2005 ISBN 1-4020-3795-3
125. J.M. Hughe, P.A.C. Raats and S. C. Cowin (eds.): *IUTAM Symposium on Physicochemical and Electromechanical Interactions in Porous Media.* 2005 ISBN 1-4020-3864-X
126. H. Ding and W. Chen: *Elasticity of Transversely Isotropic Materials.* 2005ISBN 1-4020-4033-4
127. W. Yang (ed): *IUTAM Symposium on Mechanics and Reliability of Actuating Materials.* Proceedings of the IUTAM Symposium held in Beijing, China, 1–3 September 2004. 2005 ISBN 1-4020-4131-6
128. J.P. Merlet: *Parallel Robots.* 2005 ISBN 1-4020-4132-2

Mechanics

SOLID MECHANICS AND ITS APPLICATIONS

Series Editor: G.M.L. Gladwell

129. G.E.A. Meier and K.R. Sreenivasan (eds.): *IUTAM Symposium on One Hundred Years of Boundary Layer Research.* Proceedings of the IUTAM Symposium held at DLR-Göttingen, Germany, August 12–14, 2004. 2005 ISBN 1-4020-4149-7
130. H. Ulbrich and W. Günthner (eds.): *IUTAM Symposium on Vibration Control of Nonlinear Mechanisms and Structures.* 2005 ISBN 1-4020-4160-8
131. L. Librescu and O. Song: *Thin-Walled Composite Beams.* Theory and Application. 2005 ISBN 1-4020-3457-1
132. G. Ben-Dor, A. Dubinsky and T. Elperin: *Applied High-Speed Plate Penetration Dynamics.* 2006 ISBN 1-4020-3452-0

MIX
Papier aus verantwortungsvollen Quellen
Paper from responsible sources
FSC® C105338

If you have any concerns about our products,
you can contact us on
ProductSafety@springernature.com

In case Publisher is established outside the EU,
the EU authorized representative is:
Springer Nature Customer Service Center GmbH
Europaplatz 3, 69115 Heidelberg, Germany

Printed by Libri Plureos GmbH
in Hamburg, Germany